STUDENT SOLUTIONS MANUAL

Volume 1
CHAPTERS 1—19

Edw. S. Ginsberg
UNIVERSITY OF MASSACHUSETTS BOSTON

CONTRIBUTING AUTHOR

Sen-Ben Liao
UNIVERSITY OF CALIFORNIA

Essential UNIVERSITY PHYSICS

Richard Wolfson

PEARSON

Addison
Wesley

San Francisco Boston New York
Cape Town Hong Kong London Madrid Mexico City
Montreal Munich Paris Singapore Sydney Tokyo Toronto

Vice President and Editorial Director: Adam Black
Project Editor: Martha N. Steele
Development Editor: Brad Patterson
Editorial Assistant: Kristin Rose
Managing Editor: Corinne Benson
Production Supervisor: Nancy Tabor
Production Management: Elm Street Production Services
Compositor/Illustrators: ITC
Manufacturing Buyer: Pam Augspurger
Director of Marketing: Christy Lawrence
Text and Cover Printer: Bigelow & Brown
Cover Photo Credit: Tyler Boley, Workbook

ISBN 0-8053-4009-2

1 2 3 4 5 6 7 8 9 10—B&B—10 09 08 07 06

www.aw-bc.com

CONTENTS

PREFACE

This *Student Solutions Manual to Essential University Physics,* by Richard Wolfson, is designed to increase your skill and confidence in solving physics problems—the key to success in your physics course. It coaches you through a range of helpful and effective problem-solving techniques and provides solutions to all the odd-numbered problems in your text. By working carefully through these techniques and solutions, you will master the proven four-step problem-solving approach used in the text (Interpret, Develop, Evaluate, Assess) and learn to successfully:

- interpret problems and identify the key physics concepts involved
- develop a plan and draw figures for your solution
- evaluate any mathematical expressions
- assess your solution to check that it makes sense and see how it adds to your broader understanding of physics

Do your best to solve each problem *before* reading the solution. When you need to refer to the solution, focus on the reasoning—make sure you understand how and why each step is taken. By pushing yourself, you'll develop and hone your problem-solving skills. Don't fall into the trap of just passively reading the solutions.

We have made every effort to ensure these solutions are accurate and correct. If you do find any errors, inconsistencies, or ambiguities, we would be delighted to hear from you. Please contact us at wolfson@aw.com.

DOING PHYSICS

1

EXERCISES

Section 1.2 Measurements and Units

11. **INTERPRET**   We interpret this as a problem involving the comparison of the diameter of two objects (a hydrogen and a proton) expressed in different units.

DEVELOP   Before any comparison can be made, the quantity of interest must first be expressed in the same units. From Table 1.1, we see that a nanometer is 10^{-9} m, and a femtometer (fm) is 10^{-15} m.

EVALUATE   Using the conversion factors, the diameter of a hydrogen atom is $d_H = 0.1$ nm $= 10^{-10}$ m, and the diameter of a proton is $d_p = 1$ fm $= 10^{-15}$ m. Therefore, the ratio of the diameters of a hydrogen atom and a proton (its nucleus) is

$$\frac{d_H}{d_p} = \frac{10^{-10} \text{ m}}{10^{-15} \text{ m}} = 10^5$$

ASSESS   The hydrogen atom is about 100,000 times larger than its nucleus.

13. **INTERPRET**   We interpret this as a problem involving expressing the period of cesium radiation in different units.

DEVELOP   By definition, 1 s $= 9,192,631,770$ periods of a cesium atomic clock. In addition, we know that 1 ns $= 10^{-9}$ s.

EVALUATE   One period of cesium radiation is

$$\frac{1 \text{ s}}{9,192,631,770} = 1.087827757 \times 10^{-10} \text{ s} = 0.1087827757 \text{ ns}$$

ASSESS   Since one nanosecond corresponds to about 9 periods of the cesium radiation, each period is about $\frac{1}{9}$ of a nanosecond. Note that there exists an alternative definition based on the frequency of the cesium-133 hyperfine transition, which is the reciprocal of the period.

15. **INTERPRET**   We interpret this as a problem involving expressing 1-cm line as multiples of the diameter of hydrogen atoms.

DEVELOP   We first express the quantities of interest (diameter of a hydrogen atom and 1-cm line) in the same units. Since a nanometer is 10^{-9} m (Table 1.1), we see that $d_H = 0.1$ nm $= 10^{-10}$ m. In addition, 1 cm $= 10^{-2}$ m.

EVALUATE   The desired number of atoms is the length of the line divided by the diameter of one atom:

$$N = \frac{10^{-2} \text{ m}}{10^{-10} \text{ m}} = 10^8$$

ASSESS   If 1 cm corresponds to 10^8 hydrogen atoms, then each atom would correspond to 10^{-8} cm $= 10^{-10}$ m $= 0.1$ nm.

17. **INTERPRET**   This is a problem that involves the definition of an angle subtended by a circular arc.

DEVELOP   The angle in radians is the circular arc length s divided by the radius R, or $\theta = s/R$.

EVALUATE   Using the equation above, we find the angle to be

$$\theta = \frac{s}{R} = \frac{2.1 \text{ km}}{3.4 \text{ km}} = 0.62 \text{ rad}$$

Using the conversion factor 1π rad $= 180°$, the result can be expressed as

$$\theta = 0.62 \text{ rad} = (0.62 \text{ rad})\left(\frac{180°}{\pi \text{ rad}}\right) \approx 35°$$

ASSESS   Since a complete turn corresponds to $360°$, $35°$ would be roughly 1/10 of a circle. The circumference of a circle of radius $R = 3.4$ km is $C = 2\pi(3.4 \text{ km}) = 21.4$ km. Therefore, we expect the jetliner to fly approximately 1/10 of C, or 2.1 km, in complete agreement with the problem statement.

19. **INTERPRET**   We interpret this as a problem involving expressing the mass (weight) of a letter in different units.
DEVELOP   Two different units for mass appear in the problem—ounces and grams. We must first find the conversion factor. The conversion from ounces to grams is given in Appendix C (1 oz = weight of 0.02835 kg).
EVALUATE   The maximum weight of the letter is 1 oz. Using the conversion factor above, we see that this corresponds to a weight of 0.02835 kg, or 28.35 g.
ASSESS   The conversion factor between oz and g may be obtained based on some easily remembered conversion factors between the metric and English systems (e.g., 1 lb = weight of 0.454 kg, and 1 lb = 16 oz).

21. **INTERPRET**   We interpret this as a problem involving the conversion of volume to different units.
DEVELOP   Since volume has dimension of (length)^3, the problem is equivalent to converting m to cm. The conversion factor is 1 m = 100 cm.
EVALUATE   Using the conversion factor above, we obtain

$$1 \text{ m}^3 = (1 \text{ m})^3 = (10^2 \text{ cm})^3 = 10^6 \text{ cm}^3$$

ASSESS   Another way to remember this relationship is to note that 1 m^3 = 1000 liters, and 1 liter = 1000 cm^3 = 1000 cc.

23. **INTERPRET**   We interpret this as a problem involving the conversion of both area and volume to different units.
DEVELOP   With reference to Appendix C, we write down the following conversion factors, one for the volume and one for the area:

$$1 \text{ gal} = 3.785 \times 10^{-3} \text{ m}^3 = 3.785 \text{ L}$$
$$1 \text{ ft}^2 = 9.290 \times 10^{-2} \text{ m}^2$$

EVALUATE   Combining the two conversion factors, we have

$$350 \text{ ft}^2/\text{gal} = \left(350 \frac{\text{ft}^2}{\text{gal}}\right)\left(\frac{1 \text{ gal}}{3.785 \text{ L}}\right)\left(\frac{9.290 \times 10^{-2} \text{ m}^2}{1 \text{ ft}^2}\right) = 8.59 \text{ m}^2/\text{L}$$

ASSESS   The above result implies 1 ft^2/gal = 0.0245 m^2/L.

25. **INTERPRET**   We interpret this as a problem involving the conversion of both length and time to different units.
DEVELOP   With reference to Appendix C, we write down the following conversion factors, one for length and one for time:

$$1 \text{ km} = 1000 \text{ m}$$
$$1 \text{ h} = 3600 \text{ s}$$

EVALUATE   Combining the two conversion factors, we have

$$1 \text{ m/s} = \left(1\frac{\text{m}}{\text{s}}\right)\left(\frac{3600 \text{ s}}{1 \text{ h}}\right)\left(\frac{1 \text{ km}}{1000 \text{ m}}\right) = 3.6 \text{ km/h}$$

ASSESS   The conversion factor is exact. Speed is the physical quantity with such units. A speed of 1 m/s is equivalent to 3.6 km/h. In other words, if you walk 1 m in one second, you'd be able to walk a distance of 3.6 km in one hour.

27. **INTERPRET**   This is a problem that involves the conversion of radians to degrees.
DEVELOP   The angle in radians is the circular arc length s divided by the radius R, or $\theta = s/R$. Thus, a complete (360°) revolution would correspond to $\theta = 2\pi$ rad.

EVALUATE   Using the equation above, one radian is equal to

$$1 \text{ rad} = (1 \text{ rad})\left(\frac{360°}{2\pi \text{ rad}}\right) = 57.3°$$

ASSESS   Since π rad $\leftrightarrow$ 180°, 1rad (roughly $p/3$ rad) would be equal to about 60°, which is close to the result obtained above.

Section 1.3 Working with Numbers

29.   **INTERPRET**   We interpret this as a problem involving the conversion of time to different units.

DEVELOP   With reference to Table 1.1 for SI prefixes, we have $1 \text{ ms} = 10^{-3}$ s.

EVALUATE   Using the above conversion factor, we obtain

$$\frac{4.2 \times 10^3 \text{ m/s}}{0.57 \text{ ms}} = \left(4.2 \times 10^3 \frac{\text{m}}{\text{s}}\right)\frac{1}{0.57 \times 10^{-3} \text{ s}} = 7.37 \times 10^6 \text{ m/s}^2$$

ASSESS   Acceleration is the physical quantity with such units. An average acceleration of 7.37×10^6 m/s^2 changes the speed of an object by 4.2×10^3 m/s in 0.57 ms.

31.   **INTERPRET**   This is an arithmetic problem involving taking the cube root of a number.

DEVELOP   To take a cube root of a number, N, without using a calculator, we first rewrite N as a^3. Then $N^{1/3} = a$.

EVALUATE   We may rewrite 6.4×10^{19} as

$$6.4 \times 10^{19} = 64 \times 10^{18} = (4)^3 \times (10^6)^3 = (4 \times 10^6)^3$$

Therefore, $(6.4 \times 10^{19})^{1/3} = 4 \times 10^6$.

ASSESS   The result can be readily checked by using a calculator, or raising 4×10^6 to the cubic power to recover 6.4×10^{19}.

33.   **INTERPRET**   We interpret this as a problem involving adding two lengths that are expressed in different units.

DEVELOP   Before an addition between two lengths can be made, both quantities must first be expressed in the same units. From Table 1.1, we see that $1 \text{ cm} = 10^{-2}$ m.

EVALUATE   The overall length of the airplane (of length $L_0 = 41$ m) could be increased by as much as $\Delta L = 3.6$ cm, or 0.036 m, depending on how the antenna is attached. However, to two significant figures, the overall length of the airplane is still

$$L = L_0 + \Delta L = 41 \text{ m} + 0.036 \text{ m} = 41.036 \text{ m} \approx 41 \text{ m}$$

ASSESS   To two significant figures, the result remain unchanged. That is because in this context, 41 m means a length greater than or equal to 40.5 m, but less than 41.5 m, and 41 m + 0.036 m = 41.036 m satisfies this condition.

PROBLEMS

35.   **INTERPRET**   We interpret this as a problem involving exploring the degree of numerical accuracy in evaluating $(\sqrt{3})^3$, depending on the number of significant figures used for $\sqrt{3}$.

DEVELOP   We shall carry the calculation to more digits in intermediate calculations, and then round the answers to the desired number of significant figures.

EVALUATE   (a) With $\sqrt{3} \approx 1.73$ (to 3 significant figures), we obtain $(1.73)^3 = 5.177 \approx 5.18$, to three significant figures.

(b) On the other hand, if we use $\sqrt{3} \approx 1.732$ (to 4 significant figures), we then have $(1.732)^3 \approx 5.1957$, or 5.20 to three significant figures.

ASSESS   With a calculator, one may show that $(\sqrt{3})^3 = 3^{3/2} = 5.196....$ The example shows that it is important to carry the calculation to more digits in intermediate calculations to achieve the desired accuracy. Rounding of intermediate results could affect the final answer.

37.   INTERPRET   We interpret this as a problem that calls for a rough estimate, instead of a precise numerical answer.
DEVELOP   For problems that involve rough estimates, various assumptions usually need to be made. Such assumptions must be physically motivated with reasonable order-of-magnitude estimates. We shall assume that there are 250 million people in the United States, and that each person drinks one glass of milk per day (an 8-oz glass, or half a pint, which is equivalent to half a pound, or 0.227 kg).
EVALUATE   Based on the assumption above, the amount consumed per year would be approximately

$$(250 \times 10^6)(365 \text{ d/y})(0.227 \text{ kg/d}) = 2 \times 10^{10} \text{ kg/y}$$

Dividing this by the average annual production of one cow, we estimate the number of cows needed to be

$$N = \frac{2 \times 10^{10} \text{ kg/y}}{10^4 \text{ kg/y}} = 2 \times 10^6$$

ASSESS   There are approximately 9 million milk cows in the US. Based on our estimate, we do not expect any shortage of milk supply in the near future.

39.   INTERPRET   This is a problem that calls for a rough estimate, instead of a precise numerical answer. The quantities of interest here are the total energy consumption rate (power) and the area needed for solar cells.
DEVELOP   For problems that involve rough estimates, various assumptions usually need to be made. Such assumptions must be physically motivated with reasonable order-of-magnitude estimates.
EVALUATE   The electrical power consumed by the entire population of the United States, divided by the power converted by one square meter of solar cells, is the area required by this question.
Assuming that there are 250 million people in the United States, if the electrical energy consumption rate for each person is 3 kW (a per capita average over 24 h periods of all types of weather), then the total electrical power is

$$P_{tot} = 250 \times 10^6 \times 3 \text{ kW} = 7.50 \times 10^8 \text{ kW}$$

On the other hand, for a solar cell with 20% efficiency in converting sunlight to electrical energy, the per square meter yield is $P_1/A = (0.20)(0.3 \text{ kW/m}^2) = 0.060 \text{ kW/m}^2$. Therefore, the total area needed is

$$A_{tot} = \frac{P_{tot}}{P_1/A} = \frac{7.50 \times 10^8 \text{ kW}}{0.060 \text{ kW/m}^2} = 1.25 \times 10^{10} \text{ m}^2 = 1.25 \times 10^4 \text{ km}^2$$

The land area of the United States can be approximated as the area of a rectangle the size of the distance from New York to Los Angeles by the distance from New York to Miami, or $A_{US} \approx (5000 \text{ km}) \times (2000 \text{ km}) = 10^7 \text{ km}^2$. Then the fraction of area to be covered by solar cells would be

$$\frac{A_{tot}}{A_{US}} \approx \frac{1.25 \times 10^4 \text{ km}^2}{10^7 \text{ km}^2} = 1.25 \times 10^{-3}$$

or approximately 0.13%.
ASSESS   This represents only a small fraction of the land to be used for solar cells. The area A_{tot} is comparable to the fraction of land now covered by airports.

41.   INTERPRET   We interpret this as a problem that calls for a rough estimate, instead of a precise numerical answer. The quantity of interest here is the total number of air molecules in a room.
DEVELOP   For problems that involve rough estimates, various assumptions usually need to be made. Such assumptions must be physically motivated with reasonable order-of-magnitude estimates.
Here are our assumptions: A typical dormitory single room might have dimension of 15 ft × 10 ft × 8 ft ($L \times w \times h$), with a volume of $V = 1200 \text{ ft}^3$, or approximately 34 m^3 (see Appendix C). In addition, we shall regard the air in the room as an ideal gas at standard temperature and pressure, one "mole of air" contains Avogadro's number of molecules, about 6.02×10^{23}, and occupies a volume of 22.4 liters, or $2.24 \times 10^{-2} \text{ m}^3$.
EVALUATE   Based on the above assumptions, the number of molecules in the dormitory room is about

$$N = \left(\frac{6.02 \times 10^{23}}{2.24 \times 10^{-2} \text{ m}^3} \right)(34 \text{ m}^3) = 9 \times 10^{26}$$

ASSESS   This is a fairly large number. But the result is reasonable; each cubic meter contains on the order of roughly 10^{25} molecules.

43.   **INTERPRET**   We interpret this as a problem that calls for a rough estimate, instead of a precise numerical answer. The quantity of interest here is the thickness of the bubble.

DEVELOP   For problems that involve rough estimates, various assumptions usually need to be made. Such assumptions must be physically motivated with reasonable order-of-magnitude estimates.

Here are our assumptions: The volume of gum is its mass m divided by its density r, or

$V = m/\rho = (8\text{ g})/(1\text{ g/cm}^3) = 8\text{ cm}^3$. On the other hand, the volume of the bubble (a thin spherical shell) is $4\pi R^2 d$, where R is the radius and $d \ll R$ is the thickness, and is equal to the volume of gum.

EVALUATE   With the assumptions above, the thickness of the bubble is

$$d = \frac{V}{4\pi R^2} = \frac{8\text{ cm}^3}{4\pi(5\text{ cm})^2} = 0.025\text{ cm}$$

ASSESS   The thickness of the bubble is very small. But our estimate is reasonable. Four layers of such thickness would be about 1 mm.

45.   **INTERPRET**   This is a problem that calls for a rough estimate, instead of a precise numerical answer. The quantities of interest here are the distance between the electronic components in a PC chip, and the number of calculations that can be performed each second.

DEVELOP   The area of each component (L^2) is the area of the chip divided by the number of components. The length L is the distance across each component. To estimate the number of calculations performed per second, we first find the d, the distance traveled by electrical impulses during a calculation. With the assumption that the pulse travels at the speed of light c, the time required to complete one calculation is given by $\Delta t = d/c$.

EVALUATE   **(a)** Since the area of each component is

$$A = \frac{(4\text{ mm})^2}{10^6} = 1.6 \times 10^{-11}\text{ m}^2$$

and $A = L^2$, the distance across each component is

$$L = \sqrt{A} = \sqrt{1.6 \times 10^{-11}\text{ m}^2} = 4 \times 10^{-6}\text{ m} = 4\mu\text{m}$$

(b) The distance traveled by electrical impulses during a calculation is

$$d = (10^6)(10^4)L = (10^6)(10^4)(4 \times 10^{-6}\text{ m}) = 4 \times 10^4\text{ m}$$

Traveling at the speed of light, impulses would complete one calculation in $\Delta t = \frac{d}{c} = \frac{4 \times 10^4\text{ m}}{3 \times 10^8\text{ m/s}} = 1.33 \times 10^{-4}\text{ s}$.

Therefore, in one second, the number of calculations that could be performed is

$$N = \frac{1\text{s}}{\Delta t} = \frac{1\text{s}}{1.33 \times 10^{-4}\text{ s}} = 7500$$

ASSESS   A typical PC today can execute on the order of 3 billion instructions per second. If a calculation comprises 1 million instructions, then in one second, the PC can perform on the order of 10^3 calculations. Our result is within this range.

47.   **INTERPRET**   We interpret this as a problem involving estimating uncertainty, given the number of significant figures.

DEVELOP   Since the value 3.6 can be used to represent any number between 3.55 and 3.65, rounding to two significant figures, we see that the uncertainty in the first decimal place is ±0.05. Therefore, the percent uncertainty in a one-decimal-place number, N, is

$$\Delta = 100\left(\frac{\pm 0.05}{N}\right)\%$$

This obviously decreases as N increases.

EVALUATE   For the numbers given, the percent uncertainty is

(a) $\Delta = 100(\pm 0.05/1.1)\% \approx \pm 5\%$;   **(b)** $\Delta = 100(\pm 0.05/5.0)\% \approx \pm 1\%$; and

(c) $\Delta = 100(\pm 0.05/9.9)\% \approx \pm 0.5\%$.

ASSESS   Our result indicates that, for a one-decimal place number N, while the uncertainty in the first decimal place remains the same (±0.05), independent of N, the percentage uncertainty, Δ, becomes smaller for larger N.

49.   **INTERPRET**   This problem is about converting units, and we're asked to convert a distance given in miles and yards to kilometers.

DEVELOP   We'll convert the two distances, 26 miles and 385 yards, into meters, and then add the two for our final answer. From Appendix C, one mile is 1609 meters and one yard is 0.9144 meters. We'll multiply each distance by the appropriate conversion factor and add.

EVALUATE   26 miles $\times \frac{1,609 \text{ m}}{1 \text{ mile}} = 41{,}834$ m, and 385 yards $\times \frac{0.9144 \text{ m}}{1 \text{ yard}} = 352$ m. The total distance is $41{,}834$ m $+ 352$ m $= 42{,}186$ m. We convert this to km: $42{,}186$ m $\times \frac{1 \text{ km}}{1{,}000 \text{ m}} = 42.186$ km.

ASSESS   A mile is about 1.6 km, so this answer seems reasonable.

51.   **INTERPRET**   Estimate the number of piano tuners in Chicago? Sounds impossible, but remember that this just needs to be an estimate. We'll base our answers on reasonable numbers and report it to one significant figure at most.

DEVELOP   We'll estimate the population of Chicago as 3 million. Then we'll figure that one household out of 10 has a piano, and there are 4 or 5 people per household. A piano needs to be tuned every year or two, and it takes a tuner about an hour. With travel time between pianos, we'll estimate that a full-time piano tuner might do 4 pianos a day, 300 days a year. Last, we'll calculate the number of full-time piano tuners it would take to tune every piano in Chicago, once every 2 years.

EVALUATE   3 million people $\times \frac{1 \text{ household}}{5 \text{ people}} \times \frac{1 \text{ piano}}{10 \text{ households}} = 60{,}000$ pianos

If each piano gets tuned once every 2 years, that's 30,000 pianos per year. We have estimated that a full-time piano tuner would do about 1000 pianos a year, so tuning all these pianos would take 30 full-time piano tuners.

ASSESS   This is a rough estimate only, but it's probably good to order of magnitude.

53.   **INTERPRET**   Convert kilograms to pounds to determine the price per pound.

DEVELOP   From Appendix C, 1 kg = 2.2 lb. We'll multiply the cost per kilogram by this conversion factor to find the cost per pound. Then we'll add the cost of shipping.

EVALUATE $\frac{\$8.95}{0.5 \text{ kg}} \times \frac{1 \text{ kg}}{2.20 \text{ lb}} = \frac{\$8.14}{1 \text{ lb}}$. The shipping is listed as $1.92 per bag, bringing the total price to $10.06 per bag.

ASSESS   Half a kilogram is a little more than one pound.

55.   **INTERPRET**   Convert "a half pound" to kilograms.

DEVELOP   1 kg = 2.2 lb, so 1 lb = 0.45 kg. One pound is about a half kilogram.

EVALUATE   Half a pound is about a quarter kilogram. Order accordingly.

ASSESS   Note that this is not as precise as you might expect from a physics textbook solution. That's actually fine, because the precision of the answer should not be greater than the precision of the problem.

MOTION IN A STRAIGHT LINE

EXERCISES

Section 2.1 Average Motion

13. **INTERPRET** We need to find average speed, given distance and time.

    DEVELOP Speed is distance divided by time.

    EVALUATE $v = \frac{100 \text{ m}}{9.77 \text{ s}} = 10.2$ m/s

    ASSESS His time is about 10 seconds, so his speed is about 10 m/s.

15. **INTERPRET** This is a one-dimensional kinematics problem, and we identify the bicyclist as the object of interest. His trip consists of two parts (out and back), and we are asked to compute the displacement and average velocity of each part, as well as for the trip as a whole.

    DEVELOP For motion in a straight line, the displacement, or the net change in position, is $\Delta x = x_2 - x_1$, where x_1 and x_2 are the starting and the end points, respectively. The average velocity is the displacement divided by the time interval, $\bar{v} = \Delta x / \Delta t$, as shown in Equation 2.1. In our coordinate system, we take north to be $+x$.

    EVALAUTE (a) The displacement at the end of the first 2.5 h is

    $$\Delta x_{out} = x_2 - x_1 = 24 \text{ km} - 0 \text{ km} = 24 \text{ km}$$

    (b) With $\Delta t_{out} = 2.5$ h, the average velocity over this interval is

    $$\bar{v}_{out} = \frac{\Delta x_{out}}{\Delta t_{out}} = \frac{24 \text{ km}}{2.5 \text{ h}} = 9.6 \text{ km/h} \quad \text{(north)}$$

    (c) With $\Delta t_{back} = 1.5$ h, the average velocity for the homeward leg of the trip is

    $$\bar{v}_{back} = \frac{\Delta wx_{back}}{\Delta t_{back}} = \frac{-24 \text{ km}}{1.5 \text{ h}} = -16 \text{ km/h} \quad \text{(south)}$$

    (d) Since the final position of the bicyclist is the same as his initial position, his total displacement of the trip is

    $$\Delta x_{total} = \Delta x_{out} + \Delta x_{back} = 24 \text{ km} + (-24 \text{ km}) = 0 \text{ km}.$$

    (e) Since $\Delta x_{total} = 0$, the average velocity for the entire trip is $\bar{v}_{total} = \Delta x_{total} / \Delta t_{total} = 0$.

    ASSESS Note the distinction between average velocity and average speed. The former depends only on the net displacement, while the latter takes into consideration the total distance traveled. The average speed for this trip is

    $$\text{average speed} = \frac{|\Delta x_{out}| + |\Delta x_{back}|}{\Delta t_{out} + \Delta t_{back}} = \frac{24 \text{ km} + 24 \text{ km}}{2.5 \text{ h} + 1.5 \text{ h}} = 12 \text{ km/h}$$

17. **INTERPRET** We need to find average speed, given distance and time.

    DEVELOP We're given distance in units of meters and kilometers, so we'll convert everything to meters before finding the total distance. The time is given in hours, minutes, and seconds, so we'll convert it to seconds. Then use $v = \frac{x}{t}$ to find the average speed.

    EVALUATE $x = 1500 \text{ m} + 40{,}000 \text{ m} + 10{,}000 \text{ m} = 51500 \text{ m}.$

    $$t = (1 \text{ h}) \times \frac{3600 \text{ s}}{1 \text{ h}} + (49 \text{ min}) \times \frac{60 \text{ s}}{1 \text{ min}} + (31 \text{ s}) = 6571 \text{ s}$$

    $$v = \frac{x}{t} = \frac{51{,}500 \text{ m}}{6571 \text{ s}} = 7.84 \text{ m/s}$$

    ASSESS This is faster than Olympic marathon speed (see Exercise 2.14) because of the time spent on a bike.

19.   **INTERPRET**   We interpret this as a problem involving the conversion of distance and time to different units.

DEVELOP   We shall use the following conversion factors to change from meters to miles and from seconds to hours:

$$1\text{ mi} = 1609\text{ m} \qquad \rightarrow \quad 1\text{ m} = (1/1609)\text{ mi}$$
$$1\text{h} = 60\min = 3600\text{ s} \quad \rightarrow \quad 1\text{ s} = (1/3600)\text{ h}$$

EVALUATE   Using the conversion factors above, we obtain

$$1\text{ m/s} = \left(1\frac{\text{m}}{\text{s}}\right)\cdot\left(\frac{1\text{ mi}}{1609\text{ m}}\right)\cdot\left(\frac{3600\text{ s}}{1\text{ h}}\right) = 2.24\text{ mi/h}$$

ASSESS   If you drive down the road at a speed of 22 mi/h, your car would move about 10 m in one second.

Section 2.2 Instantaneous Velocity

21.   **INTERPRET**   We interpret this as a one-dimensional kinematics problem. By making a plot of distance as a function of time, the physical quantities of interest—the average velocity and instantaneous velocity, can be deduced from the graph.

DEVELOP   Let the three cities, Houston, Des Moines, and Minneapolis be labeled as A, B, and C, respectively. With Kansas City chosen to be the origin ($x = 0$), the positions of the three cities are: $x_A = -1000$ km, $x_B = 300$ km, and $x_C = 650$ km. The negative sign in x_A indicates that Houston is *south* of Kansas City. Both trips start at the same place (Houston, point A) at time $t_A = 0$, and end at the same place (Des Moines, point B) at $t_B = 2.6$ h. They have the same overall displacement $\Delta x = x_B - x_A = 1300$ km, in the same time period, $\Delta t = t_B - t_A = 2.6$ h, and thus the same average velocity $\bar{v}_{AB} = 500$ km/h.

EVALUATE   The plot of the two trips is depicted below:

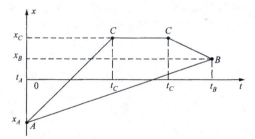

The first trip, a direct flight, is represented by a straight line AB with slope $\bar{v}_{AB}$. Thus, the position of the airplane as a function of time can be written as

$$x_1(t) = x_A + \bar{v}_{AB}t \qquad 0 \le t \le 2.6\text{ h}$$

Note that short intervals of acceleration at takeoff and landing are ignored. Since the instantaneous velocity at a particular instant t is given by the slope of $x(t)$ at that time, we have

$$v_1 = \frac{dx_1}{dt} = \bar{v}_{AB} \qquad 0 \le t \le 2.6\text{ h}$$

The second trip, using a faster plane (steeper slopes when flying), stops for a while in Minneapolis at $x_C = 650$ km (this segment is flat) and then proceeds south to Des Moines (negative velocity and slope). This trip is shown by three straight segments $ACC'B$, and is given analytically by the equations

$$x_2(t) = \begin{cases} x_A + \left(\dfrac{x_C - x_A}{t_C - 0}\right)t = x_A + \bar{v}_{AC}t & t_A = 0 \le t \le t_C \\[2ex] x_C & t_C \le t \le t_{C'} \\[2ex] x_C + \left(\dfrac{x_B - x_C}{t_B - t_{C'}}\right)(t - t_{C'}) = x_C + \bar{v}_{C'B}(t - t_{C'}) & t_{C'} \le t \le t_B \end{cases}$$

Note that each segment of the second trip has been assumed to be executed with constant velocity, ignoring takeoffs and landings. The times t_C and $t_{C'}$ and velocities $\bar{v}_{AC}$ and $\bar{v}_{C'B}$ were chosen arbitrarily. Similarly, the instantaneous velocity for this trip is

$$v_2 = \frac{dx_2}{dt} = \begin{cases} \bar{v}_{AC} & t_A = 0 \le t \le t_C \\ 0 & t_C \le t \le t_{C'} \\ \bar{v}_{C'B} & t_{C'} \le t \le t_B \end{cases}$$

ASSESS   Average velocity depends only on the net displacement between the starting and ending points. In the above situations, since $x_1(t)$ and $x_2(t)$ are linear in t (constant slope), the average velocity is equal to the instantaneous velocity.

23.   **INTERPRET**   This is a one-dimensional kinematics problem involving finding the velocity as a function of time, given position as a function of time. The object of interest is the model rocket.

DEVELOP   The instantaneous velocity $v(t)$ can be obtained by taking the derivative of $y(t)$. The derivative of a function of the form bt^n can be obtained by using Equation 2.3.

EVALUATE   (a) The instantaneous velocity is $v(t) = \frac{dy}{dt} = b - 2ct$.

(b) The velocity is zero when $b = 2ct$, or $t = \frac{b}{2c} = \frac{82 \text{ m/s}}{2(4.9 \text{ m/s}^2)} = 8.37$ s.

ASSESS   As the model rocket is launched upward with an initial velocity $b = 82$ m/s, its altitude increases and then reaches a maximum, where the instantaneous velocity is zero. The rocket then falls back to the Earth.

Section 2.3 Acceleration

25.   **INTERPRET**   The object of interest is the subway train that undergoes acceleration from rest, followed by deceleration through braking. The kinematics are one-dimensional.

DEVELOP   The average acceleration over a time interval Δt is given by Equation 2.4: $\bar{a} = \Delta v/\Delta t$.

EVALUATE   Over a time interval of $\Delta t = t_2 - t_1 = 48$ s, the velocity of the train (along a linear track) changes from $v_1 = 0$ (starting at rest) to $v_2 = 17$ m/s. The change in velocity is $\Delta v = v_2 - v_1 = 17$ m/s $-$ 0 m/s $= 17$ m/s. Thus, the average acceleration is

$$\bar{a} = \frac{\Delta v}{\Delta t} = \frac{17 \text{ m/s}}{48 \text{ s}} = 0.354 \text{ m/s}^2$$

ASSESS   We find that the average acceleration only depends on the change of velocity between the starting point and the end point; the intermediate velocity is irrelevant.

27.   **INTERPRET**   The object of interest is the egg. Its motion can be divided into two stages: (i) free fall, and (ii) stopping after striking the ground.

DEVELOP   The average acceleration over a time interval Δt is given by Equation 2.4: $\bar{a} = \Delta v/\Delta t$. We shall take downward velocity to be *negative*.

EVALUATE   While undergoing free fall, the velocity changes from $v_1 = 0$ to $v_2 = -11.0$ m/s in 1.12 s. Therefore, the average acceleration is

$$\bar{a} = \frac{\Delta v}{\Delta t} = \frac{v_2 - v_1}{t_2 - t_1} = \frac{-11.0 \text{ m/s} - 0}{1.12 \text{ s}} = -9.82 \text{ m/s}^2$$

Similarly, during the stopping process, the velocity changes from $v_2 = -11.0$ m/s to $v_3 = 0$ in 0.131 s, and the average acceleration is

$$\bar{a} = \frac{\Delta v}{\Delta t} = \frac{v_3 - v_2}{t_3 - t_2} = \frac{0 - (-11.0 \text{ m/s})}{0.131 \text{ s}} = +84.0 \text{ m/s}^2$$

ASSESS   Since acceleration is negative (downward) during free fall, the velocity becomes more negative. On the other hand, during stopping, the acceleration is positive (upward), and the velocity becomes less negative and eventually reaches zero.

29.   **INTERPRET**   The object of interest is the car, which undergoes constant acceleration. The kinematics are one-dimensional.

DEVELOP   We first convert the units km/h to m/s, using the conversion factor

$$1 \text{ km/h} = \left(1\frac{\text{km}}{\text{h}} \right)\left(\frac{1000 \text{ m}}{1 \text{ km}} \right)\left(\frac{1 \text{ h}}{3600 \text{ s}} \right) = 0.278 \text{ m/s}$$

and then use Equation 2.4, $\bar{a} = \Delta v / \Delta t$, to find the average acceleration over a time interval Δt.

EVALUATE   The speed of the car at $t = 16.0$ s is 1000 km/h, or 278 m/s. Therefore, the average acceleration is

$$\bar{a} = \frac{\Delta v}{\Delta t} = \frac{v_2 - v_1}{t_2 - t_1} = \frac{278 \text{ m/s} - 0}{16.0 \text{ s}} = 17.4 \text{ m/s}^2$$

ASSESS   The magnitude of the average acceleration is about $1.8g$, where $g = 9.8$ m/s^2 is the gravitational acceleration. An object undergoing free fall attains only a speed of 157 m/s after 16.0 s, compared to 278 m/s of the supersonic car. Given the supersonic nature of the vehicle, the value of $\bar{a}$ is completely reasonable.

Section 2.4 Constant Acceleration

31.   **INTERPRET**   The problem is to establish a connection between two mathematical equations, one for displacement and the other for velocity, in one-dimensional kinematics.

DEVELOP   By differentiating the displacement $x(t)$ given in Equation 2.10 with respect to t, we obtain the corresponding velocity $v(t)$.

EVALUATE   Using Tactic 2.1 for taking derivatives, we obtain

$$v = \frac{dx}{dt} = \frac{d}{dt}\left(x_0 + v_0 t + \frac{1}{2}at^2 \right) = 0 + v_0 + \frac{1}{2}a \cdot (2t) = v_0 + at$$

which is Equation 2.7.

ASSESS   Both Equations 2.7 and 2.10 are used to describe one-dimensional kinematics with constant acceleration a. If velocity $v(t)$ is linear in t, then the corresponding displacement $x(t)$ will be quadratic in t.

33.   **INTERPRET**   This is a one-dimensional kinematics problem with constant acceleration. The object of interest is the rocket.

DEVELOP   The three quantities, displacement, velocity, and acceleration, are related by Equation 2.11, $v^2 = v_0^2 + 2a(x - x_0)$. This is the equation we shall use to solve for the acceleration in part (a). Once the acceleration is known, the time elapsed for this ascent can be calculated by using Equation 2.7, $v = v_0 + at$.

EVALUATE   (a) In Equation 2.11 (with x positive upward) we are given that $x - x_0 = 85$ km $= 85,000$ m, $v_0 = 0$ (the rocket starts from rest), and $v = 2.8$ km/s $= 2800$ m/s. Therefore, we can solve for the acceleration:

$$a = \frac{v^2 - v_0^2}{2(x - x_0)} = \frac{(2800 \text{ m/s})^2 - 0}{2(85,000 \text{ m})} = 46.1 \text{ m/s}^2$$

(b) From Equation 2.7, the time of flight is

$$t = \frac{v - v_0}{a} = \frac{2800 \text{ m/s} - 0}{46.1 \text{ m/s}^2} = 60.7 \text{ s}$$

ASSESS   An acceleration of 46.1 m/s^2, or approximately $5g$ ($g = 9.8$ m/s^2) is typical for rocket during liftoff. This enables the rocket to reach a speed of 2.8 km/s in just about one minute.

35.   **INTERPRET**   The object of interest is the car that undergoes constant deceleration (via braking) and comes to a complete stop after traveling a certain distance.

DEVELOP   The three quantities, displacement, velocity, and deceleration (negative acceleration), are related by Equation 2.11, $v^2 = v_0^2 + 2a(x - x_0)$. This is the equation we shall use to solve for a. The initial speed can be converted as

$$v_0 = 50 \text{ mi/h} = \left(50\frac{\text{mi}}{\text{h}} \right)\left(\frac{5280 \text{ ft}}{\text{mi}} \right)\left(\frac{1 \text{ h}}{3600 \text{ s}} \right) = 73.3 \text{ ft/s}$$

EVALUATE   Since the car stops ($v = 0$) after traveling $x - x_0 = 100$ ft from an initial speed of $v_0 = 73.3$ ft/s, Equation 2.11 gives

$$a = \frac{v^2 - v_0^2}{2(x - x_0)} = \frac{0 - (73.3 \text{ ft/s})^2}{2(100 \text{ ft})} = -26.9 \text{ ft/s}^2$$

The magnitude of the deceleration is the absolute value of a: $|a| = 26.9 \text{ ft/s}^2$.

ASSESS   With this deceleration, it would take about $t = v_0/|a| = (73.3 \text{ ft/s})/(26.9 \text{ ft/s}^2) = 2.7$ s for the car to come to a complete stop. The value is in accordance with our driving experience.

37. **INTERPRET**   The object of interest is a piece of the fragments of the meteor that undergoes constant deceleration.
    DEVELOP   The three quantities, displacement, velocity, and deceleration (negative acceleration), are related by Equation 2.11, $v^2 = v_0^2 + 2a(x - x_0)$. This is the equation we shall use to solve for the speed of the fragment.
    EVALUATE   For a particular fragment that followed a straight-line path to the bottom, perpendicular to the desert surface, using Equation 2.11, the initial speed is:

    $$v_0 = \sqrt{v^2 - 2a(x - x_0)} = \sqrt{0 - 2(-4 \times 10^5 \text{ m/s}^2)(180 \text{ m})} = \sqrt{1.44 \times 10^8 \text{ m}^2/\text{s}^2} = 1.2 \times 10^4 \text{ m/s}.$$

ASSESS   With this rate of deceleration, it takes only about 0.03 s for the fragment to penetrate 180 m deep into the Earth and come to a complete stop. The impact force that created this gigantic hole was enormous!

Section 2.5 The Acceleration of Gravity

39. **INTERPRET**   The problem involves constant acceleration due to gravity. The object of interest is the rock.
    DEVELOP   If we ignore the travel time of the sound, then the depth of the well can be deduced if we know the time elapsed to hear the splash. The equation to be used is Equation 2.10, but with y as the variable. The acceleration is $a = -g = -9.8 \text{ m/s}^2$.
    EVALUATE   Neglecting the travel time of the sound, the rock fell for a duration of $t = 4.4$ s from rest ($v_0 = 0$). The height is given by

    $$y - y_0 = \frac{1}{2}at^2 = \frac{1}{2}(-g)t^2 \quad \rightarrow \quad h = y_0 - y = \frac{1}{2}(9.8 \text{ m/s}^2)(4.4 \text{ s})^2 = 94.9 \text{ m}$$

ASSESS   When the travel time of the sound is ignored, the depth of the well is quadratic in t. The longer it takes to hear the splash, the deeper the well!

41. **INTERPRET**   The problem involves constant acceleration due to gravity. The object of interest is the model rocket.
    DEVELOP   Using Equation 2.11, the altitude, the speed velocity, and the gravitational acceleration may be related as

    $$v^2 = v_0^2 + 2a(y - y_0) = v_0^2 - 2g(y - y_0)$$

    Note that the acceleration is $a = -g$, or downward. At any time t, the rocket's velocity and altitude are $v = v_0 - gt$ and $y = y_0 + v_0 t - \frac{1}{2}gt^2$.
    EVALUATE   (a) At its maximum altitude, the rocket's vertical speed is instantaneously zero, so $0 = v_0^2 - 2g(y_{max} - y_0)$, or

    $$y_{max} = y_0 + \frac{v_0^2}{2g} = 0 + \frac{(49 \text{ m/s})^2}{2(9.8 \text{ m/s}^2)} = 123 \text{ m}$$

    (b) When $t = 1.0$ s, the speed and the altitude are

    $$v = v_0 - gt = 49 \text{ m/s} - (9.8 \text{ m/s}^2)(1 \text{ s}) = 39.2 \text{ m/s}$$

    $$y = y_0 + v_0 t - \frac{1}{2}gt^2 = 0 + (49 \text{ m/s})(1 \text{ s}) - \frac{1}{2}(9.8 \text{ m/s}^2)(1 \text{ s})^2 = 44.1 \text{ m}$$

    Note that the altitude is the height above the ground, y_0, and v is positive upward.

(c) Similarly, when $t = 4$ s, we have

$$v = v_0 - gt = 49 \text{ m/s} - (9.8 \text{ m/s}^2)(4 \text{ s}) = 9.8 \text{ m/s}$$

$$y = y_0 + v_0 t - \frac{1}{2}gt^2 = 0 + (49 \text{ m/s})(4 \text{ s}) - \frac{1}{2}(9.8 \text{ m/s}^2)(4 \text{ s})^2 = 117.6 \text{ m}$$

(d) And when $t = 7$ s,

$$v = v_0 - gt = 49 \text{ m/s} - (9.8 \text{ m/s}^2)(7 \text{ s}) = -19.6 \text{ m/s}$$

$$y = y_0 + v_0 t - \frac{1}{2}gt^2 = 0 + (49 \text{ m/s})(7 \text{ s}) - \frac{1}{2}(9.8 \text{ m/s}^2)(7 \text{ s})^2 = 102.9 \text{ m}$$

ASSESS   As the rocket moves vertically upward, its velocity decreases due to gravitational acceleration, which points downward. Upon reaching its maximum height, the velocity reduces to zero. It then falls back to Earth with a negative velocity. From (c) and (d), we see that the velocities have different signs at $t = 4$ s and $t = 7$ s. Therefore, we conclude that the rocket is at a maximum height between 4 and 7 s. Further calculation using Equation 2.7 gives $t = (v - v_0)/a = (0 - 49 \text{ m/s})/(-9.8 \text{ m/s}^2) = 5.0$ s, in agreement with our expectation.

43.   **INTERPRET**   The object of interest is the rock, which travels vertically upward under the influence of gravitational acceleration.

DEVELOP   When it hits the Frisbee, the rock's velocity and height are $v = 3$ m/s and $y = 6.5$ m, while its initial velocity and height are v_0 and $y_0 = 1.3$ m. These quantities are related by Equation 2.11:

$$v^2 - v_0^2 = 2a(y - y_0) = -2g(y - y_0)$$

EVALUATE   Solving the above equation, we obtain

$$v_0 = \sqrt{v^2 + 2g(y - y_0)} = \sqrt{(3 \text{ m/s})^2 + 2(9.8 \text{ m/s}^2)(6.5 \text{ m} - 1.3 \text{ m})} = 10.5 \text{m/s}$$

ASSESS   The initial velocity v_0 must be positive since the rock is thrown upward. In addition, v_0 must be greater than the final velocity 3 m/s. These conditions are met by our result.

PROBLEMS

45.   **INTERPRET**   This is a one-dimensional problem involving two travel segments.

DEVELOP   The trip can be divided into two time intervals, t_1 and t_2 with $t = t_1 + t_2 = 40$ min (2 h/3). The total distance traveled is $x = x_1 + x_2 = 25$ mi, where x_1 and x_2 are the distances covered in each time interval.

EVALUATE   During the first time interval, $t_1 = 15$ min (or 0.25 h), with an average speed of $\bar{v}_1 = 20$ mi/h, the distance traveled is

$$x_1 = \bar{v}_1 t_1 = (20 \text{ mi/h})(0.25 \text{ h}) = 5 \text{ mi}$$

Therefore, the remaining distance $x_2 = x - x_1 = 25 \text{ mi} - 5\text{mi} = 20 \text{ mi}$ must be covered in

$$t_2 = t - t_1 = 40 \text{ min} - 15 \text{ min} = 25 \text{ min} = \frac{5}{12} \text{ h}$$

This implies an average speed of $\bar{v}_2 = \frac{x_2}{t_2} = \frac{20 \text{ mi}}{5 \text{ h}/12} = 48$ mi/h.

ASSESS   The overall average speed was pre-determined to be

$$\bar{v} = \frac{x}{t} = \frac{25 \text{ mi}}{2 \text{ h}/3} = 37.5 \text{ mi/h}$$

When you drive slower during the first segment, you make it up by driving faster during the second. In fact, the overall average speed equals the time-weighted average of the average speeds for the two parts of the trip:

$$\bar{v} = \frac{x}{t} = \frac{x_1 + x_2}{t} = \frac{\bar{v}_1 t_1 + \bar{v}_2 t_2}{t} = \left(\frac{t_1}{t}\right)\bar{v}_1 + \left(\frac{t_2}{t}\right)\bar{v}_2 = \frac{15 \text{ min}}{40 \text{ min}}(20 \text{ mi/h}) + \frac{25 \text{ min}}{40 \text{ min}}(48 \text{ mi/h}) = 37.5 \text{ mi/h}$$

47.   **INTERPRET**   This is a one-dimensional problem involving a number of evenly divided time intervals. The key concept here is the average velocity.

DEVELOP   The average velocity in a given time interval can be found by using Equation 2.1.

EVALUATE (a) We first note that each time interval consists of 2 h of driving at a velocity of 105 km/h and a 30 min (0.5 h) stop. Therefore, the distance traveled is $x_1 = (105 \text{ km/h})(2 \text{ h}) = 210 \text{ km}$, and the average velocity is

$$\bar{v}_1 = \frac{x_1}{t_1} = \frac{210 \text{ km}}{2 \text{ h} + 0.5 \text{ h}} = 84 \text{ km/h}$$

Since each 2.5 h interval covers the same distance of 210 km, we conclude that $\bar{v}_1 = 84$ km/h is also the average velocity of the entire trip. (But see **ASSESS** below)

(b) The amount of time required for this coast-to-coast trip is

$$t = \frac{x}{\bar{v}_1} = \frac{4600 \text{ km}}{84 \text{ km/h}} = 54.8 \text{ h}$$

ASSESS The values obtained above are only approximate because the exact travel time does not include a 30-min stop after the final segment. To find the total time, note that every 2.5 h you would cover a distance $x_1 = (105 \text{ km/h})(2 \text{ h}) = 210 \text{ km}$, so it would take you $21 \times 2.5 \text{ h} = 52.5 \text{ h}$ to travel $21 \times 210 \text{ km} = 4410 \text{ km}$. You could drive the remaining $4600 \text{ km} - 4410 \text{ km} = 190 \text{ km}$ in $\Delta t = 190 \text{ km}/(105 \text{ km/h}) = 1.81 \text{ h}$. Therefore, the amount of time spent for the entire trip would be $t = 52.5 \text{ h} + 1.81 \text{ h} = 54.3 \text{ h}$, which, as expected, is half an hour (or 30 min) less than that found in (b). The average velocity in this case would be

$$\bar{v} = \frac{x}{t} = \frac{4600 \text{ km}}{54.3 \text{ h}} = 84.7 \text{ km/h}$$

which is slightly higher than that obtained in (a).

49. **INTERPRET** This is a one-dimensional kinematics problem involving two jetliners taking off from the endpoints of a straight path. The key concept here is the average speed.

DEVELOP Given the average speed, the distance traveled during a time interval can be calculated using Equation 2.1: $\Delta x = \bar{v}\Delta t$. An important point here is to recognize that at the instant the planes pass each other, the sum of the total distance traveled by both is $\Delta x = 4600 \text{ km}$.

EVALUATE Suppose the two planes pass each other after a time Δt from take-off. We then have

$$\Delta x = \Delta x_1 + \Delta x_2 = \bar{v}_1 \Delta t + \bar{v}_2 \Delta t = (\bar{v}_1 + \bar{v}_2)\Delta t$$

which yields

$$\Delta t = \frac{\Delta x}{\bar{v}_1 + \bar{v}_2} = \frac{4600 \text{ km}}{1100 \text{ km/h} + 700 \text{ km/h}} = 2.56 \text{ h}$$

Thus, the encounter occurs at a point about $\Delta x_1 = \bar{v}_1 \Delta t = (1100 \text{ km/h})(2.56 \text{ h}) = 2811 \text{ km}$ from San Francisco, or $\Delta x_2 = \bar{v}_2 \Delta t = (700 \text{ km/h})(2.56 \text{ h}) = 1789 \text{ km}$ from New York.

ASSESS The point of encounter is closer to New York than San Francisco. This makes sense because the plane that leaves from New York travels at a lower speed.

51. **INTERPRET** This is a one-dimensional kinematics problem involving finding the instantaneous velocity as a function of time, given position as a function of time.

DEVELOP The instantaneous velocity $v(t)$ can be obtained by taking the derivative of $x(t)$. The derivative of a function of the form bt^n can be obtained by using Equation 2.3.

EVALUATE The instantaneous velocity is $v(t) = \frac{dx}{dt} = \frac{d}{dt}(bt^4) = 4bt^3$. On the other hand, the average velocity over the time interval from $t = 0$ to any time t can be computed by using Equation 2.1:

$$\bar{v} = \frac{\Delta x}{\Delta t} = \frac{x(t) - x(0)}{t - 0} = \frac{bt^4}{t} = bt^3$$

which is just $\frac{1}{4}$ of $v(t)$ from above.

ASSESS We note that $\bar{v}$ is not equal to the average of $v(0)$ and $v(t)$, as stated in Equation 2.8. The equation is applicable only when acceleration is constant, which is clearly not the case here.

53. **INTERPRET** This is a one-dimensional kinematics problem involving finding the instantaneous velocity, instantaneous acceleration, average velocity, and average acceleration as a function of time, given position as a function of time.

DEVELOP The instantaneous velocity $v(t)$ can be obtained by taking the derivative of $x(t)$, and by differentiating $v(t)$ with respect to t, we obtain the instantaneous acceleration $a(t)$. The derivative of a function of the form bt^n can be obtained by using Equation 2.3.

EVALUATE **(a)** Taking the first derivative of $x(t)$, the instantaneous velocity is

$$v(t) = \frac{dx}{dt} = \frac{d}{dt}(bt^3) = 3bt^2 \qquad v(2.5\text{ s}) = 3(1.5\text{ m/s}^3)(2.5\text{ s})^2 = 28.1\text{ m/s}$$

(b) Taking the first derivative of $v(t)$, or the second derivative of $x(t)$, the instantaneous acceleration is

$$a(t) = \frac{dv}{dt} = \frac{d}{dt}(3bt^2) = 6bt \qquad a(2.5\text{ s}) = 6(1.5\text{ m/s}^3)(2.5\text{ s}) = 22.5\text{ m/s}^2$$

(c) The average velocity during the first 2.5 s is

$$\bar{v} = \frac{\Delta x}{\Delta t} = \frac{x(2.5\text{ s}) - x(0)}{2.5\text{ s}} = \frac{(1.5\text{ m/s}^3)(2.5\text{ s})^3 - 0}{2.5\text{ s}} = 9.38\text{ m/s}$$

(d) Similarly, the average acceleration during the first 2.5 s is

$$\bar{a} = \frac{\Delta v}{\Delta t} = \frac{v(2.5\text{ s}) - v(0)}{2.5\text{ s}} = \frac{3(1.5\text{ m/s}^3)(2.5\text{ s})^2 - 0}{2.5\text{ s}} = 11.25\text{ m/s}^2$$

ASSESS We expect the acceleration in this problem not to be a constant, but to vary linearly in t. In a situation where the acceleration a is a constant, the displacement is quadratic function of t.

55. **INTERPRET** We interpret this as a one-dimensional problem with a car undergoing constant deceleration. The key concepts here are stopping distance and stopping time.

DEVELOP The stopping distance and the stopping time are related by Equation 2.9, for motion with constant deceleration.

EVALUATE Let v_0 be the initial velocity and $v = 0$ be the final velocity. Equation 2.9 can then be rewritten as

$$x - x_0 = \frac{1}{2}(v_0 + v)t = \frac{1}{2}v_0 t$$

Thus, we see that the stopping distance, $x - x_0$, is proportional to the stopping time, t, and both are reduced by the same amount (55%).

ASSESS Anti-lock brakes optimize the deceleration by controlling the wheels to roll just at the point of skidding. The shorter the stopping distance, the lesser the stopping time.

57. **INTERPRET** We interpret this as a one-dimensional kinematics problem with the hockey puck being the object of interest.

DEVELOP We assume the hockey puck to undergo *constant* deceleration while moving through snow. Equation 2.9, $x = x_0 + (v_0 + v)t/2$, provides the connection between the initial velocity $v_0 = 32$ m/s, the final velocity $v = 18$ m/s, the travel time t, and the distance traveled $x = 35$ cm $= 0.35$ m. For part **(b)**, to find the minimum thickness of the snow needed to stop the hockey puck entirely, we first find the acceleration a and then solve Equation 2.11, $v^2 = v_0^2 + 2a(x - x_0)$, by setting the final velocity to zero.

EVALUATE **(a)** Using Equation 2.9, the time the hockey puck spends traversing 35 cm of snow is

$$t = \frac{2(x - x_0)}{v_0 + v} = \frac{2(0.35\text{ m})}{32\text{ m/s} + 18\text{ m/s}} = 0.014\text{ s}$$

(b) Using Equation 2.7, the acceleration is

$$a = \frac{v - v_0}{t} = \frac{18\text{ m/s} - 32\text{ m/s}}{0.014\text{ s}} = -1000\text{ m/s}^2$$

The negative sign means that the puck is decelerating. Now, substituting the value into Equation 2.11 with $v = 0$, the minimum thickness of the snow wall is

$$(x - x_0) = \frac{v^2 - v_0^2}{2a} = \frac{v_0^2}{2(-a)} = \frac{(32 \text{ m/s})^2}{2(1000 \text{ m/s}^2)} = 0.512 \text{ m}$$

Thus, any wall of snow thicker than 0.512 m would stop the hockey puck.

ASSESS   We find the minimum thickness to be proportional to v_0^2 and inversely proportional to the deceleration $-a$. This agrees with our intuition: The greater the speed of the puck, the thicker the snow needed to bring it to a stop; similarly, less snow would be needed with increasing deceleration.

59.   **INTERPRET**   We interpret this as a one-dimensional kinematics problem with constant deceleration. The jetliner is the object of interest.

DEVELOP   Equation 2.9, $x = x_0 + (v_0 + v)t/2$, relates distance, initial velocity, and final velocity. The equation can be used to solve for the shortest runway.

EVALUATE   With $t = 29 \text{ s} = (29/3600)$ h, the final velocity v set to zero, Equation 2.9 gives

$$x - x_0 = \frac{(v_0 + v)t}{2} = \frac{(220 \text{ km/h} + 0)(29 \text{ h}/3600)}{2} = 886 \text{ m}$$

ASSESS   The length is a bit short compared to the typical minimum landing runway length of about 1.5 km for full-size jetliners.

61.   **INTERPRET**   This is a one-dimensional kinematics problem involving constant acceleration. The object of interest is the racing car moving along a straight path. The key concepts here are the average speed and distance.

DEVELOP   Equation 2.9, $x = x_0 + (v_0 + v)t/2$, relates distance, initial speed, and final speed. This is the equation we'll use to solve for the initial speed. To solve part (**b**), we first find the acceleration and then make use of Equation 2.11, $v^2 = v_0^2 + 2a(x - x_0)$.

EVALUATE   (**a**) Solving Equation 2.9, we find the initial speed to be

$$v_0 = \frac{2(x - x_0)}{t} - v = \frac{2(140 \text{ m})}{3.6 \text{ s}} - 53 \text{ m/s} = 24.8 \text{ m/s}$$

(**b**) From Equation 2.7, we find the acceleration to be

$$a = \frac{v - v_0}{t} = \frac{53 \text{ m/s} - 24.8 \text{ m/s}}{3.6 \text{ s}} = 7.84 \text{ m/s}^2$$

Upon substituting the result into Equation 2.11, the distance traveled starting from rest ($v_0 = 0$) while reaching a velocity $v = 53$ m/s is

$$x - x_0 = \frac{v^2 - v_0^2}{2a} = \frac{(53 \text{ m/s})^2 - 0}{2(7.84 \text{ m/s}^2)} = 179 \text{ m}$$

ASSESS   Comparing parts (**a**) and (**b**), the car travels a distance of 179 m *from rest* to the end of the 140-m distance mark. Using Equation 2.11, we can show that the additional 39 m (=179 m – 140 m) is the distance traveled to bring the car from rest to an initial speed of $v_0 = 24.8$ m/s:

$$\Delta x = \frac{v_0^2}{2a} = \frac{(24.8 \text{ m/s})^2}{2(7.84 \text{ m/s}^2)} = 39 \text{ m}$$

63.   **INTERPRET**   We interpret this as *two* problems involving one-dimensional kinematics with constant acceleration. The two objects of interest are you and the leader in a race.

DEVELOP   While the leader (runner B) maintains a constant speed, you (runner A) are trying to catch up to him with a constant acceleration. Position as a function of time is given by Equation 2.10, $x = x_0 + v_0 t + at^2/2$. We'll write down two equations, one for x_A and one for x_B. The condition that both runners finish simultaneously may be expressed as $x_A = x_B$.

EVALUATE Taking $x_0 = 0$ and $t = 0$ at the 9 km point (and assuming a straight path to the finish), we can express your position (runner A) and that of the leader (runner B) as

$$x_A = v_0 t + \frac{1}{2} a t^2$$

$$x_B = x_{B0} + v_0 t = 100 \text{ m} + v_0 t$$

Since runner B's speed is constant $v_0 = \frac{\Delta x}{\Delta t} = \frac{(9 \text{ km} + 100 \text{ m})}{35 \text{ min}} = \frac{9.1 \text{ km}}{35 \text{ min}} = 0.26$ km/min, for the remaining 0.9 km, it would take him $t = 0.9$ km/(0.26 km/min) $= 3.46$ min to finish. The condition that both runners finish simultaneously would be

$$x_A = v_0 t + \frac{1}{2} a t^2 \rightarrow 1 \text{ km} = (0.26 \text{ km/min})(3.46 \text{ min}) + \frac{1}{2} a (3.46 \text{ min})^2$$

which gives $a = 0.017$ km/min$^2 = 4.64 \times 10^{-3}$ m/s^2.

ASSESS In order for runner A to catch up to B, he must run faster than the speed he was running initially. At the instant A crosses the finish line, his speed would be $v_A = v_0 + at = (0.26 \text{ km/min}) + (0.017 \text{ km/min}^2)$ $(3.46 \text{ min}) = 0.32$ km/min. The longer he trails behind B, the greater his acceleration must be to catch up.

65. **INTERPRET** We interpret this as a one-dimensional kinematics problem. The object of interest is the Mars rover Spirit. The key concepts here are impact speed and vertical distance.

DEVELOP Equation 2.11, $v^2 = v_0^2 + 2a(y - y_0)$, can be used to describe the vertical motion of the Mars rover Spirit. After rebounding with a vertical impact speed v_0 from the surface, the spacecraft attains a maximum height when $v = 0$. Note that the gravitational acceleration of Mars is $g_{\text{Mars}} = 3.74$ m/s^2.

EVALUATE Solving Equation 2.11 with $a = -g_{\text{Mars}} = -3.74$ m/s^2, the impact speed is found to be

$$v_0 = \sqrt{v^2 - 2a(y - y_0)} = \sqrt{2 g_{\text{Mars}} (y - y_0)} = \sqrt{2(3.74 \text{ m/s}^2)(15 \text{ m})} = 10.59 \text{ m/s}$$

ASSESS We find the impact speed to be proportional to $(y - y_0)^{1/2}$, the square-root of the rebound height. This agrees with our expectation that the greater the impact speed, the higher the rover will rebound.

67. **INTERPRET** This is a one-dimensional kinematics problem that involves finding the vertical distance of an object as a function of time.

DEVELOP Using Equation 2.10, the vertical position of the object falling from y_0 as a function of time may be written as $(v_0 = 0)$

$$y(t) = y_0 + v_0 t + \frac{1}{2} a t^2 = y_0 - \frac{1}{2} g t^2$$

Note that the acceleration is $a = -g$, which points downward.

EVALUATE The vertical displacement (*negative* for downward) during the last second (an interval from $t - 1$ s to t) is

$$y(t) - y(t - 1) = \left(y_0 - \frac{1}{2} g t^2 \right) - \left(y_0 - \frac{1}{2} g(t-1)^2 \right) = \frac{1}{2} g(t-1)^2 - \frac{1}{2} g t^2 = \frac{1}{2} g(1 - 2t)$$

On the other hand, the vertical displacement from $t = 0$ to t is $y(t) - y_0 = -\frac{1}{2} g t^2$. From the problem statement, the former is one-fourth of the latter:

$$\frac{1}{4} = \frac{g(1 - 2t)/2}{-g t^2/2} = \frac{2t - 1}{t^2} \quad \rightarrow \quad t^2 - 8t + 4 = 0$$

Solving the quadratic equation for t, we obtain $t = (4 \pm 2\sqrt{3})$ s. (We discarded the negative square root because t is obviously greater than 1 s.) Substituting this value of $t = (4 + 2\sqrt{3})$ s $= 7.46$ s into the equation above, we find

$$y(t) - y_0 = -\frac{1}{2} g t^2 = -\frac{1}{2}(9.8 \text{ m/s}^2)(7.46 \text{ s})^2 = -273 \text{ m}$$

Since $y(t = 7.46 \text{ s}) = 0$, we conclude that the object must be dropped from a height of $y_0 = 273$ m.

ASSESS During a free fall, the vertical distance traveled is proportional to t^2. Therefore, we expect the object to travel a greater distance during the latter time interval. In general, we must also take into consideration air resistance.

69. **INTERPRET**   We interpret this as *two* problems involving one-dimensional kinematics with constant acceleration due to gravity. The objects of interest are the two divers.

DEVELOP   Let A be the diver who jumps upward with 1.80 m/s, and B be the one who steps off the platform. The velocity of diver A as he passes B on his way down is $v_0 = -1.8$ m/s. Using Equation 2.11, the positions and the speeds of the divers may be written as:

$$v_A^2 = v_0^2 - 2g(y - y_0)$$
$$v_B^2 = -2g(y - y_0)$$

Note that the acceleration is $a = -g$, which points downward. For part (**b**), we may use Equation 2.10 to write down the vertical position of the divers as a function of time.

EVALUATE   (**a**) At the water's surface, $y = 0$, the speeds of the divers are

$$v_A = -\sqrt{v_0^2 - 2g(y - y_0)} = \sqrt{(-1.8 \text{ m/s})^2 - 2(9.8 \text{ m/s}^2)(0 - 3 \text{ m})} = -7.88 \text{ m/s}$$
$$v_B = -\sqrt{-2g(y - y_0)} = \sqrt{-2(9.8 \text{ m/s}^2)(0 - 3 \text{ m})} = -7.67 \text{ m/s}$$

Note that we have chosen the negative square roots for v_A and v_B since the divers are moving *downward*.

(**b**) The vertical position of the divers as a function of time may be written as

$$y_A(t) = y_0 + v_0 t + \frac{1}{2}at^2 = (3.00 \text{ m}) + (-1.8 \text{ m/s})t - \frac{1}{2}(9.8 \text{ m/s}^2)t^2$$

$$y_B(t) = y_0 + \frac{1}{2}at^2 = (3.00 \text{ m}) - \frac{1}{2}(9.8 \text{ m/s}^2)t^2$$

The divers hit the water when $y(t) = 0$. Solving the equations above, we find $t_A = 0.62$ s and $t_B = \sqrt{2(3.00 \text{ m})/(9.8 \text{ m/s}^2)} = 0.782$ s. Therefore, we see that diver A hits about $\Delta t = t_B - t_A = 0.782 \text{ s} - 0.62 \text{ s} = 0.162$ s before diver B.

ASSESS   We expect diver A to spend less time in the air and hit the water first because he has a non-zero downward velocity compared to B.

71. **INTERPRET**   The object of interest is the spacecraft, which undergoes free fall under the influence of gravitational acceleration of the Moon.

DEVELOP   Using Equation 2.10, the vertical position of the spacecraft falling from y_0 as a function of time may be written as ($v_0 = 0$)

$$y(t) = y_0 + v_0 t + \frac{1}{2}at^2 = y_0 - \frac{1}{2}g_{\text{Moon}}t^2$$

The gravitational acceleration of the Moon is $g_{\text{Moon}} = 1.62 \text{ m/s}^2$, about 1/6 of that of the Earth.

EVALUATE   With $a = -g_{\text{Moon}} = -1.62 \text{ m/s}^2$, the amount of time it takes to drop 12 m from rest is

$$t = \sqrt{\frac{2(y_0 - y)}{g_{\text{Moon}}}} = \sqrt{\frac{2(12 \text{ m})}{1.62 \text{ m/s}^2}} = 3.85 \text{ s}$$

The velocity at impact is $v = -g_{\text{Moon}}t = -(1.62 \text{ m/s}^2)(3.85 \text{ s}) = -6.24$ m/s. The impact speed is therefore equal to $|v| = 6.24$ m/s.

ASSESS   Our result indicates that t is proportional to $g^{-1/2}$. Therefore, the greater the gravitational acceleration, the less time it takes for the free fall.

73. **INTERPRET**   We interpret this as *two* problems involving one-dimensional kinematics with constant acceleration. The two objects of interest are the two trains.

DEVELOP   Let the fast train be A and the slow train be B. While B maintains a constant speed, A tries to slow down to avoid collision with a constant deceleration. We take the origin $x = 0$ and $t = 0$ at the point where A begins decelerating, with positive x in the direction of motion. Position as a function of time is given by Equation 2.10, $x = x_0 + v_0 t + at^2/2$. We shall write down two equations, one for x_A and one for x_B. The condition that both trains collide may be expressed as $x_A = x_B$.

EVALUATE We first rewrite the initial speeds of the trains as

$$v_{0A} = 80 \text{ km/h} = \left(80 \, \frac{\text{km}}{\text{h}}\right)\left(\frac{1000 \text{ m}}{1 \text{ km}}\right)\left(\frac{1 \text{ h}}{3600 \text{ s}}\right) = 22.22 \text{ m/s}$$

$$v_{0B} = 25 \text{ km/h} = \left(25 \, \frac{\text{km}}{\text{h}}\right)\left(\frac{1000 \text{ m}}{1 \text{ km}}\right)\left(\frac{1 \text{ h}}{3600 \text{ s}}\right) = 6.94 \text{ m/s}$$

We can express the positions of trains A and B as

$$x_A = v_{0A}t + \frac{1}{2}at^2 = (22.22 \text{ m/s})t + \frac{1}{2}(-2.1 \text{ m/s}^2)t^2$$

$$x_B = x_{B0} + v_{0B}t = 50 \text{ m} + (6.94 \text{ m/s})t$$

When the trains collide, $x_A = x_B$. The above equations then give

$$\frac{1}{2}at^2 + (v_{0A} - v_{0B})t - x_{B0} = 0 \quad \rightarrow \quad (-1.05 \text{ m/s}^2)t^2 + (15.28 \text{ m/s})t - (50 \text{ m}) = 0$$

Using the quadratic formula to solve for the smaller root, we find

$$t = \frac{(15.28 \text{ m/s}) - \sqrt{(15.28 \text{ m/s})^2 - 4(1.05 \text{ m/s}^2)(50 \text{ m})}}{2(1.05 \text{ m/s}^2)} = 4.97 \text{ s}$$

The velocity of train A at the time of the collision is

$$v_A = v_{A0} + a_1 t = (22.22 \text{ m/s}) - (2.1 \text{ m/s}^2)(4.97 \text{ s}) = 11.78 \text{ m/s}$$

Therefore, the relative speed of impact is

$$v_{rel} = v_A - v_{B0} = 11.78 \text{ m/s} - 6.94 \text{ m/s} = 4.84 \text{ m/s}$$

or 17.4 km/h.

ASSESS The initial relative speed is $v_{rel,0} = v_{A0} - v_{B0} = 22.22 \text{ m/s} - 6.94 \text{ m/s} = 15.28 \text{ m/s}$. Braking reduces the speed of train A, and the relative speed between A and B, but apparently the deceleration $a = -2.1 \text{ m/s}^2$ is not enough to prevent collision.

75. **INTERPRET** This is a one-dimensional kinematics problem involving two travel segments. The key concept here is the average speed.

DEVELOP The average speed is the total distance divided by the total time, or $\bar{v} = \Delta x / \Delta t$. In either case, we shall find the total distance traveled and the time taken.

EVALUATE (a) Let the distances traveled during the two time intervals be L_1 and L_2. The total distance is the sum of the distances covered at each speed:

$$L = L_1 + L_2 = v_1(t/2) + v_2(t/2) = \frac{1}{2}(v_1 + v_2)t$$

Thus, the average speed is $\bar{v} = L/t = \frac{1}{2}(v_1 + v_2)$.

(b) In this case, let t_1 and t_2 be the two time intervals. The total time is the sum of the times traveled at each speed:

$$t = t_1 + t_2 = \frac{L/2}{v_1} + \frac{L/2}{v_1} = \frac{L}{2}\frac{v_1 + v_2}{v_1 v_2}$$

Therefore, the average speed is $\bar{v} = \frac{L}{t} = \frac{2v_1 v_2}{v_1 + v_2}$.

ASSESS The average speed $\bar{v}$ is the time-weighted average of the separate speeds: $\bar{v} = (t_1/t)v_1 + (t_2/t)v_2$. With this in mind, the result in part (a) may be rewritten as

$$\bar{v} = (1/2)v_1 + (1/2)v_2$$

and for part (b),

$$\bar{v} = \left(\frac{t_1}{t}\right)v_1 + \left(\frac{t_1}{t}\right)v_2 = \left(\frac{v_2}{v_1 + v_2}\right)v_1 + \left(\frac{v_1}{v_1 + v_2}\right)v_2 = \frac{2v_1 v_2}{v_1 + v_2}$$

77. **INTERPRET**   We interpret this as a one-dimensional kinematics problem that involves finding the vertical position of a person as a function of time.

    DEVELOP   Using Equation 2.10, the vertical position of a person as a function of time may be written as (setting $y_0 = 0$)

    $$y(t) = y_0 + v_0 t - \frac{1}{2}gt^2 \quad \rightarrow \quad \frac{1}{2}gt^2 - v_0 t + y = 0$$

    Note that the acceleration is $a = -g$, which points downward. The quadratic formula gives two times when the leaper passes a particular height:

    $$t_{\pm} = \frac{v_0 \pm \sqrt{v_0^2 - 2gy}}{g}$$

    The smaller value, t_-, corresponds to the time for going up and the larger, t_+, for going down. Therefore, the time spent above that height is just

    $$\Delta t(y) = t_+ - t_- = \frac{v_0 + \sqrt{v_0^2 - 2gy}}{g} - \frac{v_0 - \sqrt{v_0^2 - 2gy}}{g} = \frac{2\sqrt{v_0^2 - 2gy}}{g}$$

    Using Equation 2.11, $v^2 = v_0^2 + 2ay$, we find that in order to reach a maximum height h, the initial velocity must be $v_0 = \sqrt{2gh}$. The above expression for $\Delta t(y)$ may be simplified as

    $$\Delta t(y) = \frac{2\sqrt{2g(h-y)}}{g}$$

    EVALUATE   The total time spent in the air is the time spent above the ground. Setting $y = 0$, we have

    $$\Delta t(0) = \frac{2\sqrt{2gh}}{g} = 2\sqrt{\frac{2h}{g}}$$

    Similarly, the time spent in the upper half, above $y = h/2$, is

    $$\Delta t(h/2) = \frac{2\sqrt{2g(h/2)}}{g} = 2\sqrt{\frac{h}{g}}$$

    Therefore,

    $$\frac{\Delta t(h/2)}{\Delta t(0)} = \frac{2\sqrt{h/g}}{2\sqrt{2h/g}} = \frac{1}{\sqrt{2}} = 0.707$$

    or 70.7%.

    ASSESS   Our result indicates that while in the air, a person spends 70.7% of the time on the upper half of the height. Such a large fraction of time is what gives the illusion of "hanging" almost motionless near the top of the leap.

79. **INTERPRET**   This is a one-dimensional kinematics problem involving constant deceleration. The object of interest is the motorist moving along a straight path. The key concepts here are the average speed and the distance traveled.

    DEVELOP   The speed of radar waves $(3 \times 10^5 \text{ km/s})$ is so great compared to the speed of a motor vehicle, we can neglect any motion of the car during the travel times of the radar signals. Equation 2.11, $v^2 = v_0^2 + 2a(x - x_0)$ relates the distance traveled to the initial speed, the final speed, and the deceleration. This is the equation we shall use to solve for the deceleration.

    EVALUATE   The motorist has 0.9 km to slow down from $v_0 = 110$ km/h to $v = 70$ km/h. This requires a constant acceleration of

    $$a = \frac{v^2 - v_0^2}{2(x - x_0)} = \frac{(70 \text{ km/h})^2 - (110 \text{ km/h})^2}{2(0.9 \text{ km})} = -4000 \text{ km/h}^2 = -1.11 \text{ km/h/s} = -0.309 \text{ m/s}^2$$

    Thus, the deceleration must be at least $|a| = 0.309$ m/s^2 to avoid getting a ticket.

    ASSESS   The result means that the speed must be decreased by 1.11 km/h in each second. So, in 36 seconds, the speed is decreased from 110 km/h to 70 km/h.

81.   **INTERPRET**   We're asked to find equations for instantaneous velocity and position for a *non-constant* acceleration, given an equation for acceleration.

DEVELOP   We can't use the constant-acceleration equations, obviously, but we can still use our definitions for velocity ($v \equiv \frac{dx}{dt}$) and acceleration ($a \equiv \frac{dv}{dt}$) and work backward to get the equations we need. For example:

$$a(t) = \frac{dv}{dt} \rightarrow a(t)dt = dv \rightarrow \int dv = \int a(t)dt \rightarrow v(t) = \int a(t)dt$$

$a = a_o + bt$. The initial position (at $t = 0$) is x_o, and the initial velocity is v_o.

EVALUATE $v(t) = \int a dt = \int (a_o + bt)dt = a_o t + \frac{1}{2}bt^2 + C_1$. To find the constant C_1, note that $v(t = 0) = C_1$. We are told in the problem statement that the initial velocity is v_o, so $C_1 = v_o$ and $v(t) = v_o + a_o t + \frac{1}{2}bt^2$.

For position, use the same procedure: $x(t) = \int v(t)dt = \int (v_o + a_o t + \frac{1}{2}bt^2)dt = v_o t + \frac{1}{2}a_o t^2 + \frac{1}{6}bt^3 + C_2$. The position at $t = 0$ is x_o, so $C_2 = x_o$ and $x(t) = x_o + v_o t + \frac{1}{2}a_o t^2 + \frac{1}{6}bt^3$.

ASSESS   Note that the derivative of $a(t)$ for this problem is a constant. The derivative of acceleration is called *jerk*, so what we've just derived in these equations are the equations for constant-jerk motion.

83.   **INTERPRET**   This problem, like Example 2.6, involves constant acceleration of a ball due to gravity. We want to find the final speed and the time.

DEVELOP   The ball in Example 2.6 starts at a height of 1.5 meters, with an initial upward speed of 7.3 m/s. The second ball starts at the same height with the same speed, but downward. We're asked to find the speed of both balls just before they hit the floor and the time that the second ball hits. We can use the constant-acceleration equations, since the only acceleration is due to gravity. Start with $v^2 = v_o^2 + 2a(x - x_o)$ to find the final velocities, then use $x = x_o + v_o t + \frac{1}{2}at^2$ to find the time.

EVALUATE   **(a)** $v^2 = v_o^2 + 2a(x - x_o) = (7.3 \text{ m/s})^2 + 2(-9.8 \text{ m/s}^2)(0 \text{ m} - 1.5 \text{ m}) = 9.1 \text{ m/s}$

(b) $v^2 = v_o^2 + 2a(x - x_o) = (-7.3 \text{ m/s})^2 + 2(-9.8 \text{ m/s}^2)(0 \text{ m} - 1.5 \text{ m}) = 9.1 \text{ m/s}$

(c) $x = x_o + v_o t + \frac{1}{2}at^2 \rightarrow 0 = x_o + v_o t - \frac{1}{2}gt^2 \rightarrow t = \frac{-v_o \pm \sqrt{v_o^2 + 2x_o g}}{g}$

$$t = \frac{7.3 \text{ m/s} \pm \sqrt{(7.3 \text{ m/s})^2 + 2(1.5 \text{ m})(9.8 \text{ m/s}^2)}}{-9.8 \text{ m/s}^2} = \{-1.7 \text{ s}, 0.18 \text{ s}\}$$

Take the positive solution: The ball hits in 0.18 s.

ASSESS   Note that the answers to parts **(a)** and **(b)** are the same. This makes sense, since in the example problem the speed of the ball when it comes back down to hand-level is the same as the initial speed of the ball in part **(b)**.

85.   **INTERPRET**   We can use constant acceleration on this problem, again, since the only acceleration is that of gravity. Our goal is to find the number of drops per second.

DEVELOP   There are exactly three drops falling at any time: two halfway down and one either hitting or just leaving. So if we find the time it takes one drop to fall, and divide that time by three, we have the time between drops. We can use the constant-acceleration motion equations: the most useful one in this case would be $x = x_o + v_o t + \frac{1}{2}at^2$, with $x = 0$, $x_o = 19.6 \text{ cm} = 0.196 \text{ m}$, $v_o = 0$, and $a = -g$. The question asks for drops per second, so convert seconds per drop to drops per second for the final answer.

EVALUATE   Find the time it takes one drop to fall:

$$x = x_o + v_o t + \frac{1}{2}at^2 \rightarrow 0 = x_o - \frac{1}{2}gt^2 \rightarrow t = \sqrt{\frac{2x_o}{g}} = 0.2 \text{ s}$$

There are three drops in that time, so the time between drops is $\frac{0.2 \text{ s}}{3 \text{ drops}} = 0.067$ s/drop. The drops per second is the reciprocal of the time between drops: $\frac{3 \text{ drops}}{0.2 \text{ s}} = 15$ drops/s.

ASSESS   This is pretty fast for a leaky faucet, but the time looks about right for the distance involved.

87.   **INTERPRET**   We want to find time, given distance and speed. Make sure your units are correct!

DEVELOP   Walking speed is 3 mph, cycling speed is $25 \text{ km/h} \times \frac{1 \text{ mile}}{1.61 \text{ km}} = 15.53 \text{ mph}$, and driving speed is 55 mph. The distance is 2800 miles, so find the time for each mode of transportation. Use $t = \frac{\text{distance}}{\text{speed}}$.

EVALUATE   Walking: $t = \frac{2800 \text{ miles}}{3 \text{ mph}} = 933$ hours ≈ 39 days, or about five-and-a-half weeks. Cycling, it would take about $t = \frac{2800 \text{ miles}}{15.53 \text{ mph}} = 180 \text{ h} \approx 7.5$ days. Driving, $t = \frac{2800 \text{ miles}}{55 \text{ mph}} = 51$ hours ≈ 2.1 days.

ASSESS   The record for the cycling RAAM (Race Across America) was set in 1992 by Rob Kish with a time of 8 days, 3 hours, and 11 minutes. That's a bit slower than our 7.5 day estimate, but we did not allow for time spent eating or sleeping in our calculation!

MOTION IN TWO AND THREE DIMENSIONS

EXERCISES

Section 3.1 Vectors

17.   **INTERPRET**   We interpret this as a problem involving finding the magnitude and direction of a (displacement) vector in two dimensions.

   DEVELOP   In two dimensions, a displacement vector can generally be written as, in unit vector notation, $\Delta \vec{r} = \Delta r_x \hat{i} + \Delta r_y \hat{j}$, where Δr_x and Δr_y are the x- and y-components of the displacements, respectively. The magnitude of $\Delta \vec{r}$ is $\Delta r = \sqrt{(\Delta r_x)^2 + (\Delta r_y)^2}$, and the angle $\Delta \vec{r}$ makes with the $+x$ axis is

$$\theta = \tan^{-1}\left(\frac{\Delta r_y}{\Delta r_x}\right)$$

   We choose $+x$ direction to correspond to east and $+y$ for north.

   EVALUATE   With the coordinate system established above, the components of the displacements are $\Delta r_x = -220$ m (220 m, $-x$) and $\Delta r_y = +150$ m (150 m, $+y$). Therefore, the magnitude of the displacement $\Delta \vec{r}$ is

$$\Delta r = \sqrt{(\Delta r_x)^2 + (\Delta r_y)^2} = \sqrt{(-220 \text{ m})^2 + (150 \text{ m})^2} = 266 \text{ m}$$

   The direction of $\Delta \vec{r}$ is

$$\theta = \tan^{-1}\left(\frac{\Delta r_y}{\Delta r_x}\right) = \tan^{-1}\left(\frac{150 \text{ m}}{-220 \text{ m}}\right) = 145.7°$$

   ASSESS   The displacement vector $\Delta \vec{r}$ lies in the second quadrant. It makes an angle of 145.7° with the $+x$ axis. Alternatively, the direction of $\Delta \vec{r}$ can be specified as 34.3° N of W, or 55.7° W of N, or by the azimuth 304.3° (CW from N), etc.

19.   **INTERPRET**   We interpret this as a problem involving the addition of two displacement vectors in two dimensions and finding the magnitude and direction of the resultant vector. The object of interest is the migrating whale.

   DEVELOP   Using Equation 3.1, we see that in two dimensions, a vector $\vec{A}$ can be written as, in unit vector notation,

$$\vec{A} = A_x \hat{i} + A_y \hat{j} = A(\cos\theta_A \hat{i} + \sin\theta_A \hat{j})$$

   where $A = \sqrt{A_x^2 + A_y^2}$ and $\theta_A = \tan^{-1}(A_y / A_x)$. Similarly, we express a second vector $\vec{B}$ as $\vec{B} = B_x \hat{i} + B_y \hat{j} = B(\cos\theta_B \hat{i} + \sin\theta_B \hat{j})$. The resultant vector $\vec{C}$ is

$$\vec{C} = \vec{A} + \vec{B} = (A_x + B_x)\hat{i} + (A_y + B_y)\hat{j} = (A\cos\theta_A + B\cos\theta_B)\hat{i} + (A\sin\theta_A + B\sin\theta_B)\hat{j} = C_x \hat{i} + C_y \hat{j}$$

   EVALUATE   From the problem statement, the first segment of the travel can be written in unit-vector notation as (with $A = 360$ km and $\theta_A = 135°$)

$$\vec{A} = A(\cos\theta_A \hat{i} + \sin\theta_A \hat{j}) = (360 \text{ km})(\cos 135° \hat{i} + \sin 135° \hat{j}) = (-254.6 \text{ km})\hat{i} + (254.6 \text{ km})\hat{j}$$

   Similarly, the second segment of the travel can be expressed as (with $B = 400$ km and $\theta_B = 90°$)

$$\vec{B} = B(\cos\theta_B \hat{i} + \sin\theta_B \hat{j}) = (400 \text{ km})\hat{j}$$

Thus, the resultant displacement vector is

$$\vec{C} = \vec{A} + \vec{B} = C_x\hat{i} + C_y\hat{j} = (-254.6 \text{ km})\hat{i} + [(254.6 \text{ km}) + (400 \text{ km})]\hat{j} = (-254.6 \text{ km})\hat{i} + (654.6 \text{ km})\hat{j}$$

The magnitude of $\vec{C}$ is

$$C = \sqrt{C_x^2 + C_y^2} = \sqrt{(-254.6 \text{ km})^2 + (654.6 \text{ km})^2} = 702.4 \text{ km}$$

and its direction is

$$\theta = \tan^{-1}\left(\frac{C_y}{C_x}\right) = \tan^{-1}\left(\frac{654.6 \text{ km}}{-254.6 \text{ km}}\right) = -68.75°, \text{or } 111°$$

We choose the latter solution (111°) since the vector (with $C_x < 0$ and $C_y > 0$) lies in the second quadrant.

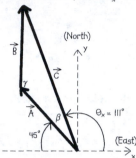

ASSESS As depicted in the figure, the resultant displacement vector $\vec{C}$ lies in the second quadrant. The direction of $\vec{C}$ can be specified as 111° CCW from the x-axis (east), or 45° + 23.7° = 68.7°N of W.

21. **INTERPRET** We interpret this as a problem involving the addition of two vectors in two dimensions and finding the magnitude and direction of the resultant vector.

DEVELOP Using Equation 3.1, we see that in two dimensions, a vector $\vec{A}$ can be written as, in unit vector notation, $\vec{A} = A_x\hat{i} + A_y\hat{j} = A(\cos\theta_A\hat{i} + \sin\theta_A\hat{j})$, where $A = \sqrt{A_x^2 + A_y^2}$ and $\theta_A = \tan^{-1}(A_y/A_x)$. Similarly, we express a second vector $\vec{B}$ as $\vec{B} = B_x\hat{i} + B_y\hat{j} = B(\cos\theta_B\hat{i} + \sin\theta_B\hat{j})$. To satisfy the condition, $\vec{A} + \vec{B} + \vec{C} = 0$, we simply set $\vec{C}$ to be

$$\vec{C} = -\vec{A} - \vec{B} = -(A_x + B_x)\hat{i} - (A_y + B_y)\hat{j} = C_x\hat{i} + C_y\hat{j}$$

EVALUATE Let +x-direction correspond to the right and +y correspond to the vertically upward direction. Then, $\vec{A} = (3.0 \text{ m})\hat{i}$, $\vec{B} = (4.0 \text{ m})\hat{j}$. Therefore,

$$\vec{C} = C_x\hat{i} + C_y\hat{j} = -(\vec{A} + \vec{B}) = (-3.0 \text{ m})\hat{i} + (-4.0 \text{ m})\hat{j}$$

The magnitude and direction of $\vec{C}$ are $C = \sqrt{C_x^2 + C_y^2} = \sqrt{(-3.0 \text{ m})^2 + (-4.0 \text{ m})^2} = 5.0 \text{ m}$, and

$$\theta = \tan^{-1}\left(\frac{C_y}{C_x}\right) = \tan^{-1}\left(\frac{-4 \text{ m}}{-3 \text{ m}}\right) = 53.1°, \text{ or } 233°$$

We choose the latter solution (233°, measured CCW from the +x-axis) since the vector (with $C_x < 0$ and $C_y < 0$) lies in the third quadrant. The angle of $\vec{C}$ could also be specified as −127°, measured CW from the x-axis.

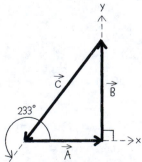

ASSESS The vectors $\vec{A}$, $\vec{B}$, and $\vec{C}$ form a 3-4-5 right triangle, as shown in the sketch. Therefore, $C = 5$ m, and the direction of $\vec{C}$, measured CCW from the direction of $\vec{A}$, is $\theta = 180° + \tan^{-1}(B/A) = 180° + 53.1° = 233°$.

23.   **INTERPRET**   We interpret this as a problem involving finding the magnitude and direction of a vector in two dimensions.

DEVELOP   Using Equation 3.1, we see that in two dimensions, a vector $\vec{A}$ can be written as, in unit vector notation, $\vec{A} = A_x \hat{i} + A_y \hat{j}$, where $A = \sqrt{A_x^2 + A_y^2}$ and $\theta = \tan^{-1}(A_y/A_x)$ is the angle $\vec{A}$ makes with the $+x$ axis.

EVALUATE   From the problem statement, we find that $A_x = 34$ m and $A_y = 13$ m. Therefore, the magnitude of $\vec{A}$ is

$$A = \sqrt{A_x^2 + A_y^2} = \sqrt{(34 \text{ m})^2 + (13 \text{ m})^2} = 36.4 \text{ m}$$

and the angle is $\theta = \tan^{-1}\left(\frac{A_y}{A_x}\right) = \tan^{-1}\left(\frac{13\,\text{m}}{34\,\text{m}} -\right) = 20.9°$ or $200.9°$. We choose the former solution ($20.9°$, measured CCW from the $+x$ axis) since the vector (with $A_x > 0$ and $A_y > 0$) lies in the first quadrant.

ASSESS   Since $A_y = 13$ m $< A_x = 34$ m (i.e., the x component of $\vec{A}$ is greater than the y component), we expect the angle to be less than $45°$. Our result indeed confirms this.

Section 3.2 Velocity and Acceleration Vectors

25.   **INTERPRET**   This problem gives us an initial velocity, time, and final velocity. We need to find average acceleration.

DEVELOP   Acceleration is $\vec{a} = \frac{d\vec{v}}{dt}$. We need the average acceleration, so use $\vec{a} = \frac{\Delta\vec{v}}{\Delta t}$. The change in velocity $\Delta\vec{v}$ is the difference in velocity before and after the rocket fires, so it will be helpful to put the velocities in coordinate form. We will also need to change all values to SI units.

Start by sketching a diagram showing the initial and final velocities, as shown in the figure later.

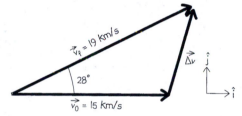

Break the vectors into components, and find the change in velocity $\Delta\vec{v}$. Divide this change by the time to obtain the average acceleration.

EVALUATE   First convert the speeds to units of m/s. $v_o = 15$ km/s $= 15{,}000$ m/s, $v_f = 19$ km/s $= 19{,}000$ m/s. The time is 10 minutes, or $t = 600$ s.

Next, express each vector in component form: $\vec{v}_o = (15{,}000 \text{ m/s})\hat{i}$,

$$\vec{v}_f = v_f \cos(28°)\hat{i} + v_f \sin(28°)\hat{j} = (16{,}800 \text{ m/s})\hat{i} + (8900 \text{ m/s})\hat{j}.$$

The change in velocity is the difference between the two:

$$\Delta\vec{v} = \vec{v}_f - \vec{v}_i = (16{,}800 \text{ m/s} - 15{,}000 \text{ m/s})\hat{i} + (8{,}900 \text{ m/s})\hat{j} = (1800 \text{ m/s})\hat{i} + (8{,}900 \text{ m/s})\hat{j}.$$

Now find acceleration: $\vec{a} = \dfrac{\Delta v}{t} = \dfrac{(1800 \text{ m/s})\hat{i} + (8900 \text{ m/s})\hat{j}}{600 \text{ s}} = (3.0 \text{ m/s}^2)\hat{i} + (15 \text{ m/s}^2)\hat{j}$

ASSESS   It's hard to estimate whether this acceleration is reasonable or not without knowing the mass of the asteroid. Large rocket engines accelerate the space shuttle at about 30 m/s^2, though, so it's probably about right.

27.   **INTERPRET**   We interpret this as a problem involving the addition of three displacements in two dimensions and finding the magnitude and direction of the resultant vector. The key concepts here are displacement and average velocity.

DEVELOP   Using Equation 3.1, we see that in two dimensions, a displacement vector $\Delta\vec{r}_1$ can be written as, in unit vector notation,

$$\Delta\vec{r}_1 = \Delta r_{1x}\hat{i} + \Delta r_{1y}\hat{j} = \Delta r_1(\cos\theta_1\hat{i} + \sin\theta_1\hat{j})$$

where $\Delta r_1 = \sqrt{(\Delta r_{1x})^2 + (\Delta r_{1y})^2}$ and $\theta_1 = \tan^{-1}(\Delta r_{1y}/\Delta r_{1x})$. One may write down a similar expression for $\Delta \vec{r}_2$ and $\Delta \vec{r}_3$. The displacement vector $\Delta \vec{r}_1$ is related to the velocity vector $\vec{v}_1$ by $\Delta \vec{r}_1 = \vec{v}_1 \Delta t$. We shall take a coordinate system with x axis east, y axis north, and origin at the starting point.

EVALUATE   (a) The first segment of the trip which last, for $\Delta t_1 = 10$ min $= (1/6)$ h, can be written, in unit-vector notation, as

$$\Delta \vec{r}_1 = \vec{v}_1 \Delta t_1 = (40 \text{ mi/h})\left(\frac{10}{60}\text{h}\right)\hat{j} = (6.67 \text{ mi})\hat{j}$$

Similarly, we have $\Delta \vec{r}_2 = (5.0 \text{ mi})\hat{i}$, and the time spent on this segment is
$\Delta t_2 = \Delta r_2/v_2 = (5.0 \text{ mi})/(60 \text{ mi/h}) = 0.083$ h $= 5$ min. Finally, the third segment has length
$\Delta r_3 = v_3 \Delta t_3 = (30 \text{ mi/h})(6\text{h}/60) = 3.0$ mi. A unit vector in the southwest direction is

$$\cos 225°\hat{i} + \sin 225°\hat{j} = -\frac{1}{\sqrt{2}}(\hat{i} + \hat{j})$$

Therefore, $\Delta \vec{r}_3 = -(3.0 \text{ mi})\frac{1}{\sqrt{2}}(\hat{i} + \hat{j}) = (-2.12 \text{ mi})\hat{i} + (-2.12 \text{ mi})\hat{j}$. These displacements and their sum are shown in the figure. The total displacement is

$$\Delta \vec{r}_{tot} = \Delta \vec{r}_1 + \Delta \vec{r}_2 + \Delta \vec{r}_3 = \vec{v}_1 \Delta t_1 = (6.67 \text{ mi})\hat{j} + (5.0 \text{ mi})\hat{i} + (-2.12 \text{ mi})\hat{i} + (-2.12 \text{ mi})\hat{j} = (2.88 \text{ mi})\hat{i} + (4.55 \text{ mi})\hat{j}$$

(b) Since the total travel time is $\Delta t = 10$ min $+ 5$ min $+ 6$min $= 21$ min $= (21/60)$ h, the average velocity for the trip is

$$\bar{\vec{v}} = \frac{\Delta \vec{r}_{tot}}{\Delta t} = \frac{(2.88 \text{ mi})\hat{i} + (4.55 \text{ mi})\hat{j}}{(21/60) \text{ h}} = (8.22 \text{ mi/h})\hat{i} + (13.0 \text{ mi/h})\hat{j}$$

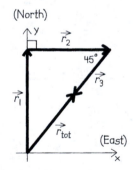

(North)

45°

(East)

ASSESS   We expect both $\Delta \vec{r}_{tot}$ and $\bar{\vec{v}}$ to be in the first quadrant since their components are all positive. Instead of unit-vector notation, $\Delta \vec{r}_{tot}$ and $\bar{\vec{v}}$ could be specified by their magnitudes $\Delta r_{tot} = \sqrt{(2.88 \text{ mi})^2 + (4.55 \text{ mi})^2} = 5.38$ mi and $\bar{v} = 15.4$ mi/h, respectively, and common direction, $\theta = \tan^{-1}[(4.55 \text{ mi})/(2.88 \text{ mi})] = 57.7°$ N of E.

29.   **INTERPRET**   We interpret this as a problem involving finding the change of velocity and the average acceleration.
DEVELOP   The average acceleration is given by Equation 3.5, $\bar{\vec{a}} = \Delta \vec{v}/\Delta t$. Therefore, the direction of $\bar{\vec{a}}$ is the same as $\Delta \vec{v}$, the change of velocity.

EVALUATE   Since the speed v is constant, we write the initial and final velocities as $\vec{v}_1 = v\hat{i}$ and
$\vec{v}_2 = v(-\hat{j}) = -v\hat{j}$, where $\hat{i}$ is east and $\hat{j}$ is north. The change in velocity for the $90°$ turn is
$$\Delta \vec{v} = -v\hat{j} - (v\hat{i}) = -v(\hat{i} + \hat{j})$$
Thus, the direction of the average acceleration $\bar{\vec{a}}$ is the same as that of $\Delta \vec{v}$, which is parallel to $-(\hat{i} + \hat{j})$, or southwest.
ASSESS   The angle between $\bar{\vec{a}}$ and the $+x$ axis is $\theta = \tan^{-1}[(-v)/(-v)] = 225°$, measured CCW. The acceleration must have westbound and southbound components because in order for the car to change its direction from eastbound to southbound, there must be a westbound velocity component that exactly cancels out the initial eastbound component. The remaining southbound component then determines the final direction of the car.

31.   **INTERPRET**   We interpret this as a velocity addition problem that involves a constant acceleration $\vec{a}$.

DEVELOP   Let $\vec{v}_0$ be the initial velocity, and $\vec{v}$ be the final velocity. As depicted in the figure, the relationship between $\vec{v}$ and $\vec{v}_0$ is given by

$$\vec{v} = \vec{v}_0 + \Delta \vec{v} = \vec{v}_0 + \vec{a}\Delta t$$

The angle between $\vec{a}$ and $\vec{v}_0$ can be found by using the law of cosine.

EVALUATE   The law of cosines gives

$$v^2 = v_0^2 + (a\Delta t)^2 - 2v_0(a\Delta t)\cos(180° - \theta_0) = v_0^2 + (a\Delta t)^2 + 2v_0(a\Delta t)\cos\theta_0$$

When the given magnitudes are substituted, one can solve for θ_0:

$$\theta_0 = \cos^{-1}\left(\frac{v^2 - v_0^2 - a^2(\Delta t)^2}{2v_0 a\Delta t}\right) = \cos^{-1}\left(\frac{(5.7 \text{ m/s})^2 - (2.4 \text{ m/s})^2 - (1.1 \text{ m/s}^2)^2(3.0 \text{ s})^2}{2(2.4 \text{ m/s})(1.1 \text{ m/s}^2)(3.0 \text{ s})}\right) = \cos^{-1}(1.00) = 0°$$

That is, the vectors $\vec{a}$ and $\vec{v}_0$ are collinear.

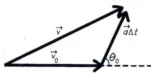

ASSESS   In two dimensions, acceleration and velocity vectors generally need not be collinear. But when they are (either parallel or anti-parallel), the change in speed would be maximal, as can be readily demonstrated graphically.

Section 3.3 Relative Motion

33.   **INTERPRET**   This is a problem involving relative velocities. The quantity of interest is the wind velocity relative to the ground.

DEVELOP   Equation 3.7, $\vec{v} = \vec{v}' + \vec{V}$, is what we shall use to find the wind velocity $\vec{V}$. Here $\vec{v}$ is the velocity of the jetliner relative to the ground, and $\vec{v}'$ is the velocity of the jetliner relative to the air. We shall take a coordinate system with x axis east, y axis north.

EVALUATE   From the problem statement, the velocity of the jetliner relative to the ground is

$$\vec{v} = \left(\frac{1500 \text{ km}}{100 \text{ min}}\right)(-\hat{j}) = \left(\frac{1500 \text{ km}}{(100/60)\text{h}}\right)(-\hat{j}) = -(900 \text{ km/h})\hat{j}$$

Similarly, using the fact that the unit vector in the direction 15° west of south (255° CCW from the +x axis) is $\cos 255°\hat{i} + \sin 255°\hat{j}$, the velocity of the jetliner relative to the air is

$$\vec{v}' = (1000 \text{ km/h})(\cos 255°\hat{i} + \sin 255°\hat{j}) = (-259 \text{ km/h})\hat{i} + (-965.9 \text{ km/h})\hat{j}$$

Thus, the wind velocity is

$$\vec{V} = \vec{v} - \vec{v}' = (259 \text{ km/h})\hat{i} + [(-900 \text{ km/h}) - (-965.9 \text{ km/h})]\hat{j} = (259 \text{ km/h})\hat{i} + (65.9 \text{ km/h})\hat{j}$$

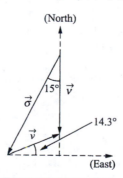

ASSESS   The wind speed is $V = \sqrt{(259 \text{ km/h})^2 + (65.9 \text{ km/h})^2} = 267 \text{ km/h},$ and the angle $\vec{V}$ makes with the x-axis is $\theta = \tan^{-1}[(65.9 \text{ km/h})/(259 \text{ km/h})] = 14.3°$ (north of east). The wind direction, by convention, is the direction facing the wind, in this case 14.3° S of W.

35.   **INTERPRET**   This is a problem involving relative velocities. The quantity of interest is the velocity of the jet
stream relative to the ground.

DEVELOP   Equation 3.7, $\vec{v} = \vec{v}' + \vec{V}$, is what we shall use to find $\vec{V}$, the velocity of the jet stream. Here $\vec{v}$ is the
velocity of the airplane relative to the ground, and $\vec{v}'$ is the velocity of the airplane relative to the air. We shall take
a coordinate system with x axis east, y axis north. The relationship between $\vec{v}$, $\vec{v}'$, and $\vec{V}$ is shown in the figure.

EVALUATE   We are given that the triangle is a right triangle, with $\vec{v} \perp \vec{V}$, and that the angle between the airspeed
and groundspeed is 32°, and the hypotenuse (magnitude of airspeed) is 370 km/h. From trigonometry, the
magnitude of the jet stream speed is

$$V = v' \sin\theta = (370 \text{ km/h}) \sin 32° = 196 \text{ km/h}$$

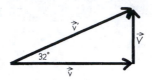

ASSESS   The speed of the airplane relative to the ground is

$$v = v' \cos\theta = (370 \text{ km/h})(\cos 32°) = 314 \text{ km/h}$$

The plane's heading of 32° north of east is a reasonable compensation for the southward wind blowing at a speed
of 196 km/h.

Section 3.4 Constant Acceleration

37.   **INTERPRET**   We interpret this as a problem involving finding the acceleration vectors in two dimensions, with
position vector given.

DEVELOP   The acceleration can be found by taking the second derivative of the position, $\vec{a} = \frac{d^2\vec{r}}{dt^2}$.

EVALUATE   The second derivative of $\vec{r} = (3.2t + 1.8t^2)\hat{i} + (1.7t - 2.4t^2)\hat{j}$ is

$$\vec{a} = \frac{d^2\vec{r}}{dt^2} = \frac{d}{dt}\frac{d}{dt}[(3.2t + 1.8t^2)\hat{i} + (1.7t - 2.4t^2)\hat{j}] = \frac{d}{dt}[(3.2 + 3.6t)\hat{i} + (1.7 - 4.8t)\hat{j}]$$
$$= (3.6)\hat{i} + (-4.8)\hat{j}$$

in units of m/s². The magnitude of $\vec{a}$ is $a = \sqrt{(3.6 \text{ m/s}^2)^2 + (-4.8 \text{ m/s}^2)^2} = 4.49 \text{ m/s}^2$, and the direction is
$\theta = \tan^{-1}(-4.8/3.6) = -53.1°$, or 53.1° CW from the x-axis.

ASSESS   We find the acceleration to be a constant vector. In general, a position vector in two dimensions that is
quadratic in t can be related to the velocity and acceleration vectors as

$$\vec{r} = r_x\hat{i} + r_y\hat{j} = \left(r_{x0} + v_{0x}t + \frac{1}{2}a_{0x}t^2\right)\hat{i} + \left(r_{y0} + v_{0y}t + \frac{1}{2}a_{0y}t^2\right)\hat{j}$$
$$= (r_{x0}\hat{i} + r_{y0}\hat{j}) + (v_{0x}\hat{i} + v_{0y}\hat{j})t + \frac{1}{2}(a_{0x}\hat{i} + a_{0y}\hat{j})t^2$$
$$= \vec{r}_0 + \vec{v}_0 t + \frac{1}{2}\vec{a}_0 t^2$$

From the above expression, we see that the constant acceleration vector $\vec{a}$ is equal to twice the coefficient of the
t^2 term.

Section 3.5 Projectile Motion

39.   **INTERPRET**   This is a problem involving projectile motion. The objects under consideration are the apple and the
peach, and the quantity of interest is the total flight time.

DEVELOP   We first note that the horizontal and vertical motions are independent of each other. The time of flight
t for either projectile is determined from the vertical component of the motion, which is the same for both, since
$v_{0y} = 0$. With $y_0 = 2.6$ m, we can calculate t by solving Equation 3.13, $y = y_0 + v_{y0}t - gt^2/2$.

EVALUATE   From the above equation, the total flight time is found to be

$$t = \sqrt{\frac{2(y_0 - y)}{g}} = \sqrt{\frac{2(2.6 \text{ m} - 0)}{9.8 \text{ m/s}^2}} = 0.728 \text{ s}$$

ASSESS   The apple and the peach both reach the ground at the same time. This is to be expected, since the total flight time is determined by the equation of motion in the vertical direction. The non-vanishing horizontal component of the velocity for the apple only makes it move further away from you.

41.   **INTERPRET**   This is a problem involving projectile motion. The object under consideration is an arrow, and the quantity of interest is the initial height from which it is fired.

DEVELOP   The horizontal and vertical motions of the arrows are independent of each other, and we can consider them separately. The time of flight t of the arrow can be determined from its range (horizontal motion, Equation 3.12). Once t is found, we can substitute the value into the equation of motion for the vertical direction (Equation 3.13) to determine the initial height.

EVALUATE   From Equation 3.12, we find the total flight time of the arrow to be

$$t = \frac{x - x_0}{v_{0x}} = \frac{23 \text{ m}}{41 \text{ m/s}} = 0.561 \text{ s}$$

Substituting the value into Equation 3.13, noting that $v_{y0} = 0$, we find the height to be

$$y_0 = y + \frac{1}{2}gt^2 = 0 + \frac{1}{2}(9.8 \text{ m/s}^2)(0.561 \text{ s})^2 = 1.54 \text{ m}$$

ASSESS   Dropping a height of 1.5 m in half a second is reasonable for free fall. We may relate y_0 to x as

$$y_0 = \frac{1}{2}gt^2 = \frac{1}{2}g\left(\frac{x - x_0}{v_{0x}}\right)^2$$

From the equation, it is clear that the greater the value of y_0, the longer it takes for the arrow to reach the ground, and the greater the horizontal distance traveled.

43.   **INTERPRET**   This is a problem involving projectile motion. The object under consideration is a proton, and the quantity of interest is the average speed.

DEVELOP   The horizontal and vertical distances in projectile motion (range and drop) are related by the trajectory Equation 3.14:

$$y = x \tan\theta_0 - \frac{g}{2v_0^2 \cos^2\theta_0}x^2$$

This is the equation we shall solve to find the initial speed v_0.

EVALUATE   With $\theta_0 = 0$, the above equation can be simplified to $y = -\dfrac{g}{2v_0^2}x^2$, which gives

$$v_0 = x\sqrt{\frac{-g}{2y}} = (1.7 \times 10^3 \text{ m})\sqrt{\frac{(-9.8 \text{ m/s}^2)}{2(-1.2 \times 10^{-6} \text{ m})}} = 3.44 \times 10^6 \text{ m/s}$$

This is the x-component of the velocity and it remains unchanged during the course of flight. The y-component of the velocity is

$$v_y = -gt = -g\frac{(x - x_0)}{v_0} = -(9.8 \text{ m/s}^2)\frac{1.7 \times 10^3 \text{ m}}{3.44 \times 10^6 \text{ m/s}} = -4.85 \times 10^{-3}\text{m/s}$$

Since $v_y \ll v_0$, it is negligible and v_0 is the approximate average speed.

ASSESS   Traveling a distance of 1700 m with a drop of only 1.2×10^{-6} m indicates that the horizontal component of the speed must be much greater than the vertical component. With $v_y \ll v_0$, the average speed can be approximated as $v = \sqrt{v_0^2 + v_y^2} \approx v_0$.

Section 3.6 Uniform Circular Motion

45.   **INTERPRET**   This is a problem about uniform circular motion. The object of interest is the car and the key concept involved here is centripetal acceleration.

DEVELOP   Given the radius and the acceleration, we may use Equation 3.16, $a = v^2/r$, to solve for the speed.

EVALUATE   Using Equation 3.16, and setting $a = g = 9.8$ m/s², the speed of the car is

$$v = \sqrt{ar} = \sqrt{(9.8 \text{ m/s}^2)(75 \text{ m})} = 27.1 \text{ m/s} = 97.6 \text{ km/h} = 60.7 \text{ mi/h}$$

ASSESS   If the radius of the turn is kept fixed, then the only means to attain a higher centripetal acceleration is to increase the speed. A centripetal acceleration of 1g is just within the capability of autocross tires.

47.   **INTERPRET**   We are asked to find the time it takes to go around a circle, knowing the centripetal acceleration and the radius.

DEVELOP   We use $a_c = \frac{v^2}{r}$ to find the speed, and $C = 2\pi r$ for the distance. We find the time by using $t = \frac{\text{distance}}{\text{speed}}$. The acceleration a_c is provided by gravity, which is 5.8% of the surface value, 9.8 m/s². We have to be careful to note that the radius r is the Earth's radius *plus* the altitude, not just the altitude.

EVALUATE $v = \sqrt{a_c r} \rightarrow T = \dfrac{2\pi r}{\sqrt{a_c r}} = 2\pi \sqrt{\dfrac{r}{a_c}}$

$$a_c = (9.8 \text{ m/s}^2) \times 0.058 = 0.568 \text{ m/s} \quad r = (6.38 \times 10^6 \text{ m}) + (20000 \times 10^3 \text{ m}) = 26.4 \times 10^6 \text{ m}$$

$$T = 2\pi \sqrt{\frac{r}{a_c}} = 2\pi \sqrt{\frac{26.4 \times 10^6 \text{m}}{0.568 \text{ m/s}^2}} = (42.8 \times 10^3 \text{ s}) \times \frac{1 \text{ h}}{3600 \text{ s}} = 11.9 \text{ h}$$

ASSESS   Part of the expense of placing satellites in orbit is the difficulty in not only getting them to altitude, but getting them moving fast enough so that they can stay in orbit.

PROBLEMS

49.   **INTERPRET**   We interpret this as a problem involving adding two vectors in two dimensions to produce a resultant vector which points in a certain direction.

DEVELOP   Using Equation 3.1, we see that in two dimensions, a vector $\vec{A}$ can be written as, in unit vector notation, $\vec{A} = A_x \hat{i} + A_y \hat{j} = A(\cos\theta_A \hat{i} + \sin\theta_A \hat{j})$, where $A = \sqrt{A_x^2 + A_y^2}$ and $\theta_A = \tan^{-1}(A_y/A_x)$. Similarly, we express a second vector $\vec{B}$ as $\vec{B} = B_x \hat{i} + B_y \hat{j} = B(\cos\theta_B \hat{i} + \sin\theta_B \hat{j})$. We choose $+x$ direction to correspond to the right and $+y$ to be the vertical direction. The resultant vector is

$$\vec{C} = \vec{A} + \vec{B} = (A_x + B_x)\hat{i} + (A_y + B_y)\hat{j} = (A\cos\theta_A + B\cos\theta_B)\hat{i} + (A\sin\theta_A + B\sin\theta_B)\hat{j} = C_y \hat{j}$$

which points in the vertical direction.

EVALUATE   Using the fact that $\vec{C}$ is vertical with $C_x = 0$, we find

$$0 = C_x = A\cos\theta_A + B\cos\theta_B$$

or

$$\cos\theta_B = -\frac{A\cos\theta_A}{B} = -\frac{(1.0 \text{ m})\cos(-35°)}{1.8 \text{ m}} = -0.455$$

The two possible solutions are $\theta_B = 117°$, or 243°. The former corresponds to $B_y > 0$, while the latter corresponds to $B_y < 0$. The results are depicted in the figure on the right.

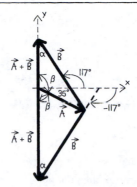

ASSESS   Since $|\cos\theta_B| \leq 1$, solutions are possible only if $B \geq A$. In the case where $B = A$, the above condition reduces to $\cos\theta_B = -\cos\theta_A = \cos(\theta_A + \pi)$, or $\theta_B = \theta_A + \pi$. This result can be readily verified with graphical constructions.

51. **INTERPRET**   We interpret this as a problem involving finding the magnitude and direction of the average velocity vector.

DEVELOP   Given two vectors $\vec{r}_1$ and $\vec{r}_2$, the displacement vector is given by $\Delta\vec{r} = \vec{r}_2 - \vec{r}_1$. The average velocity vector is $\bar{\vec{v}} = \Delta\vec{r}/\Delta t$. The magnitude of $\bar{\vec{v}}$ is

$$\bar{v} = \sqrt{(\bar{v}_x)^2 + (\bar{v}_y)^2 + (\bar{v}_z)^2}$$

EVALUATE   The displacement is

$$\Delta\vec{r} = \vec{r}_2 - \vec{r}_1 = (4.6\hat{i} + 1.9\hat{k})\mu m - (2.2\hat{i} + 3.7\hat{j} - 1.2\hat{k})\mu m = (2.4\hat{i} - 3.7\hat{j} + 3.1\hat{k})\mu m$$

With $\Delta t = 6.2$ s, the average velocity is

$$\bar{\vec{v}} = \frac{\Delta\vec{r}}{\Delta t} = \frac{(2.4\hat{i} - 3.7\hat{j} + 3.1\hat{k})\mu m}{6.2 \text{ s}} = (0.387\mu m/s)\hat{i} - (0.597\mu m/s)\hat{j} + (0.50\mu m/s)\hat{k}$$

The magnitude of $\bar{\vec{v}}$ is

$$\bar{v} = \sqrt{(\bar{v}_x)^2 + (\bar{v}_y)^2 + (\bar{v}_z)^2} = \sqrt{(0.387\mu m/s)^2 + (-0.597\mu m/s)^2 + (0.50\mu m/s)^2} = 0.869 \ \mu m/s$$

ASSESS   The typical size of bacteria is on the order of 1 micron, or $1 \ \mu m$. In this example, the bacterium has an average speed of $0.869 \ \mu m/s$. This means that in one second it moves a distance which is comparable to its own length. This is a reasonable result.

53. **INTERPRET**   We interpret this as a problem involving finding the change of velocity and the average acceleration. The object of interest is the car.

DEVELOP   The acceleration of the car is opposite to the direction of the skid, since it comes to a stop with final velocity $\vec{v} = 0$. We shall use Equation 3.5, $\bar{\vec{a}} = \Delta\vec{v}/\Delta t$, to calculate the average acceleration. Note that the direction of $\bar{\vec{a}}$ is the same as $\Delta\vec{v}$, the change of velocity.

EVALUATE   The initial speed of the car when skidding begins is

$$v_0 = 80 \text{ km/h} = \left(80\frac{\text{km}}{\text{h}}\right)\left(\frac{1000 \text{ m}}{1 \text{ km}}\right)\left(\frac{1 \text{ h}}{3600 \text{ s}}\right) = 22.22 \text{ m/s}$$

In unit-vector notation, the initial velocity can be written as

$$\vec{v}_0 = v_0(\cos\theta_0\hat{i} + \sin\theta_0\hat{j}) = (22.22 \text{ m/s})(\cos 30°\hat{i} + \sin 30°\hat{j}) = (19.2 \text{ m/s})\hat{i} + (11.1 \text{ m/s})\hat{j}$$

Note that $\vec{v}_0$, the velocity at the start of the skid, is not in the direction of the initial motion before the skid. The magnitude of the average acceleration is

$$\bar{a} = \frac{|\Delta v|}{\Delta t} = \frac{22.22 \text{ m/s}}{3.9 \text{ s}} = 5.70 \text{ m/s}^2$$

The direction of $\bar{\vec{a}}$ is in the opposite direction of $\vec{v}_0$. Therefore, in unit-vector notation, we can express $\bar{\vec{a}}$ as

$$\bar{\vec{a}} = \bar{a} \ [\cos(\theta_0 + 180°)\hat{i} + \sin(\theta_0 + 180°)\hat{j}] = (5.70 \text{ m/s}^2)(\cos 210°\hat{i} + \sin 210°\hat{j})$$

$$= (-4.93 \text{ m/s}^2)\hat{i} + (-2.85 \text{ m/s}^2)\hat{j}$$

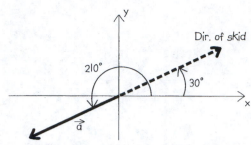

ASSESS The average acceleration only depends on the change of velocity between the starting point and the end point. In this case, it must point in the opposite direction as $\vec{v}_0$ in order to bring the car to a complete stop.

55. **INTERPRET** This is a problem about uniform circular motion. The object of interest is the second hand of the clock and the key concepts involved here are the average acceleration and average velocity.

DEVELOP We first find the displacement vector $\Delta\vec{r}$. We can then use Equation 3.3, $\bar{\vec{v}} = \Delta\vec{r}/\Delta t$, to obtain the average velocity vector. The average acceleration can be calculated by using Equation 3.5, $\bar{\vec{a}} = \Delta\vec{v}/\Delta t$, where $\Delta\vec{v}$ is the change of velocity.

EVALUATE (a) The distance moved by the sweep-second hand in 60 seconds (one complete revolution) is $s = 2\pi R$, where R is the radius of the circle. Therefore, during a five-second interval the distance traveled is

$$\Delta r = 2\pi R/12 = \pi R/6 = \pi(3.1 \text{ cm})/6 = 1.62 \text{ cm}$$

The position vectors (from the center hub) of the tip at the beginning and end of the interval, $\vec{r}_1$ and $\vec{r}_2$, form the sides of an isosceles triangle whose base is Δr, the magnitude of the displacement, and whose base angle is $(180° - 30°)/2 = 75°$ (see figure). Thus, the average velocity $\bar{\vec{v}}$ has a magnitude $\bar{v} = \Delta r/\Delta t = (1.62 \text{ cm})/(5.0 \text{ s}) = 0.325 \text{ cm/s}$, and the direction is $\theta_{\bar{v}} = 180° - 75° = 105°$ measured CW from $\vec{r}_1$.

(b) The instantaneous speed of the tip of the second-hand is a constant and equal to the circumference divided by 60 s, or $\bar{v} = \Delta r/\Delta t = (1.62 \text{ cm})/(5.0 \text{ s}) = 0.325 \text{ cm/s}$. The direction of the velocity of the tip is tangent to the circumference, or perpendicular to the radius, in the direction of motion (CW). The angle between the two tangents is the same as the angle between the two corresponding radii, so that $\vec{v}_1$, $\vec{v}_2$, and $\Delta\vec{v}$ form an isosceles triangle similar to the one in part (a).
Thus

$$|\Delta v| = 2 \, |v| \sin\frac{1}{2}\theta = 2(0.325 \text{ cm/s})\sin\left(\frac{1}{2} \times 30°\right) = 0.168 \text{ cm/s}$$

The magnitude of the average acceleration $\bar{\vec{a}}$ is

$$\bar{a} = \Delta v/\Delta t = (0.168 \text{ cm/s})/(5.0 \text{ s}) = 0.034 \text{ cm/s}^2$$

and its direction is $105°$ CW from the direction of $\vec{v}_1$, or $195°$ CW from the direction of $\vec{r}_1$.
(c) The angle between $\bar{\vec{a}}$ and $\bar{\vec{v}}$, from parts (a) and (b), is $195° - 105° = 90°$.

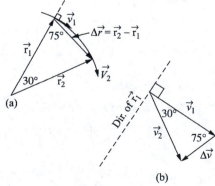

ASSESS The two vectors $\bar{\vec{a}}$ and $\bar{\vec{v}}$ are always perpendicular to each other. While $\bar{\vec{v}}$ is tangential to the circular path, the direction of $\bar{\vec{a}}$ is toward the center of the circle. Thus, $\bar{\vec{a}}$ is the centripetal acceleration of the circular motion. In the above, the magnitude of the average acceleration can also be calculated as

$$\bar{a} = v^2/R = (0.325 \text{ cm/s})^2/(3.1 \text{ cm}) = 0.034 \text{ cm/s}^2$$

57.   **INTERPRET**   We interpret this as a problem involving comparing the magnitudes of two vectors.

DEVELOP   Using Equation 3.1, we see that in two dimensions, a vector $\vec{A}$ can be written as, in unit vector notation, $\vec{A} = A_x\hat{i} + A_y\hat{j}$, where $A = \sqrt{A_x^2 + A_y^2}$ and $\theta_A = \tan^{-1}(A_y/A_x)$. Similarly, we express vector $\vec{B}$ as $\vec{B} = B_x\hat{i} + B_y\hat{j}$. Let $\vec{C}$ be the sum of the two vectors:

$$\vec{C} = \vec{A} + \vec{B} = (A_x + B_x)\hat{i} + (A_y + B_y)\hat{j} = C_x\hat{i} + C_y\hat{j}$$

and $\vec{D}$ be the difference of the two vectors:

$$\vec{D} = \vec{A} - \vec{B} = (A_x - B_x)\hat{i} + (A_y - B_y)\hat{j} = D_x\hat{i} + D_y\hat{j}$$

Since $\vec{C}$ and $\vec{D}$ are perpendicular to each other, for simplicity, we take $\vec{C} = C_x\hat{i}$ and $\vec{D} = D_y\hat{j}$.

EVALUATE   The conditions we have set for $\vec{C}$ and $\vec{D}$ imply $C_y = 0$ and $D_x = 0$, or

$$C_y = A_y + B_y = 0 \quad \rightarrow \quad A_y = -B_y$$
$$D_x = A_x - B_x = 0 \quad \rightarrow \quad A_x = B_x$$

With the results obtained above, we can readily show that the magnitudes of $\vec{A}$ and $\vec{B}$ are equal:

$$A = \sqrt{A_x^2 + A_y^2} = \sqrt{B_x^2 + (-B_y)^2} = \sqrt{B_x^2 + B_y^2} = B$$

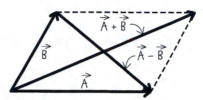

ASSESS   An alternative way to establish the equality between A and B is to note that the vectors $\vec{A} + \vec{B}$ and $\vec{A} - \vec{B}$ are the two diagonals of the parallelogram formed by sides $\vec{A}$ and $\vec{B}$. If the diagonals are perpendicular, the parallelogram is a rhombus; hence $A = B$.

59.   **INTERPRET**   This is a problem involving motion in two dimensions. The physical quantity of interest is the magnitude of acceleration.

DEVELOP   We first note that the motions in x and y directions are independent of each other. The x component of the displacement is due only to the initial velocity, $\Delta x = v_0\Delta t$. On the other hand, the y component is just due to the acceleration, $\Delta y = a(\Delta t)^2/2$.

EVALUATE   The condition that $\Delta x = \Delta y$ when $\Delta t = 18$ s implies

$$v_0\Delta t = a(\Delta t)^2/2 \quad \rightarrow \quad a = \frac{2v_0}{\Delta t} = \frac{2(4.5 \text{ m/s})}{18 \text{ s}} = 0.5 \text{ m/s}^2$$

ASSESS   The answer can be readily checked by substituting the value of a into Δy:

$$\Delta y = \frac{1}{2}a(\Delta t)^2 = \frac{1}{2}(0.5 \text{ m/s}^2)(18 \text{ s})^2 = 81 \text{ m}$$

This is equal to $\Delta x = v_0\Delta t = (4.5 \text{ m/s})(18 \text{ s}) = 81$ m.

61.   **INTERPRET**   This is a problem involving projectile motion. The physical quantity of interest is the initial speed of water.

DEVELOP   We first note that the projectile motion can be decomposed into horizontal and vertical motions that are independent of each other. We are given the initial height of the water, $y_0 = 1.6$ m, the final height, $y = 0.93$ m, and the range $x = 2.1$ m. These quantities are related by the trajectory Equation 3.14:

$$y - y_0 = x\tan\theta_0 - \frac{g}{2v_0^2\cos^2\theta_0}x^2$$

This is the equation we shall solve to find the initial speed v_0.

EVALUATE   With $\theta_0 = 0$, the above equation can be simplified to $y - y_0 = -\frac{g}{2v_0^2}x^2$, which gives

$$v_0 = x\sqrt{\frac{-g}{2(y - y_0)}} = (2.1 \text{ m})\sqrt{\frac{(-9.8 \text{ m/s}^2)}{2(0.93 \text{ m} - 1.6 \text{ m})}} = 5.68 \text{ m/s}$$

ASSESS We check the answer by solving the problem in a different way: Since the water was fired horizontally ($v_{0y} = 0$), the time it takes to fall from $y_0 = 1.6$ m to $y = 0.93$ m is given by Equation 3.13:

$$t = \sqrt{2(y_0 - y)/g} = \sqrt{2(1.6 \text{ m} - 0.93 \text{ m})/(9.8 \text{m/s}^2)} = 0.370 \text{ s.}$$

Its initial speed, $v_0 = v_{0x}$, can be found from Equation 3.12,

$$v_0 = (x - x_0)/t = (2.1 \text{ m})/(0.370 \text{ s}) = 5.68 \text{ m/s}$$

Both approaches lead to the same final answer for v_0.

63. **INTERPRET** This is a problem involving projectile motion. The physical quantities of interest are the initial speed and angle of the package.

DEVELOP We first note that the projectile motion can be decomposed into horizontal and vertical motions that are independent of each other. We interpret "just barely" to mean that the maximum height of the package equals the height of the window sill. Therefore, when the package reaches the sill (in the coordinate system shown), $v_y = 0$.

EVALUATE Using Equation 2.11, $v_y^2 = v_{0y}^2 - 2gy$, with $v_y = 0$ when the package reaches the sill, y-component of the initial velocity is

$$v_{0y} = \sqrt{2(9.8 \text{ m/s}^2)(2.7 \text{ m})} = 7.27 \text{m/s}$$

Since $v_y = 0 = v_{0y} - gt$, the time of flight is $t = \frac{v_{0y}}{g} = \frac{7.27 \text{ m/s}}{9.8 \text{ m/s}^2} = 0.742$ s. Therefore, the x-component of the initial velocity is

$$v_{0x} = \frac{x}{t} = \frac{3.0 \text{ m}}{0.742 \text{ s}} = 4.04 \text{ m/s}$$

From these components, we find the magnitude and direction of $\vec{v}_0$ to be

$$v_0 = \sqrt{v_{0x}^2 + v_{0y}^2} = \sqrt{(4.04 \text{ m/s})^2 + (7.27 \text{ m/s})^2} = 8.32 \text{ m/s,}$$

and $\theta = \tan^{-1}\left(\frac{v_{0y}}{v_{0x}}\right) = 60.9°.$

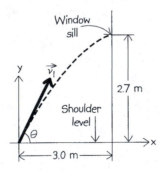

ASSESS Checking the answer using Equation 3.14, we find

$$y = x\tan\theta_0 - \frac{g}{2v_0^2 \cos^2\theta_0}x^2 = (3.0 \text{ m})\tan 60.9° - \frac{(9.8 \text{ m/s}^2)(3.0 \text{ m})^2}{2(8.32 \text{ m/s})^2 \cos^2 60.9°} = 2.70 \text{ m}$$

in agreement with the problem statement.

65. **INTERPRET** This is a problem involving projectile motion. We are interested in the dependence of total flight time of a projectile on its launch angle.

DEVELOP Suppose the projectile is launched at an angle θ_0 with an initial velocity $\vec{v}_0$. The x and y components of the initial velocities are then equal to

$$v_{0x} = v_0 \cos\theta_0, \quad v_{0y} = v_0\sin\theta_0$$

To find the total flight time, we note that $v_y = 0$ at the maximum height y_{max}, and the amount of time it takes for the projectile to reach y_{max} is given by Equation 3.11:

$$0 = v_y = v_0 \sin\theta_0 - gt \ \rightarrow \ t = \frac{v_0 \sin\theta_0}{g}$$

Since the time of ascent is equal to the time of descent, the total flight time is

$$t_{tot} = 2t = \frac{2v_0 \sin\theta_0}{g}$$

EVALUATE   Using the equation above, the ratio of the total flight time when $\theta_0 = 30°$ and $60°$ is

$$\frac{t_{tot}(60°)}{t_{tot}(30°)} = \frac{\sin 60°}{\sin 30°} = \sqrt{3}$$

In Figure 3.19, $v_0 = 50$ m/s, so $t(30°) = \dfrac{2(50 \text{ m/s}) \sin 30°}{9.8 \text{ m/s}^2} = 5.10$ s, and $t(60°) = \sqrt{3} \ t(30°) = 8.84$ s.

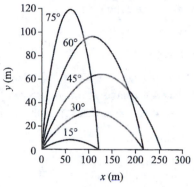

ASSESS   The total flight time is longest when $\sin\theta_0 = 1$, or $\theta_0 = 90°(v_{0y} = v_0)$, and shortest when $\sin\theta_0 = 0$, or $\theta_0 = 0$ $(v_{0y} = 0)$.

67.   **INTERPRET**   This is a problem involving projectile motion. We want to prove that a projectile launched on level ground reaches its maximum height midway along its trajectory.

DEVELOP   The total flight time can be found by using Equation 3.13, $y = y_0 + v_{0y}t - \frac{1}{2}gt^2$ and setting $y = y_0$ (on level ground). Similarly, to find the amount of time it takes for the projectile to reach its maximum height y_{max}, we note that $v_y = 0$ at y_{max} and apply Equation 3.11.

EVALUATE   From Equation 3.13, we find the total flight time to be

$$0 = y - y_0 = v_{0y}t_{tot} - \frac{1}{2}gt_{tot}^2 \ \rightarrow \ t_{tot} = \frac{2v_{0y}}{g}$$

On the other hand, solving Equation 3.11 for t', the amount of time it takes to reach y_{max}, we obtain

$$0 = v_y = v_{0y} - gt' \ \rightarrow \ t' = \frac{v_{0y}}{g}$$

Comparing the two expressions, we see that $t_{tot} = 2t'$. Thus, a projectile launched on level ground reaches its maximum height midway along its trajectory.

ASSESS   The result shows that the time of ascent is equal to the time of descent, as expected. An alternative proof could be carried out by differentiating Equation 3.14 with respect to x:

$$\frac{dy}{dx} = \frac{d}{dx}\left(x\tan\theta_0 - \frac{g}{2v_0^2 \cos^2\theta_0}x^2 \right) = \tan\theta_0 - \frac{g}{v_0^2 \cos^2\theta_0}x$$

We see that $y = y_{max}$, when $x' = \frac{v_0^2 \sin\theta_0 \cos\theta_0}{g} = \frac{v_0^2 \sin 2\theta_0}{2g}$. But this is only half of the horizontal range $x = \frac{v_0^2 \sin 2\theta_0}{g}$.

69. **INTERPRET** This is a problem involving projectile motion. The physical quantity of interest is the initial speed of the cyclist.

DEVELOP Suppose the projectile is launched at an angle θ_0 with an initial velocity $\vec{v}_0$. The x and y components of the initial velocities are then equal to

$$v_{0x} = v_0 \cos\theta_0 \quad v_{0y} = v_0 \sin\theta_0$$

From Equation 3.13, we find the total flight time to be

$$0 = y - y_0 = v_{0y}t_{tot} - \frac{1}{2}gt_{tot}^2 \quad \rightarrow \quad t_{tot} = \frac{2v_{0y}}{g}$$

The range of the projectile is

$$x = v_{0x}t_{tot} = (v_0 \cos\theta_0)\left(\frac{2v_0 \sin\theta_0}{g}\right) = \frac{2v_0^2 \cos\theta_0 \sin\theta_0}{g} = \frac{v_0^2 \sin 2\theta_0}{g}$$

EVALUATE If the motorcyclist was deflected upward from the road at an angle of 45°, the horizontal range formula found above implies a minimum initial speed of

$$v_0 = \sqrt{xg} = \sqrt{(39 \text{ m})(9.8 \text{ m/s}^2)} = 19.5 \text{ m/s} = 70.4 \text{ km/h}.$$

In fact, some speed would be lost during impact with the car, so the cyclist probably was speeding.

ASSESS The greater the range, the larger the initial speed. If the cyclist were traveling at a speed of 60 km/h, or 16.67 m/s, he would have landed at a distance of about 28 m from his bike.

71. **INTERPRET** This is a problem involving projectile motion. The physical quantity of interest is the launch angle of the basketball.

DEVELOP We are given the initial height of the ball, $y_0 = 8.2$ ft, the final height, $y = 10$ ft, the initial speed $v_0 = 26$ ft/s, and the range $x = 15$ ft. These quantities are related by the trajectory Equation 3.14:

$$y - y_0 = x \tan\theta_0 - \frac{g}{2v_0^2 \cos^2\theta_0}x^2$$

This is the equation we shall solve to find the launch angle θ_0.

EVALUATE With origin at the point from which the ball is thrown, the equation of the trajectory, evaluated at the basket, becomes

$$y - y_0 = (10 - 8.2)\text{ft} = (15 \text{ ft}) \tan\theta_0 - \frac{(32 \text{ ft/s}^2)(15 \text{ ft})^2}{2(26 \text{ ft/s})^2 \cos^2\theta_0}$$

or $1.8 \text{ ft} = (15 \text{ ft}) \tan\theta_0 - \frac{5.33 \text{ ft}}{\cos^2\theta_0}$

Using the trigonometric identity $1 + \tan^2\theta_0 = 1 = \cos^2\theta_0$, we can convert this equation into a quadratic in $\tan\theta_0$:

$$7.13 - 15\tan\theta_0 + 5.33\tan^2\theta_0 = 0$$

The answers are

$$\theta_0 = \tan^{-1}\left[\frac{15 \pm \sqrt{15^2 - 4(5.33)(7.13)}}{2(5.33)}\right] = 31.2° \quad \text{or} \quad 65.7°$$

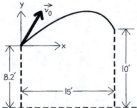

ASSESS Like the horizontal range formula for given v_0, there are two launch angles whose trajectories pass through the basket, although in this case they are not symmetrically placed about 45°.

73. **INTERPRET** This is a problem involving projectile motion. The physical quantity of interest is the launch speed.

DEVELOP If we take the origin of coordinates at the slingshot and the stranded climbers at $x = 390$ m and $y = 270$ m, we can use Equation 3.14 for the trajectory to solve for v_0.

EVALUATE Using Equation 3.14 to solve for v_0, we obtain

$$v_0 = \frac{x}{\cos\theta_0}\sqrt{\frac{g}{2(x\tan\theta_0 - y)}} = \frac{390 \text{ m}}{\cos 70°}\sqrt{\frac{9.8 \text{ m/s}^2}{2(390 \text{ m}\cdot\tan 70° - 270 \text{ m})}} = 89.2 \text{ m/s}$$

ASSESS As expected, the initial speed v_0 increases with y. The projectile would have a much greater range had it not been obstructed by the mountain.

75. **INTERPRET** This is a problem involving projectile motion. The physical quantities of interest are the launch speed and the maximum height.

DEVELOP From the expression for the horizontal range, we see that with a given initial speed v_0, the maximum range is attained with a launch angle of $\theta_0 = 45°(\sin 2\theta_0 = 1)$, and $x_{max} = v_0^2/g$. On the other hand, the projectile reaches its maximum height when $v_y = 0$. Using Equation 2.11, $v_y^2 = v_{0y}^2 - 2g(y - y_0)$, we have

$$y_{max} = y_0 + \frac{v_{0y}^2}{2g}$$

EVALUATE A maximum range of $x_{max} = 180$ km $= 1.8\times 10^5$ m implies

$$v_0 = \sqrt{x_{max}g} = \sqrt{(1.8\times 10^5 \text{ m})(9.8 \text{ m/s}^2)} = 1328 \text{ m/s}$$

This in turn gives $(y_0 = 0)$

$$y_{max} = \frac{v_{0y}^2}{2g} = \frac{(1328 \text{ m/s})^2}{2(9.8 \text{ m/s}^2)} = 90,000 \text{ m} = 90 \text{ km}$$

ASSESS We find the maximum height to be half the maximum range. To see why, we note that the maximum height of a projectile is $y_{max} = \frac{v_0^2 \sin^2\theta_0}{g}$. Thus,

$$\frac{y_{max}}{x} = \frac{v_0^2 \sin^2\theta_0/g}{v_0^2 \sin 2\theta_0/g} = \frac{\sin^2\theta_0}{\sin 2\theta_0} = \frac{1}{2\tan\theta_0}$$

When choosing $\theta_0 = 45°$ to give $x = x_{max}$, $\tan 45° = 1$ and we have $\frac{y_{max}}{x_{max}} = \frac{1}{2}$.

77. **INTERPRET** This is a problem involving projectile motion. The physical quantities of interest are the initial speed and angle of the diver.

DEVELOP We first note that the projectile motion can be decomposed into horizontal and vertical motions that are independent of each other. Since we are given the maximum height (at which point $v_y = 0$), Equation 2.11 can be used to find the y component of the diver's initial velocity. The x component of v_0 can be found from Equation 3.12, once the time of flight is known.

EVALUATE Solving Equation 2.11 gives

$$0 = v_{0y}^2 - 2g(y_{max} - y_0) \rightarrow v_{0y} = \sqrt{2(9.8 \text{ m/s}^2)(2.5 \text{ m})} = 7.00 \text{ m/s}$$

To find the total flight time, we solve the quadratic equation (Equation 3.13):

$$y - y_0 = v_{0y}t - \frac{1}{2}gt^2$$

With $y_0 - y = 3$ m (a 3-m board is 3 m above the water level), we obtain

$$t = \frac{v_{0y} + \sqrt{v_{0y}^2 + 2g(y_0 - y)}}{g} = \frac{7 \text{ m/s} + \sqrt{(7 \text{ m/s})^2 + 2(9.8 \text{ m/s}^2)(3 \text{ m})}}{9.8 \text{ m/s}^2} = 1.77 \text{ s}$$

We take the positive square root because the diver springs upward off the board. Thus, the x component of the velocity is

$$v_{0x} = \frac{x - x_0}{t} = \frac{2.8 \text{ m}}{1.77 \text{ s}} = 1.58 \text{ m/s}$$

From v_{0x} and v_{0y}, we find the magnitude of $\vec{v}_0$ to be

$$v_0 = \sqrt{v_{0x}^2 + v_{0y}^2} = \sqrt{(1.58 \text{ m/s})^2 + (7.0 \text{ m/s})^2}$$
$$= 7.18 \text{ m/s}$$

and direction $\theta_0 = \tan^{-1}\left(\dfrac{v_{0y}}{v_{0x}}\right) = \tan^{-1}\left(\dfrac{7.0 \text{ m/s}}{1.58 \text{ m/s}}\right) = 77.3°$

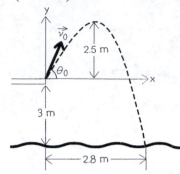

ASSESS   It takes the diver 1.77 s to complete the dive. The result is reasonable. The greater the value of θ_0, the closer the diver will be to the diving board.

79.   **INTERPRET**   This is a problem involving projectile motion. The physical quantity of interest is the slope of the ground in the landing zone of the ski jump.

DEVELOP   We first note that the direction of the skier's velocity is $\theta = \tan^{-1}(v_y/v_x)$, where angles are measured CCW from the x axis, chosen horizontal to the right in Fig. 3.25 with the y axis upward.

In the landing zone, θ is in the fourth quadrant, which can be represented by a negative angle below the x axis. The slope of the ground at this point can be represented by a similar angle θ_g, and for the safety of ski jumpers, $\theta_g - \theta = 3.0°$.

EVALUATE   The slope $\dfrac{v_y}{v_x} = \dfrac{dy/dt}{dx/dt} = \dfrac{dy}{dx}$ can be calculated by using Equations 3.10, 3.11, and 3.12. Thus,

$v_x = v_{0x} = v_0 \cos\theta_0$, and $v_y = v_{0y} - gt = v_0 \sin\theta_0 - gt$. Since $x - x_0 = v_{0x}t = 55$ m is given, the time of flight can be eliminated, and we obtain

$$\frac{v_y}{v_x} = \frac{v_{0y}}{v_{0x}} - \frac{g(x - x_0)}{v_{0x}^2} = \tan\theta_0 - \frac{g(x - x_0)}{v_0^2 \cos^2\theta_0} = \tan(-9.5°) - \frac{(9.8 \text{ m/s}^2)(55 \text{ m})}{(28 \text{ m/s})^2[\cos(-9.5°)]^2} = -0.874$$

Thus, $\theta = \tan^{-1}(v_y/v_x) = \tan^{-1}(-0.874) = -41.2°$, and $\theta_g = \theta + 3.0° = -38.2°$

ASSESS   An angle of 38.2° below the x axis is a reasonable value. A typical value is between 35°–40°.

81.   **INTERPRET**   We need to find a general equation for the initial speed of a projectile, given the range R and maximum height h.

DEVELOP   If the trajectory is over level ground, then it reaches the maximum height at a time halfway through the flight. The time for the entire trajectory is twice the time it takes to fall from height h, so we use this time and the range to find the x-component of the initial velocity. We also use the height to find the y-component of the initial velocity.

We use $y = y_0 + v_{y0}t - \frac{1}{2}gt^2$, starting at height h, to find the time.

We then use $R = v_{x0}t$ to find the x-component of initial velocity.

We use $v_y = v_{y0} + a_y t$ to find the y-component of initial velocity.

The initial speed is the magnitude of the initial velocity: $v = \sqrt{v_{x0}^2 + v_{y0}^2}$

EVALUATE

$$y = y_0 + v_{y0}t - \frac{1}{2}gt^2 \rightarrow 0 = h - \frac{1}{2}gt^2 \rightarrow t = \sqrt{\frac{2h}{g}}$$

$$R = v_{x0}t \rightarrow v_{x0} = \frac{R}{t} = R\sqrt{\frac{g}{2h}}$$

$$v_y = v_{y0} + a_y t \rightarrow 0 = v_{y0} - gt \rightarrow v_{y0} = gt = g\sqrt{\frac{2h}{g}} = \sqrt{2gh}$$

$$v = \sqrt{v_{x0}^2 + v_{y0}^2} = \sqrt{R^2\frac{g}{2h} + 2gh}$$

ASSESS The problem does not specify that the ground must be level, but it's not possible to solve unless you make this assumption.

83. **INTERPRET** We find whether a certain acceleration on a meteor will give the desired result. We can use the equations for constant-acceleration motion.

DEVELOP There are two things to check. We want to know if the acceleration will give the desired displacement in 4 minutes, and we want to know if the new velocity will be at the desired angle. We use $v_x = a_x t$ to find the component of velocity perpendicular to the original direction, then check the angle by noting that $\theta = \tan^{-1}\left(\frac{v_x}{v_0}\right)$ in the figure below.

We find the displacement x using $x = x_o + v_o t + \frac{1}{2}at^2$.

The acceleration is $a_x = 0.035$ km/s^2. The time is 4 minutes, and the initial velocity is $v_0 = 21$ km/s. The new direction of the velocity should be 22.6° degrees from the original velocity, and the desired displacement is 5.36×10^3 km.

EVALUATE We first convert time to seconds: $t = (4 \text{ min}) \times \frac{60 \text{ s}}{1 \text{ min}} = 240$ s

$v_x = a_x t = (0.035 \text{ km/s}^2)(240 \text{ s}) = 8.4$ km/s. The angle is $\theta = \tan^{-1}\left(\frac{v_x}{v_0}\right) = \tan^{-1}\left(\frac{8.4 \text{ km/s}}{21 \text{ km/s}}\right) = 21.8°$, which is a bit short of the desired angle. The displacement is $x = x_o + v_o t + \frac{1}{2}at^2 = \frac{1}{2}a_x t^2 = \frac{1}{2}(0.035 \text{ km/s}^2)(240 \text{ s})^2 = 1.01 \times 10^3$ km. This is far short of the desired result!

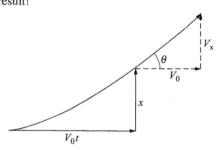

ASSESS Note that we did not convert our units of km to m. Since all distances are in km, and the speeds are given in km/s, this is okay: the units are self-consistent.

85. **INTERPRET** What is the actual best launch angle for maximum range with a ball? Why would range change with altitude?

DEVELOP The problem suggests searching the internet for this information.

EVALUATE A quick check using any search engine reveals that the exact angle depends on the speed, but in general it is lower than 45°. At higher altitudes, there is less air resistance and thus greater range.

ASSESS The hardest part of any internet search is trying to determine which pages are actually useful. It is also possible to write a simple program, or spreadsheet, to estimate the effect of air resistance. Ask your instructor whether there is a computational physics course available on your campus!

4

FORCE AND MOTION

EXERCISES

Section 4.2 Newton's First and Second Laws

13. **INTERPRET** We interpret this as a problem involving the application of Newton's second law. The object under consideration is the train and the physical quantity of interest is the net force acting on it.

DEVELOP The net force can be found by using Equation 4.3, $\vec{F}_{net} = m\vec{a}$.

EVALUATE Using Equation 4.3, the magnitude of the force acting on the train is found to be

$$F_{net} = ma = (1.5 \times 10^6 \text{ kg})(2.5 \text{ m/s}^2) = 3.75 \times 10^6 \text{ N}$$

ASSESS The result is reasonable, since by definition, one newton is the force required to accelerate a 1-kg mass at the rate of 1 m/s².

15. **INTERPRET** We interpret this as a problem involving the application of Newton's second law. The object under consideration is the airplane and the physical quantity of interest is the plane's mass.

DEVELOP We shall assume that the runway is horizontal (so that the vertical force of gravity and the normal force of the surface cancel) and neglect aerodynamic forces (which are small just after the plane begins to move). Then the net force equals the engine's thrust and is parallel to the acceleration. The plane's mass can be found by using Equation 4.3, $\vec{F}_{net} = m\vec{a}$.

EVALUATE Using Equation 4.3, the mass of the plane is found to be

$$m = \frac{F_{net}}{a} = \frac{1.1 \times 10^4 \text{ N}}{7.2 \text{ m/s}^2} = 1.53 \times 10^3 \text{ kg}$$

ASSESS First, the units are consistent since $1 \text{ N} = 1 \text{ kg} \cdot \text{m/s}^2$. The result is reasonable, since by definition, one newton is the force required to accelerate a 1-kg mass at the rate of 1 m/s².

17. **INTERPRET** In this problem we want to find the relationship between the initial speed of the car and the force required to stop it.

DEVELOP From Equation 4.3, we see that the net force on a car of given mass is proportional to the acceleration, $F_{net} \sim a$. We can then relate the three quantities, displacement, velocity, and acceleration, by Equation 2.11, $v^2 = v_0^2 + 2a(x - x_0)$.

EVALUATE To stop a car in a given distance, $(x - x_0)$, the acceleration is $a = \frac{v^2 - v_0^2}{2(x - x_0)} = \frac{-v_0^2}{2(x - x_0)}$.

Therefore, we see that $F_{net} \sim v_0^2$. Doubling v_0 quadruples the magnitude of F_{net}.

ASSESS The conclusion that $F_{net} = v_0^2$ is an important fact to remember when driving at high speeds.

19. **INTERPRET** In this problem we want to find the force that's required to accelerate a car to cover a certain distance within a given time interval.

DEVELOP The displacement of the car as a function of time is given by Equation 2.10, $x = x_0 + v_0 t + \frac{1}{2}at^2$. The equation can be used to solve for the acceleration a. Also, from Equation 4.3, we see that the net force on a car of given mass is proportional to the acceleration, $F_{net} \sim a$.

EVALUATE Using Equation 2.10 with $v_0 = 0$, the acceleration of the car is

$$a = \frac{2(x - x_0)}{t^2} = \frac{2(400 \text{ m})}{(4.95 \text{ s})^2} = 32.6 \text{ m/s}^2$$

Newton's second law gives the average net force on the car as

$$F_{av} = ma = (940 \text{ kg})(32.6 \text{ m/s}^2) = 3.07 \times 10^4 \text{ N}$$

The force acts in the direction of the motion.

ASSESS Our answer for the acceleration a can be checked by using other kinematic equations. The speed of the car after 4.95 s is $v = at = (32.6 \text{ m/s}^2)(4.95 \text{ s}) = 161 \text{ m/s}$. Using Equation 2.11, $v^2 = v_0^2 + 2a(x - x_0)$, we find the distance traveled to be

$$x - x_0 = \frac{v^2 - v_0^2}{2a} = \frac{(161 \text{ m/s})^2 - 0}{2(32.6 \text{ m/s}^2)} = 400 \text{ m}$$

in agreement with the value given in the problem statement.

21. **INTERPRET** We interpret this as a problem involving the application of Newton's second law. The object under consideration is the car and the physical quantity of interest is the bumper deformation to withstand the impact force and avoid damage.

    DEVELOP In order that the force on the bumper does not exceed the stated limit, the magnitude of the deceleration should satisfy $a_{av} \leq \frac{F_{max}}{m}$. The distance moved by the bumper can then be calculated from Equation 2.11, $v^2 = v_0^2 + 2a(x - x_0)$.

    EVALUATE From the reasoning above, the magnitude of the maximum acceleration is

    $$a_{av} = \frac{F_{max}}{m} = \frac{65,000 \text{N}}{1300 \text{ kg}} = 50 \text{ m/s}^2$$

    With an initial speed of $v_0 = 10 \text{ km/h} = 2.78 \text{ m/s}$, the minimum bumper deformation is

    $$x - x_0 = \frac{|v^2 - v_0^2|}{2a} = \frac{|0 - (2.78 \text{ m/s})^2|}{2(50 \text{ m/s}^2)} = 0.0772 \text{ m} = 7.72 \text{ cm}$$

    ASSESS A bumper is typically allowed to deform up to a maximum of 12.5 cm before stopping the car.

Section 4.4 The Force of Gravity

23. **INTERPRET** The problem here is to identify the planet the person is in based on the gravitational force experienced.

    DEVELOP If the mass and weight are known, then the gravitational acceleration of the planet can be obtained by using Equation 4.5, $\vec{w} = m\vec{g}$.

    EVALUATE The surface gravity of the planet is

    $$g = \frac{w}{m} = \frac{532 \text{ N}}{60 \text{ kg}} = 8.87 \text{ m/s}^2$$

    precisely the value for Venus in Appendix E.

    ASSESS The gravitational acceleration of Venus is lower than that of the Earth. Therefore, the person's weight is less on Venus. The mass, however, remains unchanged.

25. **INTERPRET** In this problem we are asked about the actual weight of a cereal box in SI units and in ounces.

    DEVELOP In many contexts, the phrase "net weight" actually refers to the mass, rather than the actual weight of the object.

    EVALUATE (a) The actual weight (equal to the force of gravity at rest on the surface of the Earth) is

    $$w = mg = (0.340 \text{ kg})(9.81 \text{ m/s}^2) = 3.33 \text{ N}$$

    (b) With reference to Appendix C, 1 ounce is equal to the weight of 0.02835 kg, or $(0.02835 \text{ kg})(9.81 \text{ m/s}^2) = 0.2778 \text{ N}$. Therefore,

    $$w = (3.33 \text{ N})\left(\frac{1 \text{ oz}}{0.2778 \text{ N}}\right) = 12.0 \text{ oz}$$

ASSESS The word "net" in net weight means just the weight of the contents; gross weight includes the weight of the container, etc. This may be compared with the use of the word in net force, which means the sum of all the forces or the resultant force. A net weight, profit, or amount is the resultant after all corrections have been taken into account.

27. **INTERPRET** The problem is to find the weight of an object, given its mass and the magnitude of gravitational acceleration.

DEVELOP If the mass and the gravitational acceleration are known, the weight can be obtained by using Equation 4.5, $\vec{w} = m\vec{g}$.

EVALUATE The magnitude of the Earth's gravitational force on the astronaut at this altitude is

$$w = mg(r) = (68 \text{ kg})(0.93 \times 9.8 \text{ m/s}^2) = 620 \text{ N}$$

where $g(r)$ is the gravitational acceleration at the appropriate distance, r, from the center of the Earth.

ASSESS If the astronaut is orbiting with the shuttle, i.e., in free fall around the Earth, his or her "weight" is zero. An operational definition of "weight" is the force read on a scale at rest relative to the object being weighed.

Section 4.5 Using Newton's Second Law

29. **INTERPRET** In this problem there are two forces acting on the motorboat, the thrust and the drag. We are asked to find the drag force.

DEVELOP The horizontal forces acting on the boat are the thrust and the oppositely directed drag force (take the positive x axis in the direction of the acceleration). The net force on the motorboat is the vector sum of the two forces.

EVALUATE The horizontal component of Newton's second law gives

$$F_{net,x} = F_{thrust,x} + F_{drag,x} = 3900 \text{ N} + F_{drag,x} = ma_x = (930 \text{ kg})(2.3 \text{ m/s}^2) = 2139 \text{ N} \text{ from which one finds}$$

$$F_{drag,x} = 2139 \text{ N} - 3900 \text{ N} = -1761 \text{ N}$$

ASSESS As expected, the horizontal component of the drag force is negative, i.e., opposite to the direction of the acceleration.

31. **INTERPRET** In this problem there are two forces acting on the airplane, the downward gravitational force and the upward lift force. We are asked to find the lift force under different conditions.

DEVELOP The vertical forces on the airplane are lift and gravity, so with the y axis upward, $F_{net,y} = F_{lift} - mg = ma_y$, or

$$F_{lift} = m(g + a_y)$$

EVALUATE (a) At constant altitude, $a_y = 0$, and

$$F_{lift} = mg = (4.5 \times 10^5 \text{ kg})(9.8 \text{ m/s}^2) = 4.41 \times 10^6 \text{ N}$$

which is equal to the weight of the airplane.

(b) If $a_y = 1.1 \text{ m/s}^2$, then

$$F_{lift} = m(g + a_y) = (4.5 \times 10^5 \text{ kg})(9.8 \text{ m/s}^2 + 1.1 \text{ m/s}^2) = 4.91 \times 10^6 \text{ N}$$

ASSESS As expected, the greater the upward acceleration, the greater the upward lift force is required.

33. **INTERPRET** The problem is to find the apparent weight of a person riding an elevator that is accelerating.

DEVELOP Your apparent weight w_{app} is the force you exert on the floor of the elevator, equal and opposite to the upward force the floor exerts on you. The latter and gravitational force, which equals your actual weight w, are the two forces that determine your vertical acceleration, a_y. Take the positive direction to be vertically upward. The apparent weight is given by $w_{app} - w = ma_y$ or

$$w_{app} = w + ma_y = mg + ma_y = m(g + a_y)$$

Thus, $w_{app} > w = mg$ for upward acceleration $a_y > 0$. However, $w_{app} < w$ for downward acceleration.

EVALUATE In our case, the acceleration of the elevator is

$$a_y = \frac{\Delta v}{\Delta t} = -\frac{9.2 \text{ m/s}}{2.1 \text{ s}} = -4.38 \text{ m/s}^2$$

(negative sign for downward). Therefore, the apparent weight is

$$w_{app} = m(g + a_y) = mg\left(1 + \frac{a_y}{g}\right) = w\left(1 + \frac{-4.38 \text{ m/s}^2}{9.8 \text{ m/s}^2}\right) = 0.553w$$

That is, w_{app} is only about 55.3% of the actual weight.

ASSESS To see that our expression for w_{app} makes sense, let's consider the case of free fall. In this limit, $a_y = -g$, and we have the expected weightless situation, $w_{app} = 0$.

Section 4.6 Newton's Third Law

35. **INTERPRET** This problem deals with interaction between a pair of objects—the Earth and the person. The key concept involved here is Newton's third law.

    DEVELOP By Newton's third law, the force the Earth exerts on you is equal to the force you exert on the Earth, but in the opposite direction:

    $$\vec{F}_{\text{on you by E}} = -\vec{F}_{\text{on E by you}}$$

    However, the force exerted on you by Earth is simply equal to the weight force: $\vec{F}_{\text{on you by E}} = \vec{w} = m\vec{g}$. Therefore, as you fall toward the Earth, your acceleration would be

    $$\vec{a} = \frac{\vec{F}_{\text{on you by E}}}{m} = \vec{g}$$

    On the other hand, for the Earth with mass M_E, its acceleration is

    $$\vec{a}_E = \frac{\vec{F}_{\text{on E by you}}}{M_E} = \frac{-\vec{F}_{\text{on you by E}}}{M_E} = -\frac{m\vec{g}}{M_E}$$

    Clearly, $\vec{a}$ and $\vec{a}_E$ point in the opposite directions. If you and the Earth both start from rest, the distance moved by each would be

    $$d = \frac{1}{2}|a|t^2 = \frac{1}{2}gt^2 \quad \text{(down)}$$

    $$d_E = \frac{1}{2}|a_E|t^2 = \frac{1}{2}\frac{m}{M_E}gt^2 = \left(\frac{m}{M_E}\right)d \quad \text{(up)}$$

    with $d + d_E = 1.2$ m.

    EVALUATE Solving for d_E using the equations obtained above gives

    $$d_E = \frac{1.2 \text{ m}}{1 + M_E/m}$$

    Since $M_E/m = (5.97 \times 10^{24} \text{ kg})/(65 \text{ kg}) = 9.18 \times 10^{22}$, the distance moved by the Earth is

    $$d_E \approx \frac{1.2 \text{ m}}{9.18 \times 10^{22}} = 1.3 \times 10^{-23} \text{ m}$$

    ASSESS The value of d_E is too small to be noticeable; it is about 10^5 times smaller than the smallest physically meaningful distances studied to date.

37. **INTERPRET** This is a problem that involves applying Hooke's law to a spring that is being stretched.

    DEVELOP Given the spring force F_{sp} and the spring constant k, the length stretched can be calculated by using Hooke's law in Equations 4.9: $F_{sp} = -kx$.

EVALUATE From Hooke's law, a force of magnitude 35 N produces a stretch of

$$|x| = \frac{|F_{sp}|}{k} = \frac{35 \text{ N}}{220 \text{ N/m}} = 0.159 \text{ m} = 15.9 \text{ cm}$$

ASSESS A spring constant of $k = 220$ N/m $= 2.20$ N/cm means that a force of 2.20 N must be applied to stretch (or compress) the spring by 1.0 cm. Therefore, to stretch the spring by 15.9 cm would require a force of 35 N.

PROBLEMS

39. **INTERPRET** This problem is about finding the force that changes the velocity of a moving object.

DEVELOP Newton's second law, shown in Equation 4.2, says that the average force acting on an object is equal to the rate of change of momentum: $\vec{F}_{av} = m\frac{\Delta \vec{v}}{\Delta t}$. Once $\Delta \vec{v}$, the change of velocity is known, $\vec{F}_{av}$ can be readily calculated.

EVALUATE The initial velocity is $\vec{v}_1 = (17.4 \text{ m/s})i$ m/s, and the final velocity is

$$\vec{v}_2 = (26.8 \text{ m/s})(\cos 34° \hat{i} + \sin 34° \hat{j}) = (22.2 \text{ m/s})\hat{i} + (15 \text{ m/s})\hat{j}$$

so $\Delta \vec{v} = \vec{v}_2 - \vec{v}_1 = [(22.2 \text{ m/s}) - (17.4 \text{ m/s})]\hat{i} + (15 \text{ m/s})\hat{j} = (4.8 \text{ m/s})\hat{i} + (15 \text{ m/s})\hat{j}$

Therefore, the average force is

$$\vec{F}_{av} = m\frac{\Delta \vec{v}}{\Delta t} = (1.25 \text{ kg})\frac{(4.8 \text{ m/s})\hat{i} + (15 \text{ m/s})\hat{j}}{3.41 \text{ s}} = (1.77 \text{ N})\hat{i} + (5.49 \text{ N})\hat{j}$$

The magnitude of $\vec{F}_{av}$ is $F_{av} = \sqrt{(1.77 \text{ N})^2 + (5.49 \text{ N})^2} = 5.77$ N, and the direction is

$$\theta = \tan^{-1}\left(\frac{F_{av,y}}{F_{av,x}}\right) = \tan^{-1}\left(\frac{5.49 \text{ N}}{1.77 \text{ N}}\right) = 72°$$

measured CCW from the x axis.

ASSESS Note that the force $\vec{F}_{av}$ is in the same direction as $\Delta \vec{v}$, not the final velocity $\vec{v}_2$. The latter holds only when the object is initially at rest so that $\Delta \vec{v} = \vec{v}_2$.

41. **INTERPRET** The problem involves the application of Newton's second law. Two forces are acting on the tree surgeon, the gravitational force $\vec{F}_g$ and the normal force $\vec{n}$ of the bucket. We are asked to calculate $\vec{n}$ under various conditions.

DEVELOP We shall assume that the only vertical forces acting on the tree surgeon are those given, namely, the force of gravity, $\vec{F}_g = m\vec{g}$ acting downward, and the normal force of the bucket, $\vec{n}$ acting upward. Taking the upward direction to be positive, the net force is $F_{net,y} = -mg + n = ma_y$, which gives

$$n = m(g + a_y)$$

EVALUATE For (a), (b), and (c), the acceleration of the tree surgeon is zero, the normal force is equal (in magnitude) to the weight,

$$n = mg = (74 \text{ kg})(9.8 \text{ m/s}^2) = 725 \text{ N}$$

(d) If $a_y = +1.7$ m/s^2, $n = (74 \text{ kg})(9.8 \text{ m/s}^2 + 1.7 \text{ m/s}^2) = 851$ N.
(e) Similarly, for $a_y = -1.7$ m/s^2, we have $n = (74 \text{ kg})(9.8 \text{ m/s}^2 - 1.7 \text{ m/s}^2) = 599$ N.

ASSESS The upward normal force exerted by the bucket is greatest when the lift is moving upward with a non-zero acceleration. The tree surgeon's feet will feel "heavy."

43. **INTERPRET** The problem involves the application of Newton's second law. The object of interest is the passenger riding the elevator, and we are asked to compute the minimum stopping time the elevator can have so that the passengers are to remain on the floor.

DEVELOP We shall take the y axis positive upward. There are two vertical forces on a passenger, gravity $\vec{F}_g = m\vec{g}$ downward (magnitude equal to his or her weight), and the upward normal force of the floor $\vec{n}$. The latter is a contact force and always acts in a direction away from the surface of contact. (Otherwise, we would be dealing with an adhesive force.)

EVALUATE   Using the *y* component of Newton's second law $n - mg = ma_y$, we may express the condition for contact as

$$n = m(a_y + g) \geq 0 \quad \rightarrow \quad a_y \geq -g$$

since *m* is positive. In words, the passenger remains in contact with the floor as long as the vertical acceleration of the elevator is either upward, or downward with magnitude less than *g*. Using Equation 3.8, the time required the elevator come to a stop $(v_y = 0)$ from an initial upward velocity $(v_{0y} = 5.2 \text{ m/s})$ is

$$t = \frac{v_y - v_{0y}}{a_y} = \frac{-v_{0y}}{a_y}$$

Therefore, $t \geq \frac{v_{0y}}{g} = \frac{5.2 \text{ m/s}}{9.8 \text{ m/s}^2} = 0.531$ s is the condition for the passengers to stay on the floor.

ASSESS   Half a second is a reasonable value. The condition $n = 0$ is the limit that the person and the floor remain in contact. As long as the passenger is in contact with the floor, his or her vertical acceleration is the same as that of the floor and the elevator.

45.   **INTERPRET**   This problem deals with interaction between different pairs of objects. The key concepts involved here are Newton's second and third laws.

DEVELOP   Let the three masses be denoted, from left to right, as m_1, m_2, and m_3, respectively. We take the right direction to be $+x$. Assume that the table surface is horizontal and frictionless so that the only horizontal forces are the applied force and the contact forces between the blocks. For example, $\vec{F}_{12}$ denotes the force exerted by block 1 on block 2. Since the blocks are in contact, they all have the same acceleration *a*, to the right. Newton's second law can be applied to each block separately:

$$\vec{F}_{app} + \vec{F}_{21} = m_1 \vec{a}$$
$$\vec{F}_{12} + \vec{F}_{32} = m_2 \vec{a}$$
$$\vec{F}_{23} = m_3 \vec{a}$$

EVALUATE   Adding all three equations and using Newton's third law ($\vec{F}_{12} + \vec{F}_{21} = 0$, etc.), one finds

$$\vec{a} = \frac{\vec{F}_{app}}{m_1 + m_2 + m_3} = \frac{12 \text{ N}}{1 \text{kg} + 2 \text{kg} + 3 \text{kg}} = 2 \text{ m/s}^2 \quad \text{(to the right)}$$

Thus, the force block 2 exerts on block 3 is $F_{23} = m_3 a = (3 \text{ kg})(2.0 \text{ m/s}^2) = 6.0 \text{ N}$ (to the right).

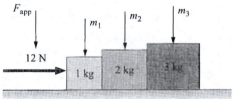

ASSESS   If we draw a free-body diagram for m_3, we will see that F_{23} is the only force acting on it (ignoring friction). Thus, the force that accelerates m_3 comes entirely from F_{23}.

47.   **INTERPRET**   The objects of interest are the airplane and the two gliders, connected by ropes. The problem involves applying Newton's second law to find horizontal thrust of the plane's propeller, the tension forces in each rope, and the net force on the first glider.

DEVELOP   Let the masses of the airplane and the two gliders be denoted, from right to left, as m_1, m_2, and m_3, respectively (see figure). We take the right direction to be $+x$. Assume that the runway surface is horizontal and frictionless so that the only horizontal forces are the thrust and the tensions in the ropes.

We may then write the horizontal component (positive in direction of the acceleration) of the equations of motion (Newton's second law) for the three planes (all assumed to have the same acceleration) as follows:

$$F_{th} - T_1 = m_1 a \quad \text{(airplane)}$$
$$T_1 - T_2 = m_2 a \quad \text{(first glider)}$$
$$T_2 = m_3 a \quad \text{(second glider)}$$

Note: The tension has the same magnitude at every point in a rope of negligible mass.

EVALUATE **(a)** Add all the equations of motion (the tensions cancel in pairs due to Newton's third law), the thrust of the airplane is

$$F_{th} = (m_1 + m_2 + m_3)a = (2200\,\text{kg} + 310\,\text{kg} + 260\,\text{kg})(1.9\ \text{m/s}^2) = 5.26 \times 10^3\,\text{N}$$

(b) The tension in the first rope is

$$T_1 = F_{th} - m_1 a = (m_2 + m_3)a = (310\ \text{kg} + 260\,\text{kg})(1.9\ \text{m/s}^2) = 1.08 \times 10^3\ \text{N}$$

(c) The tension in the second rope is $T_2 = m_3 a = (260\ \text{kg})(1.9\ \text{m/s}^2) = 494\ \text{N}$.

(d) The net force on the first glider is $F_{net} = m_2 a = (310\ \text{kg})(1.9\ \text{m/s}^2) = 589\ \text{N}$.

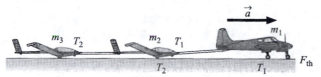

ASSESS The tension in the first rope provides the force of acceleration for m_2 and m_3, while the tension in the second force accelerates only m_3.

49. **INTERPRET** The object of interest is the rat. The problem involves applying Newton's second law to find the mass of the rat.

DEVELOP Using Newton's second law stated in Equation 4.3, the mass of a system undergoing acceleration can be written as $m = F_{net}/a$. According to the scale, a force of 0.46 N applied to the cage and rat produces an acceleration of 0.40 m/s^2:

$$F_{net} = (m_{cage} + m_{rat})a \quad \rightarrow \quad m_{cage} + m_{rat} = \frac{F_{net}}{a} = \frac{0.46\ \text{N}}{0.40\ \text{m/s}^2} = 1.15\ \text{kg}$$

Since the mass of the cage is $m_{cage} = 0.32\,\text{kg}$, the mass of the rat is

$$m_{rat} = 1.15\ \text{kg} - m_{cage} = 1.15\ \text{kg} - 0.32\,\text{kg} = 0.83\,\text{kg}\ \ (\text{or } 830\ \text{g})$$

ASSESS The answer can be checked by asking the force required to provide an acceleration of 0.40 m/s^2 to the cage and the rat. We find

$$F_{net} = (m_{cage} + m_{rat})a = (0.83\,\text{kg} + 0.32\,\text{kg})(0.40\ \text{m/s}^2) = 0.46\ \text{N}$$

which is the value given in the problem statement.

51. **INTERPRET** This is a problem that involves applying Newton's second law and Hooke's law to a spring that is being stretched.

DEVELOP Given the spring force F_{sp} and the spring constant k, the length stretched can be calculated by using Hooke's law in Equation 4.9: $F_{sp} = -kx$ (the negative sign means that the spring force opposes the distortion). The spring stretches until the acceleration of both masses is the same (positive in the direction of the applied force in Fig. 4.23). Since the spring is assumed massless, the tension in it is the same at both ends. If this were not so, there would be a non-zero net horizontal force on the spring.

EVALUATE The magnitude of the spring tension is given by Hooke's law, $F_{sp} = k\,|x|$ where $|x|$ is the stretch. The horizontal component of Newton's second law applied to each mass is

$$F_{app} - F_{sp} = m_3 a$$
$$F_{sp} = m_2 a$$

as indicated in the sketch. Adding the two equations, the acceleration of the entire system is

$$a = \frac{F_{app}}{m_2 + m_3} = \frac{15\ \text{N}}{2.0\,\text{kg} + 3.0\,\text{kg}} = 3.0\ \text{m/s}^2$$

The spring force is

$$F_{sp} = m_2 a = (2.0\,\text{kg})(3.0\ \text{m/s}^2) = 6.0\ \text{N}$$

Applying Hooke's law, the amount of stretching of the spring is

$$|x| = \frac{F_{sp}}{k} = \frac{6.0\ \text{N}}{140\ \text{N/m}} = 0.0429\ \text{m} = 4.29\ \text{cm}$$

ASSESS   The spring force may be rewritten as

$$F_{sp} = m_2 a = \left(\frac{m_2}{m_2 + m_3} \right) F_{app}$$

In the limit where $m_2 \gg m_3$, $F_{sp} \approx F_{app}$. On the other hand, if the mass m_2 is negligible, then $F_{sp} \approx 0$, as expected.

53. **INTERPRET**   The problem involves finding the force exerted on the air by the blade of a helicopter under various conditions. The key concept involved here is Newton's third law.

    DEVELOP   Using Newton's third law, we see that the downward force exerted on the air by the helicopter is equal and opposite to the upward force on the helicopter (the engine's thrust). If we neglect air resistance, the thrust and gravity are the only vertical forces acting, so Newton's second law for the helicopter (positive component up) is

    $$F_{th} - mg = ma \quad \rightarrow \quad F_{th} = m(g + a)$$

    EVALUATE   (a) Hovering means zero acceleration, $a = 0$ (also $v = 0$, but v doesn't enter the equation of motion if air resistance is neglected). Therefore, the downward force on the air is

    $$F_{th} = mg = (4300 \text{ kg}) (9.8 \text{ m/s}^2) = 42.1 \times 10^3 \text{ N}$$

    (b) If v is decreasing downward, then the acceleration must be $a = 3.2 \text{ m/s}^2$ upward, and

    $$F_{th} = m(g + a) = (4300 \text{ kg})(9.8 \text{ m/s}^2 + 3.2 \text{ m/s}^2) = 55.9 \times 10^3 \text{ N}$$

    (c) In this case, the acceleration a is the same as in part (b), $a = 3.2 \text{ m/s}^2$ upward , and so is F_{th}:

    $$F_{th} = m(g + a) = (4300 \text{ kg})(9.8 \text{ m/s}^2 + 3.2 \text{ m/s}^2) = 55.9 \times 10^3 \text{ N}$$

    (d) If the speed v is constant, then $a = 0$, and F_{th} is the same as the hovering case in (a):

    $$F_{th} = mg = (4300 \text{ kg})(9.8 \text{ m/s}^2) = 42.1 \times 10^3 \text{ N}$$

    (e) If v is decreasing upward, then a is downward, and

    $$F_{th} = (4300 \text{ kg})(9.8 \text{ m/s}^2 - 3.2 \text{ m/s}^2) = 28.4 \times 10^3 \text{ N}$$

    ASSESS   The thrust force from the engine is greatest when the helicopter is either moving upward and accelerating, or moving downward and decelerating. In this case, the force exerted on the air is also the greatest, by Newton's third law.

55. **INTERPRET**   In this problem we are asked to find out how many 65-kg passengers an elevator can accommodate within the guideline of safety standards. The forces involved here are the downward gravitational force $\vec{F}_g$ and the upward cable tension $\vec{T}$.

    DEVELOP   Assume that the only forces involved are $\vec{F}_g$ and $\vec{T}$ in the vertical direction. Newton's second law gives $\vec{F}_{net} = \vec{T} + \vec{F}_g = M\vec{a}$, where M is the total mass of the elevator and its passengers. Taking $+y$ to point upward, the equation in component form is $T - Mg = Ma_y$, which implies

    $$T = M(g + a_y)$$

    EVALUATE   For safety's sake, we require that

    $$T \leq \frac{2}{3} T_{max} = \frac{2}{3}(19,500 \text{ N}) = 13,000 \text{ N}$$

    Therefore, if the acceleration is $a = 2.24 \text{ m/s}^2$, then the upper mass limit (combined mass of elevator and passenger) is

    $$M = \frac{T}{g + a_y} \leq \frac{13000 \text{ N}}{9.8 \text{ m/s}^2 + 2.24 \text{ m/s}^2} = 1080 \text{ kg}$$

    Since M equals the sum of elevator mass $m_{elevator} = 490 \text{ kg}$ and the mass of N passengers, each with $m = 65 \text{ kg}$, we have $N < \frac{M - m_{elevator}}{m} = \frac{1080 \text{ kg} - 490 \text{ kg}}{65 \text{ kg}} = 9.07$ or $n \leq 9$.

ASSESS   An elevator that accommodates 9 passengers, with a total mass of 590 kg sounds reasonable. Many passenger elevators, depending on their size, can accommodate up to about 2500 kg.

57.   INTERPRET   This problem involves computing the total force exerted on two springs (of spring constants k_1 and k_2) that are connected side-by-side or end-to-end.

DEVELOP   Two springs are connected side-by-side (in "parallel"), $F = F_1 + F_2$, and $x = x_1 = x_2$, where F and x are the (magnitude of the) force and stretch of the combination, and subscripts 1 and 2 refer to the individual springs. On the other hand, when connected end-to-end (in "series"), the tension is the same in both springs, $F = F_1 = F_2$ (true for "massless" springs), while the total stretch is the sum of the individual stretches, $x = x_1 + x_2$.

EVALUATE   (a) For the "series" combination, from Hooke's law, $F_1 = k_1 x_1$, and $F_2 = k_2 x_2$. Therefore, the total force is

$$F = k_1 x_1 + k_2 x_2 = (k_1 + k_2)x$$

(b) In the "series" combination, again using Hooke's Law, we find

$$x = x_1 + x_2 = \frac{F_1}{k_1} + \frac{F_2}{k_2} = F\left(\frac{1}{k_1} + \frac{1}{k_2}\right) = F\left(\frac{k_1 + k_2}{k_1 k_2}\right)$$

or

$$F = \left(\frac{k_1 k_2}{k_1 + k_2}\right)x$$

ASSESS   For a system with many springs, we may define an effective spring constant as $k_{eff} = F/x$. In the parallel case, we have $k_p = k_1 + k_2$, and in the series case, $k_s = k_1 k_2/(k_1 + k_2)$. Common experience tells us that the parallel combination is stiffer than the series combination, and thus requires a greater amount of force to stretch by the same amount. One can readily see this by considering the simple case where $k_1 = k_2 = k$. The above formulae give $k_p = 2k$ and $k_s = k/2$.

59.   INTERPRET   We use Newton's second law, in the most general form, to find the force necessary to keep a constant velocity while mass changes. We must find the force applied by the engine on the railroad car.

DEVELOP   We use the result of Problem 58: $F = ma + v\frac{dm}{dt}$. The velocity v is a constant 2.0 m/s, so the acceleration a is zero. The car gains mass at a rate of $\frac{dm}{dt} = 450$ kg/s.

EVALUATE $F = ma + v\frac{dm}{dt} = (2.0 \text{ m/s}) \times (250 \text{ kg/s}) = 500 \text{ kg m/s}^2 = 500 \text{ N}$

ASSESS   Note that no force is required to keep the railroad car itself moving: The ma term is zero. This 500-N force is the force needed to accelerate the grain so that it is moving at the same speed as the car.

61.   INTERPRET   This is a Newton's second law problem: The mass being accelerated is the bucket of concrete, and the forces involved are the tension in the crane cable and the force of gravity. We want to find the tension.

DEVELOP   We draw a free-body diagram and find the net force on the bucket (see the figure).

Use $F = ma$ and the given acceleration of $a = 2.6$ m/s^2 to find the tension in the cable.

EVALUATE   The net force is $\vec{F} = \vec{T} + \vec{F}_g \rightarrow F = T - mg$.

$F = ma \rightarrow T - mg = ma \rightarrow T = m(a + g) = (1200 \text{ kg}) \times (2.6 \text{ m/s}^2 + 9.8 \text{ m/s}^2) = 14,900 \text{ N}$

ASSESS   *Check:* Is this force more than the weight of the bucket? The weight is $mg = 11,800$ N, which is good. In our equation, what happens if $a = 0$? We get the weight of the bucket again. Everything checks out ok, then.

63.   INTERPRET   We are asked to find the force necessary to stop a jet in a given distance. We know the mass and the speed of the jet.

DEVELOP   We can use the distance and the speed in the equations for motion in constant acceleration. Starting with $v^2 = v_o^2 + 2a(x - x_o)$, find the necessary acceleration a. Then we apply Newton's second law, $F = ma$, to find the force.

EVALUATE
$$v^2 = v_0^2 + 2a(x - x_0) \to 0 = v_0^2 + 2ax \to a = -\frac{v_0^2}{2x}$$

$$F = ma = -m\frac{v_0^2}{2x} = -(55 \times 10^3 \text{ kg})\frac{(36 \text{ m/s})^2}{2(120 \text{ m})} = -280 \text{ kN}$$

ASSESS   The force is negative, since we took our direction to be positive.

65. **INTERPRET**   We use Newton's second law, and a spring and a mass, to find the acceleration of an elevator and the force on a person in the elevator. The force and stretch of the spring are related by Hooke's law.
DEVELOP   (a) From Hooke's law, $F = -kx$, we can find the force applied by the spring on the mass with a spring stretch of $x = 0.05$ m and a spring constant of $k = 1080$ N/m. We compare this force with the weight of the mass to find whether the elevator is accelerating, and how much, using $F = ma$.
(b) From the acceleration found in part (a), and $F = ma$, find the force of the scale on the person. Remember that the person is holding an extra 5 kg mass.
EVALUATE   (a) $F = -kx = -(1080 \text{ N/m}) \times (-0.05 \text{ m}) = 54$ N. The actual weight of the mass is
$F_g = -mg = -49$ N. We can see that the spring force is larger than what is needed to support the mass, so the mass is accelerating upward. We find the actual acceleration from the excess force:
$$\Delta F = 5 \text{ N} = ma = (5 \text{ kg})a \to a = 1.0 \text{ m/s}^2 \text{ upward}$$

(b) The scale must provide enough force to counteract gravity and accelerate the person at 1.0 m/s², so
$$F = ma = (70 \text{ kg})(9.8 \text{ m/s}^2 + 1.0 \text{ m/s}^2) = 756 \text{ N}$$

ASSESS   This problem, *like most force problems,* is easier if you draw a free-body diagram first.

67. **INTERPRET**   We relate force and acceleration using the general form of Newton's second law and relativistic momentum.
DEVELOP   The general form of Newton's second law is $F = \frac{dp}{dt}$, and relativistic momentum is $p = \dfrac{mu}{\sqrt{1 - \frac{u^2}{c^2}}}$.

Acceleration is $a = \dfrac{du}{dt}$. We'll start by substituting the relativistic momentum into the force law and taking the derivative.
EVALUATE

$$F = \frac{dp}{dt} = \frac{d}{dt}\left[\frac{mu}{\sqrt{1 - \frac{u^2}{c^2}}}\right] = \frac{m}{\sqrt{1 - \frac{u^2}{c^2}}}\left(\frac{du}{dt}\right) - \frac{1}{2}\frac{mu}{\left(1 - \frac{u^2}{c^2}\right)^{\frac{3}{2}}}\left(-2\frac{u}{c^2}\right)\left(\frac{du}{dt}\right)$$

$$F = \frac{m}{\left(1 - \frac{u^2}{c^2}\right)^{\frac{1}{2}}}(a) + \frac{m}{\left(1 - \frac{u^2}{c^2}\right)^{\frac{3}{2}}}\left(\frac{u^2}{c^2}\right)(a)$$

$$F = ma\left[\frac{\left(1 - \frac{u^2}{c^2}\right) + \frac{u^2}{c^2}}{\left(1 - \frac{u^2}{c^2}\right)^{\frac{3}{2}}}\right] = ma\left[\frac{1}{\left(1 - \frac{u^2}{c^2}\right)^{\frac{3}{2}}}\right]$$

ASSESS   This looks more complex than it really is: We're just using the chain rule and the product rule for differentiation. The term in square brackets in our final answer is unitless, so the units work out fine. At "normal" speeds ($u = c$), $\frac{u^2}{c^2} \approx 0 \to F \approx ma$, so we can continue to use the equation we've been using as long as our speeds stay "slow."

69. **INTERPRET**   We need to find the acceleration required to stop a car with the given speed in just under a meter, and express our answer in g's. This problem does not involve force, just acceleration; so we use the equations of motion for constant acceleration.
DEVELOP   We can relate acceleration, speed, and distance with $v^2 = v_0^2 + 2a(x - x_0)$, after converting our speed to units of m/s. Once we solve the equation for a, divide by g to find the answer in g's.

EVALUATE $v_0 = \frac{70 \text{ km}}{1 \text{ h}} \times \frac{1 \text{ h}}{3600 \text{ s}} \times \frac{1000 \text{ m}}{1 \text{ km}} = 19.4$ m/s.

We solve for a:

$$v^2 = v_0^2 + 2a(x - x_0) \rightarrow 0 = v_0^2 + 2ax \rightarrow a = -\frac{v_0^2}{2x}$$

Next we find the answer, in units of g:

$$\left| \frac{a}{g} \right| = \frac{v_0^2}{2gx} = 2.1 \ g's$$

ASSESS   This is hard on the car, but quite survivable for the passengers. Note that for a given speed, having the car crumple further *decreases* the magnitude of the acceleration of the passengers. Modern cars are designed specifically to crumple on impact for just this reason.

5

USING NEWTON'S LAWS

EXERCISES

Section 5.1 Using Newton's Second Law

13. **INTERPRET** In this problem, two forces are exerted on the object of interest and produce an acceleration. With the mass of the object and one force given, we are asked to find the other force.

DEVELOP Newton's second law for this mass says $\vec{F}_{net} = \vec{F}_1 + \vec{F}_2 = m\vec{a}$, where we assume no other significant forces are acting. The second force is given by

$$\vec{F}_2 = m\vec{a} - \vec{F}_1$$

EVALUATE Substituting the expressions given in the problem statement, we obtain

$$\vec{F}_2 = m\vec{a} - \vec{F}_1 = (3.1 \text{ kg})[(0.91 \text{ m/s}^2)\hat{i} - (0.27 \text{ m/s}^2)\hat{j}] - [(-1.2 \text{ N})\hat{i} - (2.5 \text{ N})\hat{j}] = (4.02 \text{ N})\hat{i} + (1.66 \text{ N})\hat{j}$$

ASSESS To show that the answer is correct, let's add the two forces together. The net force is

$$\vec{F}_{net} = \vec{F}_1 + \vec{F}_2 = [(-1.2 \text{ N})\hat{i} - (2.5 \text{ N})\hat{j}] + [(4.02 \text{ N})\hat{i} + (1.66 \text{ N})\hat{j}]$$
$$= (2.82 \text{ N})\hat{i} + (-0.84 \text{ N})\hat{j} = (3.1)[(0.91 \text{ N})\hat{i} + (-0.27 \text{ N})\hat{j}]$$

From the expression, it is clear that $\vec{F}_{net}$ points in the same direction as $\vec{a}$.

15. **INTERPRET** In this problem, we are asked to find the tilt angle of the table such that the acceleration of an object sliding on the surface of the table is the same as the gravitational acceleration near the surface of the Moon.

DEVELOP In Example 5.1 it was shown that the acceleration down an incline is $a = g\sin\theta$. This equation is what we shall use to solve for the tilt angle θ.

EVALUATE Given that the gravitational acceleration near the surface of the Moon is $g = 1.6 \text{ m/s}^2$, the angle of tilt should be

$$\theta = \sin^{-1}\left(\frac{a}{g}\right) = \sin^{-1}\left(\frac{1.6 \text{ m/s}^2}{9.8 \text{ m/s}^2}\right) = 9.40°$$

ASSESS A tilt angle of 9.40° would give an acceleration that is the same as that of the Moon's surface. On the other hand, if we want to simulate the motion near Earth's surface, then $a = g$, and $\theta = 90°$. The greater the value of a, the larger the tilt angle.

17. **INTERPRET** In this problem, the physical quantity of interest is the tension force in the cable. To compute the tension, we analyze the motion of the car.

DEVELOP We note that the only force on the car with a horizontal component (in the direction of the acceleration) is the tension. In addition, the acceleration of the car is the same as that of the tow truck.

EVALUATE Applying Newton's second law to the motion in the horizontal direction, we obtain $T\cos\theta = ma$, or

$$T = \frac{ma}{\cos\theta} = \frac{(1400 \text{ kg})(0.57 \text{ m/s}^2)}{\cos 25°} = 880 \text{ N}$$

Note that the force of static friction acts between the tires and the road with magnitude sufficient to keep the wheels turning, but is assumed to be negligible.

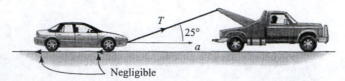

Negligible

ASSESS To see that our expression for T makes sense, let's consider the limiting case where the cable is completely horizontal with $\theta = 0$. The tension in this case would simply be $T = ma$, which is the force required to accelerate the car.

Section 5.2 Multiple Objects

19. **INTERPRET** In this problem two masses that rest on slopes of unequal angles are being connected by a rope passing over a pulley. We are asked to find out the ratio of the two masses if they both remain at rest.

DEVELOP Let the masses on the right and on the left be denoted as m_r and m_L, respectively. The free-body diagrams for m_r and m_L are shown in the sketch, where there is only a normal contact force since each slope is frictionless, and we indicate separate parallel and perpendicular x-y axes.

If the masses don't slide, the net force on each must be zero.

EVALUATE Applying Newton's second law to the force component that is parallel to the surface, we have

$$T_L - m_L g \sin \theta_L = 0$$
$$m_r g \sin \theta_r - T_r = 0$$

Now, if the masses of the string and pulley are negligible and there is no friction, then the tension must be the same throughout the entire string: $T_L = T_r$.

Finally, adding the two force equations, we find $m_r g \sin \theta_r - m_L g \sin \theta_L = 0$. With $\theta_L = 60°$ and $\theta_r = 20°$, the mass ratio must be $\frac{m_r}{m_L} = \frac{\sin \theta_L}{\sin \theta_r} = \frac{\sin 60°}{\sin 20°} = 2.53$ for no motion.

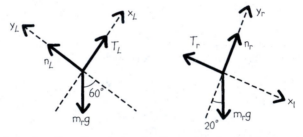

ASSESS We conclude that when $\theta_L > \theta_r$, $m_r > m_\ell$ for the system to remain at rest. This makes sense because when the angle θ_L on the left-hand side increases, there is a greater tendency for m_L to slide down. To counter this, m_r, the mass on the right-hand-side must go up! We may also consider the extreme case where $\theta_r = 0$. The situation corresponds to having a mass m_r on a frictionless horizontal surface, with m_L hanging over the edge and connected to m_r via a string. For this system to remain stationary, m_r must be infinitely massive compared to m_L!

21. **INTERPRET** This problem is about two climbers, tied together by a rope, sliding down the icy mountainside. The physical quantities of interest are their acceleration, and the force required to bring them to a complete stop.

DEVELOP We choose a coordinate system with x axis parallel to the slope and y axis perpendicular to it (see Example 5.1). Assume that the icy surface is frictionless and that the climbers move together as a unit, with the same magnitude of down-slope acceleration a, then the net force acting on them is the sum of the down-slope components of gravity on each. The force components on the upper and lower climbers are $F_u = m_u g \sin \theta_u$ and $F_l = m_l g \sin \theta_l$, respectively.

EVALUATE The net force in the x direction (downward positive) is

$$F_{net} = F_u + F_l = m_u g \sin \theta_u + m_l g \sin \theta_l = (75 \text{ kg}) g \sin 12° + (63 \text{ kg}) g \sin 38° = 533 \text{ N}$$

Therefore, the acceleration of the pair is

$$a = \frac{F_{net}}{m_u + m_l} = \frac{533 \text{ N}}{75 \text{ kg} + 63 \text{ kg}} = 3.86 \text{ m/s}^2$$

(b) After they have stopped, the force of the ice ax against the ice must balance the down-slope components of gravity calculated in part **(a)**. That is, the force exerted by the ax must be $F_{ax} = 533$ N up the slope.

ASSESS   If the two climbers were not roped together, then their acceleration would have been

$$a_u = \frac{F_u}{m_u} = g \sin\theta_u = (9.8 \text{ m/s}^2)\sin 12° = 2.04 \text{ m/s}^2$$

$$a_l = \frac{F_l}{m_l} = g \sin\theta_l = (9.8 \text{ m/s}^2)\sin 38° = 6.03 \text{ m/s}^2$$

The acceleration of the pair is the mass-weighted average of the individual accelerations:

$$a = \frac{F_{net}}{m_u + m_l} = \frac{m_u g \sin\theta_u + m_l g \sin\theta_l}{m_u + m_l} = \left(\frac{m_u}{m_u + m_l}\right)g\sin\theta_u + \left(\frac{m_l}{m_u + m_l}\right)g\sin\theta_l = \left(\frac{m_u}{m_u + m_l}\right)a_u + \left(\frac{m_l}{m_u + m_l}\right)a_l$$

Section 5.3 Circular Motion

23.   **INTERPRET**   In this problem we are asked to show that the force required to keep a mass m in a circular path of radius r with period T is $4\pi^2 mr/T^2$.

DEVELOP   To derive the formula, we first note that for an object of mass m in uniform circular motion, the magnitude of the net force is given by Equation 5.1: $F = ma = mv^2/r$. Next, we make use of the fact that the period of the motion (time for one revolution) is $T = 2\pi r/v$.

EVALUATE   Combining the two expressions, the force can be rewritten as

$$F = \frac{mv^2}{r} = \frac{m}{r}\left(\frac{2\pi r}{T}\right)^2 = \frac{4\pi^2 mr}{T^2}$$

ASSESS   Our result indicates that for a fixed radius r, the centripetal force is inversely proportional to T^2. For example, if T is very large (i.e., it takes a very long time for the mass to complete one revolution), then the speed v is very small and the centripetal force F is also very small.

25.   **INTERPRET**   This problem involves circular motion and Newton's second law. By analyzing the force acting on the unused strap inside the train, the radius of the circular turn can be determined.

DEVELOP   The net force on the unused strap is the vector sum of the tension in the strap (acting along its length at $15°$ to the vertical) and its weight. This must equal the mass times the horizontal centripetal acceleration. The free-body diagram for the strap is shown on the right. The component equations are

$$x: \quad T\sin\theta = \frac{mv^2}{r}$$
$$y: \quad T\cos\theta = mg$$

EVALUATE   Dividing the two equations allows us to eliminate T and obtain

$$\frac{T\sin\theta}{T\cos\theta} = \frac{mv^2/r}{mg} \quad \rightarrow \quad \tan\theta = \frac{v^2}{rg}$$

With $v = 67$ km/h $= 18.6$ m/s, we solve for the radius of the turn and find

$$r = \frac{v^2}{g\tan\theta} = \frac{(18.6 \text{ m/s})^2}{(9.8 \text{ m/s}^2)\tan 15°} = 132 \text{ m}$$

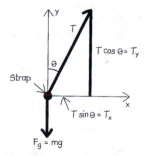

ASSESS   The result indicates that the radius r is inversely proportional to $\tan\theta$. If the speed v is kept fixed, a smaller radius (sharper turn) would result in a larger angle. This agrees with our experience.

27.   **INTERPRET**   This problem involves circular motion and Newton's second law. The object of interest is the airplane. By analyzing the force acting on the plane while it travels a circular path, its speed can be determined.

DEVELOP   We follow the strategy outlined in Example 5.6. There are two forces acting on the plane: gravitational force $\vec{F}_g$ and the normal force $\vec{n}$. Applying Newton's second law gives $\vec{F}_g + \vec{n} = m\vec{a}$. In terms of the x and y components of the force, we have

$$x: \quad n\sin\theta = \frac{mv^2}{r}$$
$$y: \quad n\cos\theta = mg$$

EVALUATE   Dividing the two equations allows us to eliminate n and obtain

$$\frac{n\sin\theta}{n\cos\theta} = \frac{mv^2/r}{mg} \quad \rightarrow \quad \tan\theta = \frac{v^2}{rg}$$

With $\theta = 28°$ and $r = 3.6$ km $= 3600$ m, the speed of the plane is

$$v = \sqrt{gr\tan\theta} = \sqrt{(9.8 \text{ m/s}^2)(3600 \text{ m})\tan 28°} = 137 \text{ m/s} = 493 \text{ km/h}$$

ASSESS   The result shows that the speed of the airplane is proportional to $\sqrt{\tan\theta}$. If we want to increase the speed v while the radius r is kept fixed, then the banking angle would also need to be increased. This is due to the fact that the horizontal component of the normal force $(n\sin\theta)$ is what keeps the plane in circular motion.

Section 5.4 Friction

29.   **INTERPRET**   The object of interest is the hockey puck. Three forces are involved: gravity $\vec{F}_g$, normal force $\vec{n}$, and friction $\vec{f}_k$. Friction is what causes acceleration. The physical quantity to be computed is the coefficient of kinetic friction.

DEVELOP   Newton's second law gives $\vec{F}_g + \vec{n} + \vec{f}_k = m\vec{a}$. The force of friction is the only horizontal force acting, and the normal force is vertical and equal to the puck's weight. Using the horizontal/vertical coordinate system, the equation can be rewritten in terms of the force components:

$$x: \; -\mu_k n = ma_x$$
$$y: \; -mg + n = 0$$

We take the positive direction parallel to the initial velocity, and use Equation 2.11, $v^2 = v_0^2 + 2a\Delta x$ to calculate the acceleration.

EVALUATE   Solving the y equation for n and substituting in the x equation gives

$$f_k = ma_x = -\mu_k mg \quad \rightarrow \quad a_x = -\mu_k g$$

On the other hand, Equation 2.11 gives

$$a = \frac{v^2 - v_0^2}{2\Delta x} = \frac{0 - (14 \text{ m/s})^2}{2(56 \text{ m})} = -1.75 \text{ m/s}^2$$

Thus, the coefficient of kinetic friction is

$$\mu_k = -\frac{a}{g} = \frac{-(-1.75 \text{ m/s}^2)}{9.8 \text{ m/s}^2} = 0.18$$

ASSESS   The result $a = -\mu_k g$ shows that increasing the coefficient of friction would result in a greater acceleration. This makes sense because friction is what causes the acceleration.

31.   **INTERPRET**   This problem involves circular motion and Newton's second law. The object of interest is the car. Three forces involved are: gravity $\vec{F}_g$, normal force $\vec{n}$, and friction $\vec{f}_s$. By analyzing the force acting on the car while it travels a circular path, we can determine the coefficient of friction needed to keep the car on the road.

DEVELOP   Applying Newton's second law to the car rounding a banked turn gives $\vec{F}_g + \vec{n} + \vec{f}_k = m\vec{a}$. Let R be the radius of the curve, and $v_d = \sqrt{gR\tan\theta}$ be the design speed for the proper banking angle (see Example 5.6). The acceleration is radial and has a magnitude of $a_r = v^2/R$. The forces are as shown in the figure. Note that the frictional force $\vec{f}_s$ changes direction for v greater or less than the design speed v_d.

Taking components parallel and perpendicular to the road, we find

$$\text{perpendicular:}\quad n - mg\cos\theta = \frac{mv^2}{R}\sin\theta$$

$$\text{parallel:}\quad mg\sin\theta \pm f_s = \frac{mv^2}{R}\cos\theta$$

where the upper sign is for $v > v_d$, and the lower for $v < v_d$. The car does not skid if $f_s \leq \mu_s n$. Substituting the expressions from above for f_s and n, the condition of no skidding becomes

$$\mu_s \geq \frac{f_s}{n} = \frac{\mp mg\sin\theta \pm mv^2\cos\theta/R}{mg\cos\theta + mv^2\sin\theta/R} = \frac{\mp gR\tan\theta \pm v^2}{gR(1 + v^2\tan\theta/gR)} = \frac{\mp v_d^2 \pm v^2}{gR(1 + v^2 v_d^2/g^2 R^2)}$$

Now let $\Delta v = v - v_d$ be the difference between the actual speed and the design speed. The condition on μ_s becomes:

$$\mu_s \geq \pm \frac{(v - v_d)(v + v_d)}{gR(1 + v^2 v_d^2/g^2 R^2)} = \pm \frac{\Delta v(2v_d + \Delta v)}{gR(1 + v_d^2 v^2/g^2 R^2)} = \frac{|\Delta v|(2v_d + \Delta v)}{gR(1 + v_d^2 v^2/g^2 R^2)}$$

where $|\Delta v|$ is $+\Delta v$ for $v > v_d$ and $-\Delta v$ for $v < v_d$.

EVALUATE   Substituting the values given in the problem, $R = 130$ m, $v = 40$ km/s $= 11.1$ m/s and $v_d = 60$ km/h $= 16.7$ m/s (with $\Delta v = -5.56$ m/s), we obtain

$$\mu_s \geq \frac{(5.56\text{ m/s})[2(16.7\text{ m/s}) + (-5.56\text{ m/s})]}{(9.8\text{ m/s}^2)(130\text{ m})[1 + (16.7\text{ m/s})^2(11.1\text{ m/s})^2/((9.8\text{ m/s}^2)^2(130\text{ m})^2)]} = 0.12$$

ASSESS   The general expression for the minimum coefficient of friction looks complicated. However, for $v = 0$ (car at rest) it reduces to the expected result (see Example 5.10):

$$\mu_s \geq \frac{v_d^2}{gR} = \frac{gR\tan\theta}{gR} = \tan\theta$$

PROBLEMS

33.   **INTERPRET**   The problem involves a block sliding up a frictionless ramp with an initial velocity. The quantity of interest is the distance it travels before coming to a complete stop.

DEVELOP   The problem can be reduced to a one-dimensional kinematics problem by noting that the component of the gravitational acceleration in the direction parallel to the ramp is $g\sin\theta$ (downward). The stopping distance can then be calculated by solving Equation 2.11: $v^2 = v_0^2 + 2a\Delta x$.

EVALUATE   The acceleration up the ramp is $a = -g\sin\theta$. Therefore, from Equation 2.11, the distance traveled by the block is

$$\Delta x = \frac{v^2 - v_0^2}{2a} = \frac{v^2 - v_0^2}{2(-g\sin\theta)} = \frac{0 - (2.2\text{ m/s})^2}{-2(9.8\text{ m/s}^2)\sin 35°} = 0.43\text{ m}$$

ASSESS   The result shows that the distance traveled is inversely proportional to $\sin\theta$. To see that this makes sense, let's consider the limit where $\theta \to 0$. The situation corresponds to a frictionless horizontal surface. In this case, we expect the block to travel indefinitely, with $\Delta x \to \infty$, in agreement with our expression for Δx.

35.   **INTERPRET**   The object of interest in this problem is the 10-kg mass. The forces acting on it are gravity and the tension forces in the two strings. At equilibrium, all forces cancel and the mass remains stationary.

DEVELOP   Newton's second law gives

$$\vec{F}_{\text{net}} = \vec{F}_g + \vec{T}_1 + \vec{T}_2 = m\vec{a}$$

The free-body diagram is shown on the right. Since the mass is at rest, the sum of the forces is zero, $\vec{F}_g + \vec{T}_1 + \vec{T}_2 = 0$. In terms of the x and y components, we have

$$x:\quad T_1\cos\theta - T_2 = 0$$
$$y:\quad T_1\sin\theta - mg = 0$$

where we have used $F_g = W = mg$.

EVALUATE Solving for the magnitudes of the tensions, we obtain

$$T_1 = \frac{mg}{\sin\theta} = \frac{(10\text{ kg})(9.8\text{ m/s}^2)}{\sin 45°} = 139\text{ N}$$

Similarly, the tension on the second string is

$$T_2 = T_1\cos\theta = (139\text{ N})\cos 45° = 98\text{ N}$$

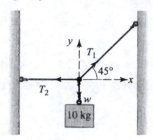

ASSESS To see that our result makes sense, let's consider some special cases. When $\theta = 90°$, the tensions in the string become

$$T_1 = \frac{mg}{\sin 90°} = mg = (10\text{ kg})(9.8\text{ m/s}^2) = 98\text{ N}$$
$$T_2 = T_1\cos\theta = (98\text{ N})\cos 90° = 0$$

This is to be expected, since in this limit, the weight of the mass is supported entirely by the first rope. On the other hand, if $\theta = 0$, the tensions would become infinite. This is equivalent to saying that it's impossible to support a weight with a purely horizontal string.

37. **INTERPRET** The key concepts in this problem are circular motion and Newton's second law. The object of interest is the mass m_1 that travels in a circular path. By analyzing the force acting on m_1, the tension in the string can be determined.

DEVELOP For each mass, there are two forces acting on it: gravitational force $\vec{F}_g$ and the tension force $\vec{T}$. The role of m_2 is only to provide the tension. On the other hand, it is the tension in the string that is providing the centripetal force for m_1. Applying Newton's second law gives:

$$m_1: \quad T = \frac{m_1 v^2}{R}$$
$$m_2: \quad T - m_2 g = 0$$

EVALUATE **(a)** Newton's second law applied to the stationary mass yields $T - m_2 g = 0$, so the tension is

$$T = m_2 g$$

(b) The tension in the string also provides the centripetal force for m_1. Let $\tau = 2\pi R/v$ be the period of the circular motion. The above equation for m_1 then gives

$$T = \frac{m_1 v^2}{R} = \frac{m_1}{R}\left(\frac{2\pi R}{\tau}\right)^2 = \frac{4\pi^2 m_1 R}{\tau^2}$$

But from **(a)**, we also have $T = m_2 g$. So, when the two equations are combined, we obtain $\frac{4\pi^2 m_1 R}{\tau^2} = m_2 g$, which gives

$$\tau = 2\pi\sqrt{\frac{m_1 R}{m_2 g}}$$

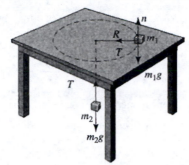

ASSESS From the expression for τ, we could draw the following conclusions: **(i)** Since $\tau \propto \sqrt{R}$, the larger the radius, the longer the period. **(ii)** With the tension (or m_2) kept fixed, increasing m_1 also leads to a longer period.

39. **INTERPRET** This problem involves circular motion and Newton's second law. The object of interest is the skater. Three forces involved are: gravity $\vec{F}_g$, normal force $\vec{n}$, and friction $\vec{f}_s$. The physical quantities of interest are the vertical and horizontal components of the force exerted on the blade as the skater makes a circular turn.

DEVELOP Using Newton's second law, the equation of motion for the skater is $\vec{F}_g + \vec{n} + \vec{f}_k = m\vec{a}$. The equation can be decomposed into two equations, one for the vertical direction and one for the horizontal. By analyzing the force acting on the skater, we can also find the maximum angle she can lean without falling over.

EVALUATE **(a)** In terms of the x and y components of the force (see free-body diagram), we have

$$x: \quad f_s = \frac{mv^2}{r}$$
$$y: \quad n = mg$$

Thus, we conclude that the normal force n (vertical) is equal to the weight mg, and static friction (horizontal) provides the centripetal acceleration:

$$n = mg = (45 \text{ kg})(9.8 \text{ m/s}^2) = 441 \text{N}$$
$$f_s = \frac{mv^2}{r} = \frac{(45 \text{ kg})(6.3 \text{ m/s})^2}{5 \text{ m}} = 357 \text{ N}$$

(b) Stability requires that the center of gravity of the skater should be along the line of action of the contact force. She should lean at an angle

$$\theta = \tan^{-1}\left(\frac{f_s}{n}\right) = \tan^{-1}\left(\frac{357 \text{ N}}{441 \text{ N}}\right) = 39°$$

relative to the vertical.

ASSESS We see that friction is what provides the centripetal force for circular motion. Without the friction, the skater would never be able to turn. In **(b)**, the angle could be written as

$$\theta = \tan^{-1}\left(\frac{f_s}{n}\right) = \tan^{-1}\left(\frac{mv^2/r}{mg}\right) = \tan^{-1}\left(\frac{v^2}{rg}\right)$$

This is precisely the expression we find (see Example 5.6) for the banking angle.

41. **INTERPRET** This problem involves circular motion and Newton's second law. The object of interest is the water in the bucket. The forces involved are gravity $\vec{F}_g$ and normal force $\vec{n}$. We would like to find the minimum speed of whirl such that the water will remain in the bucket.

DEVELOP   Newton's second law gives $\vec{F}_g + \vec{n} = m\vec{a}$. At the top of the whirl, the y component of the force equation is (taking downward to be positive)

$$mg + n = \frac{mv^2}{r}$$

EVALUATE   Solving the above equation for speed, we obtain $v = \sqrt{gr + \frac{nr}{m}}$. The minimum possible speed corresponds to the situation where the water is just about to lose contact with the bucket, with $n = 0$. Thus,

$$v_{min} = \sqrt{gr} = \sqrt{(9.8 \text{ m/s}^2)(0.85 \text{ m})} = 2.89 \text{ m/s}$$

ASSESS   With this minimum speed, the normal force vanishes just right at the top of the loop, and gravity alone provides the centripetal force which keeps the water moving in its circular path. If the whirl speed is too slow (less than v_{min}), then the water will not remain in the bucket.

43.   **INTERPRET**   This is a problem involving Newton's second law with zero acceleration. The object of interest is the mower, and the forces involved are: gravity $\vec{F}_g$, normal force $\vec{n}$, friction $\vec{f}_k$, and the applied force $\vec{F}$. The physical quantity of interest is the value of $\vec{F}$ which keeps the mower moving at a constant speed.

DEVELOP   Using Newton's second law, the force equation for the mower is $\vec{F}_{net} = \vec{F} + \vec{F}_g + \vec{n} + \vec{f}_k = m\vec{a}$. Assuming the ground is also horizontal, this equation can be decomposed into two equations, one for the vertical direction and one for the horizontal.

EVALUATE   We may depict the forces on the lawnmower as shown. At constant velocity (constant speed in a straight line) $\vec{a} = 0$, and $\vec{F}_{net} = \vec{F} + \vec{F}_g + \vec{n} + \vec{f}_k = 0$. The x and y components of this equation are

$$x: \quad F\cos\theta - f_k = 0$$
$$y: \quad n - F\sin\theta - mg = 0$$

which leads to $f_k = \mu_k n = \mu_k(F\sin\theta + mg)$. Substituting the expression for f_k back to the x equation and solving for F, we obtain $F\cos\theta - \mu_k(F\sin\theta + mg) = 0$, or $F = \frac{\mu_k mg}{\cos\theta - \mu_k \sin\theta}$ with the values given in the problem statement, the magnitude of the applied force is

$$F = \frac{\mu_k mg}{\cos\theta - \mu_k \sin\theta} = \frac{(0.68)(22 \text{ kg})(9.8 \text{ m/s}^2)}{\cos 35° - (0.68)\sin 35°} = 342 \text{ N}$$

In terms of the weight of the mower, we have $\frac{F}{mg} = \frac{\mu_k}{\cos\theta - \mu_k \sin\theta} = 1.58$.

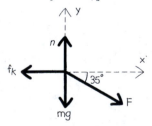

ASSESS   To show that our result makes sense, let's check some limiting cases. First, if $\mu_k = 0$, our expression gives $F = 0$. This is so because without friction, no force needs to be applied to keep the mower moving at constant speed. On the other hand, if $\theta = 0$, then we get $F = \mu_k mg$. In this limit the normal force is simply $n = mg(\vec{F}_g + \vec{n}_k = 0)$, and the friction $f_k = \mu_k n = \mu_k mg$. Thus, the force applied has the same magnitude as the friction, but points in the opposite direction: $\vec{F} = -\vec{f}_k$.

45.   **INTERPRET**   This is a problem involving Newton's second law. The object of interest is the bat, and the forces involved are gravity $\vec{F}_g$, normal force $\vec{n}$, and static friction $\vec{f}_s$. The physical quantity of interest is the minimum acceleration of the train that keeps the bat in place.

DEVELOP   Using Newton's second law, the force equation for the bat is $\vec{F}_{net} = \vec{F}_g + \vec{n} + \vec{f}_s = m\vec{a}$. This can be decomposed into two equations, one for the vertical direction and one for the horizontal:

$$\text{horizontal}: \quad n = ma$$
$$\text{vertical}: \quad f_s - mg = 0$$

EVALUATE   The normal force n is parallel to the acceleration, but perpendicular to gravity and friction, $n = ma$, and

$$f_s = mg \le \mu_s n = \mu_s ma$$

Therefore, in order to remain in place, the minimum acceleration must be

$$a_{min} = \frac{g}{\mu_s} = \frac{9.8 \text{ m/s}^2}{0.86} = 11.4 \text{ m/s}^2$$

ASSESS   The minimum acceleration is inversely proportional to μ_s, the coefficient of friction between bat and train. The smaller the value of μ_s, the greater the acceleration is needed to keep the bat in place. In the limit $\mu_s \to 0$ (frictionless surface), the acceleration would have to be infinitely large. On the other hand, for $\mu_s \to \infty$ (infinitely sticky surface), we would then expect $a_{min} \to 0$.

47.   **INTERPRET**   This is a problem involving circulation motion and Newton's second law. The object of interest is the bug, and the forces acting on it are gravity $\vec{F}_g$, normal force $\vec{n}$, and static friction $\vec{f}_s$. We want to find out how far the bug gets from the center before it begins to slip.

DEVELOP   Using Newton's second law, the force equation for the bug (at a distance r from the center of the compact disc) is $\vec{F}_{net} = \vec{F}_g + \vec{n} + \vec{f}_s = m\vec{a}$. Assume that the disc is level. This equation can be decomposed into two equations, one for the vertical direction and one for the horizontal:

$$\text{horizontal}: \quad n = ma = \frac{mv^2}{r}$$
$$\text{vertical}: \quad f_s - mg = 0$$

The period of the compact disc is $T = \frac{1 \min}{200} = \frac{60 \text{ s}}{200} = 0.3$ s.

EVALUATE   From the equations for the force components, we see that the frictional force produces the (centripetal) acceleration of the bug, and the normal force equals its weight. Thus,

$$f_s = \frac{mv^2}{r} = \frac{m}{r}\left(\frac{2\pi r}{T}\right)^2 = \frac{4\pi^2 mr}{T^2} \le \mu_s n = \mu_s mg$$

Thus, the radial distance traveled by the bug before slipping occurs is

$$r \le \frac{\mu_s g T^2}{4\pi^2} = \frac{(1.2)(9.8 \text{ m/s}^2)(0.3 \text{ s})^2}{4\pi^2} = 0.0268 \text{ m} = 2.68 \text{ cm}$$

ASSESS   Our result indicates that the larger the friction coefficient, the greater the distance the bug can travel. This makes sense because friction is what produces the centripetal force. In the limit $\mu_s \to 0$ (frictionless surface), the bug would not be able to move at all ($r \to 0$). On the other hand, if the surface of the disc is very sticky (large μ_s), we would then expect the distance traveled by the bug to be very large.

49.   **INTERPRET**   This is a problem involving Newton's second law. The object of interest is one of the children. Three forces involved are gravity $\vec{F}_g$, normal force $\vec{n}$, and friction $\vec{f}_k$. By analyzing the forces acting on the child while he travels down the hill, we can determine the child's acceleration a and his speed when he reaches the level. Subsequently, we can solve the one-dimensional kinematics problem to determine the distance he slides before coming to a complete stop.

DEVELOP   Newton's second law gives $\vec{F}_g + \vec{n} + \vec{f}_k = m\vec{a}$. In terms of force components parallel and perpendicular to the surface of the hill, we have

$$\text{perpendicular:}\quad n - mg\cos\theta = 0$$
$$\text{parallel:}\quad mg\sin\theta - f_k = ma$$

The acceleration is in the downward direction of the hill. The frictional force is $f_k = \mu_k n$.

EVALUATE   The first equation gives $n = mg\cos\theta$. Using this result in the second equation, we find the acceleration to be

$$a = g\sin\theta - \frac{f_k}{m} = g\sin\theta - \frac{\mu_k mg\cos\theta}{m} = g(\sin\theta - \mu_k\cos\theta) = (9.8\ \text{m/s}^2)(\sin25° - 0.12\cos25°) = 3.08\ \text{m/s}^2$$

Using Equation 2.11, the speed of the child at the bottom of the hill is

$$v_b = \sqrt{v_0^2 + 2as} = \sqrt{2as} = \sqrt{2(3.08\ \text{m/s}^2)(41\ \text{m})} = 15.9\ \text{m/s}$$

Now, on level ground, the deceleration is (see Example 5.8) $a' = -\frac{f_k}{m} = -\frac{\mu_k mg}{m} = -\mu_k g$. Again, using Equation 2.11, the distance traveled before stopping is

$$\Delta x = \frac{v^2 - v_b^2}{2a'} = \frac{-v_b^2}{2(-\mu_k g)} = \frac{-(15.9\ \text{m/s})^2}{-2(0.12)(9.8\ \text{m/s}^2)} = 107\ \text{m}$$

ASSESS   After traveling 41 m down the hill, the speed of the child at the bottom of the hill is 15.9 m/s, or roughly 36 mi/h! So sliding 107 m on the level ground before coming to a complete stop sounds very reasonable.

51.   **INTERPRET**   This is a problem involving Newton's second law. The object of interest is the car. The three forces involved are gravity $\vec{F}_g$, normal force $\vec{n}$, and friction $\vec{f}_s$. By analyzing the force acting on the car, we can solve the one-dimensional kinematics problem to determine its initial speed.

DEVELOP   On a level road, the acceleration of a skidding car is $a = -\mu_k g$ (see Example 5.8). Given that the final speed of the car is $v = 25\ \text{km/h} = 6.94\ \text{m/s}$, we can solve Equation 2.11, $v^2 = v_0^2 + 2a\Delta x$, to find its initial speed.

EVALUATE   Using $a = -\mu_k g$ for Equation 2.11, we have

$$v_0 = \sqrt{v^2 - 2a\Delta x} = \sqrt{(6.94\ \text{m/s})^2 + 2(0.71)(9.8\ \text{m/s})^2(47\ \text{m})} = 26.5\ \text{m/s} = 95.4\ \text{km/h} = 59.3\ \text{mi/h}$$

ASSESS   The police officer should add speeding to the traffic citation! Note that friction is what causes the acceleration.

53.   **INTERPRET**   This is a problem involving Newton's second law with zero acceleration. The object of interest is the trunk, and the forces involved are gravity $\vec{F}_g$, normal force $\vec{n}$, friction $\vec{f}_s$, and the applied force $\vec{F}_a$. We want to show that if the coefficient of static friction exceeds a certain value, the trunk won't move no matter how hard you push.

DEVELOP   Using Newton's second law, the force equation for the trunk is $\vec{F}_{net} = \vec{F}_a + \vec{F}_g + \vec{n} + \vec{f}_s = m\vec{a}$. Assuming the ground is also horizontal, this equation can be decomposed into two equations, one for the vertical direction and one for the horizontal.

EVALUATE   We may depict the forces on the trunk as shown. At constant velocity (constant speed in a straight line) $\vec{a} = 0$, and $\vec{F}_{net} = \vec{F}_a + \vec{F}_g + \vec{n} + \vec{f}_s = 0$. The x and y components of this equation are

$$x:\quad F_a\cos\theta - f_s = 0$$
$$y:\quad n - F_a\sin\theta - mg = 0$$

which leads to

$$f_s = F_a\cos\theta \le \mu_s n = \mu_s(mg + F_a\sin\theta)$$

Substituting the expression for f_s back to the x-equation and solving for F_a, we obtain $F_a\cos\theta - \mu_s(F_a\sin\theta + mg) = 0$, or

$$F_a = \frac{\mu_s mg}{\cos\theta - \mu_s\sin\theta}$$

If $\cos\theta - \mu_s \sin\theta < 0$, then the applied force F_a will become negative. This means that the trunk will not budge and remain in equilibrium. Thus, the equilibrium condition will always be satisfied if

$$\mu_s > \cot\theta = \cot 50° = 0.84$$

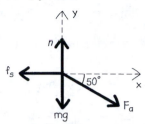

ASSESS   As the coefficient of friction increases, we must apply a greater force in order to move the trunk. However, because the applied force has a downward component $F_a \sin\theta$, this adds to the normal force and results in a greater the friction force. Obviously when the force is applied vertically downward ($\theta = 90°$), it's impossible to move the truck. On the other hand, if $\theta = 0$, then we get $F_a = \mu_s mg$. In this limit the normal force is simply $n = mg$ $(\vec{F}_g + \vec{n}_k = 0)$, and the friction $f_s = \mu_s n = \mu_s mg$. Thus, the applied force applied has the same magnitude as the friction, but points in the opposite direction: $\vec{F} = -\vec{f}_s$.

55.   **INTERPRET**   This is a problem involving Newton's second law. The object of interest is the box. The three forces involved are gravity $\vec{F}_g$, normal force $\vec{n}$, and friction $\vec{f}_k$. By analyzing the forces acting on the box as it slides down the ramp, we can determine the maximum frictional coefficient that can be tolerated.

DEVELOP   Newton's second law gives $\vec{F}_g + \vec{n} + \vec{f}_k = m\vec{a}$. In terms of force components parallel and perpendicular to the surface of the ramp, we have

$$\text{perpendicular:}\quad n - mg\cos\theta = 0$$
$$\text{parallel:}\quad mg\sin\theta - f_k = ma$$

The acceleration is in the downward direction of the ramp and the frictional force is $f_k = \mu_k n$.

EVALUATE   The first equation gives $n = mg\cos\theta$. Using this result in the second equation, we find the acceleration to be

$$a = g\sin\theta - \frac{f_k}{m} = g\sin\theta - \frac{\mu_k mg\cos\theta}{m} = g(\sin\theta - \mu_k\cos\theta)$$

Since the time required for the box to travel a distance Δx down the ramp is given by

$$\Delta x = \frac{1}{2}at^2 \quad \rightarrow \quad a = \frac{2\Delta x}{t^2}$$

equating the two expression leads to, $a = g(\sin\theta - \mu_k\cos\theta) = \frac{2\Delta x}{t^2}$, or $\mu_k = \tan\theta - \frac{2\Delta x}{gt^2\cos\theta}$.

This is the maximum coefficient that can be tolerated. Substituting the values given in the problem statement, we obtain

$$\mu_k = \tan 30° - \frac{2(5.4\ \text{m})}{(9.8\ \text{m/s}^2)(3.3\ \text{s})^2\cos 30°} = 0.46$$

ASSESS   The above expression shows that increasing t gives a greater value of μ_k. What this means is that if the coefficient of friction were greater than 0.46, the slide would take longer than 3.3 s. The value $\mu_k = \tan\theta = \tan 30° = 0.577$ corresponds to $t = \infty$. This means that the box will stay indefinitely on the top of the ramp without sliding when the coefficient of friction exceeds 0.577.

57.   **INTERPRET**   The aim of this problem is to establish a condition under which a car moving too slowly as it goes around a loop-the-loop roller coaster would leave the track.

DEVELOP   The forces involved are gravity $\vec{F}_g$ and normal force $\vec{n}$. Newton's second law gives $\vec{F}_g + \vec{n} = m\vec{a}$. The angle ϕ and the forces acting on the car are shown in the sketch. The radial component of the net force (toward the center of the track) equals the mass times the centripetal acceleration,

$$n + mg\cos\phi = \frac{mv^2}{r}$$

The tangential component is not of interest in this problem.

EVALUATE   The car leaves the track when the normal force becomes zero (no more contact):

$$n = \frac{mv^2}{r} - mg\cos\phi = 0 \quad \rightarrow \quad \cos\phi = \frac{v^2}{gr}$$

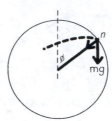

ASSESS   The result implies that the car leaves the track when the speed is too small: $v < \sqrt{gr}$; otherwise, the car never leaves the track, as in Example 5.7.

59.   **INTERPRET**   In this problem we would like to find out the difference in a person's weights measured at the north pole and at the equator.

DEVELOP   When standing on the Earth's surface, you are rotating with the Earth about its axis through the poles, with a period of 1d. The radius of your circle of rotation (your perpendicular distance to the axis) is $r = R_E \cos\theta$, where R_E is the radius of the Earth (constant if geographical variations are neglected) and θ is your latitude. Your centripetal acceleration has magnitude $a_c = \frac{v^2}{r} = \frac{(2\pi r/T)^2}{r} = \frac{4\pi^2 r}{T^2}$ and is directed toward the axis of rotation.

We assume there are only two forces acting on you, gravity, $\vec{F}_g$ (with magnitude mg approximately constant, directed towards the center of the Earth), and the force exerted by the scale, $\vec{F}_s$. Newton's second law requires that $\vec{F}_g + \vec{F}_s = m\vec{a}_c$.

EVALUATE   (a) At the north pole, $\vec{a}_c = 0$, so the magnitudes of $\vec{F}_g$ and $\vec{F}_{s, \text{pole}}$ are equal, or $F_{s, \text{pole}} = mg$. On the other hand, at the equator, $m\vec{a}_c$ has a maximum magnitude, equal to the difference in the magnitudes of $\vec{F}_g$ and $\vec{F}_{s, \text{eq}}$, or

$$F_{s, \text{eq}} = mg - ma_c = mg - \frac{4\pi^2 mR_E}{T^2}$$

Therefore $F_{s, \text{eq}}$ (your "weight") is lower at the equator than at the pole.

(b) The fractional difference of these two values is

$$\frac{F_{s, \text{pole}} - F_{s, \text{eq}}}{F_{s, \text{pole}}} = \frac{ma_c}{mg} = \frac{a_c}{g} = \frac{4\pi^2 R_E}{gT^2} = \frac{4\pi^2 (6.37 \times 10^6 \text{ m})}{(9.81 \text{ m/s}^2)(86,400 \text{ s})^2} = 0.00344 = 0.344\%$$

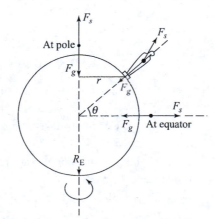

ASSESS   Our result shows that you weigh more at the north pole because the centripetal acceleration is zero there. But the difference is hardly noticeable.

61.   **INTERPRET**   The object of interest in this problem is the block which slides up an incline and slides down again. We want to show that when the coefficient of kinetic friction is $\mu_k = \frac{3}{5}\tan\theta$, the final speed of the block half its initial speed.

DEVELOP   When going up the incline, the block's acceleration (positive down the incline) is

$$a_{up} = -g(\sin\theta + \mu_k\cos\theta)$$

On the other hand, while going down, the acceleration is

$$a_{down} = g(\sin\theta - \mu_k\cos\theta)$$

Knowing the acceleration, Equation 2.11, $v^2 = v_0^2 + 2a\Delta x$, can be used to find the initial speed.

EVALUATE   Suppose the block slides up a distance L, then by using Equation 2.11, its initial speed upward was

$$v_{up} = \sqrt{v^2 - 2a\Delta x} = \sqrt{-2a_{up}L}$$

Similarly, as the block slides down the same distance, it returns to the bottom with speed

$$v_{down} = \sqrt{2a_{down}L}$$

The condition that $\frac{v_{down}}{v_{up}} = \frac{1}{2}$ implies

$$\frac{1}{4} = \left(\frac{v_{down}}{v_{up}}\right)^2 = \frac{2a_{down}L}{-2a_{up}L} = \frac{g(\sin\theta - \mu_k\cos\theta)}{g(\sin\theta + \mu_k\cos\theta)} = \frac{(\tan\theta - \mu_k)}{(\tan\theta + \mu_k)}$$

which gives $\mu_k = \frac{3}{5}\tan\theta$,

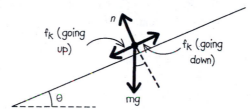

ASSESS   To see that our result makes sense, let's check some limiting cases. If the incline were frictionless with $\mu_k = 0$, then we would expect $v_{down} = v_{up}$. Our expression clearly satisfies this condition. On the other hand, if the coefficient of friction becomes too large ($\mu_k > \tan\theta$, see Example 5.10), the block will not slide down at all ($v_{down} = 0$).

63.   **INTERPRET**   We are asked to graph the tension force from Example 5.11 as a function of angle. From the graph, we can determine the minimum force necessary to move the trunk and the angle for this minimum force.

DEVELOP   The equation given in Example 5.11 is $T = \frac{\mu_k mg}{\cos\theta + \mu_k\sin\theta}$. If we plot $\frac{T}{mg}$ versus θ accurately we can read the minimum force ratio and the optimal angle directly from the plot.

EVALUATE   **(a)** From the figure below, we see that the minimum force ratio is 0.6, so the minimum force required is 60% of the weight of the trunk.

(b) The angle at which the minimum occurs is about 0.65 radians, or $\theta \approx (0.65 \text{ radian}) \times \frac{360 \text{ degrees}}{2\pi \text{ radians}} = 37°$

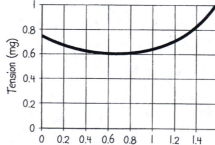

ASSESS   The shape of the graph makes sense, intuitively. As we increase the angle, we decrease the frictional force, up to a point. At too high an angle, we're spending too much effort trying to lift the trunk instead of moving it. At an angle of π (90°) the force is equal to the trunk's weight, because we're just lifting it. This graphical method of solution is not bad! The exact solution (see Problem 64) is 60% of the trunk's weight, at an angle of 36.9°.

65. **INTERPRET**   We need to find an equation for the speed of an object as a function of time, where instead of just a constant acceleration due to gravity there is also a drag force that depends on velocity. We also need to find the terminal velocity of the object.

     DEVELOP   We choose the upward direction to be positive. Instead of $F = -mg = ma$, we have $F = -mg - bv = ma$. The acceleration is $a = \frac{dv}{dt}$, so we have a differential equation $m\frac{dv}{dt} + bv + mg = 0$. The initial velocity of the object is zero. We find terminal velocity by finding the value of $v(t)$ in the limit as $t \to \infty$.

     EVALUATE

     (a) $\dfrac{dv}{dt} + \dfrac{b}{m}v + g = 0 \to \dfrac{dv}{dt} + g = -\dfrac{b}{m}v \to v(t) = Ae^{-bt/m} - g\dfrac{m}{b}$. The initial velocity is zero, so

     $v(0) = Ae^0 - \frac{mg}{b} = A - \frac{mg}{b} \to A = \frac{mg}{b}$. We then have our complete equation: $v(t) = \frac{mg}{b}(e^{-bt/m} - 1)$

     (b) $\lim\limits_{t\to\infty}[v(t)] = \frac{mg}{b}\lim\limits_{t\to\infty}[e^{-bt/m} - 1] = -\frac{mg}{b}$, so the terminal velocity is $v_t = \frac{mg}{b}$ in the downward direction.

     ASSESS   Our units check out for the equation in part (a), and the equation has a form consistent with what we would expect. We can also check our answer for part (a) by taking the derivative and seeing that it fits our initial equation. In (b) we have shown what was requested.

67. **INTERPRET**   We have a chain hanging from two supports at the same height, and are asked to find the relationship between the magnitude of the tension at one support to the magnitude of the tension at the center. We also need to find the relationship between the tension at the supports and the weight of the chain, and find a differential equation for the shape of the chain. We will use Newton's first law, the vector nature of forces, and a fair amount of calculus.

     DEVELOP   We start by drawing a diagram, as shown in the figure. We draw the diagram for *half* the chain, since the other half is the same. We can obtain the relationship between T_o (the tension at the center) and T_s (the tension at a support) from the diagram. We can derive the relationship between the force at the support and the weight mg from Newton's first law. The last part, finding the differential equation, we do by using the equation for arclength: $s = \int ds = \int \sqrt{dx^2 + dy^2}$.

     EVALUATE   From the figure, we see that $T_o = T_s \cos\theta_s$.

     From Newton's first law, and the observation that the chain is not accelerating, we can deduce that the vertical component of tension is equal to half the weight of the chain. (Half, since there is another support not shown.) So $T_v = T_s \sin\theta_s = \frac{1}{2}mg \to T_s = \frac{mg}{2\sin\theta_s}$.

     The last part is a bit trickier. First, we note that at any point P on the curve, the tension is directed along the tangent to the curve, so $\frac{dy}{dx} = \frac{T_y}{T_x}$. The tension in the y direction at point P is equal to the weight of this section of chain up to point P: $T_y = \mu s$, where $\mu \equiv \frac{m}{L}$. So $\frac{dy}{dx} = \frac{\mu s}{T_x} \to s = \frac{T_x}{\mu}\frac{T_y}{dx}$. We also have from the definition of arclength $s = \int ds = \int \sqrt{dy^2 + dx^2} = \int\left[\sqrt{\left(\frac{dy}{dx}\right)^2 + 1}\right]dx$. Equate these two terms for s:

     $$\frac{T_x}{\mu}\frac{dy}{dx} = \int\left[\sqrt{\left(\frac{dy}{dx}\right)^2 + 1}\right]dx \to \frac{dy}{dx} = \frac{\mu}{T_x}\int\left[\sqrt{\left(\frac{dy}{dx}\right)^2 + 1}\right]dx.$$

     We can see that $T_x = T_0 = T_s \cos\theta_s$ from the observation that the chain is in equilibrium: If T_x was not constant at some point P, then that point would accelerate. So we can substitute the value of T_x and take the derivative with respect to x of both sides to obtain the desired differential equation: $\frac{d^2y}{dx^2} = \frac{m}{LT_s\cos\theta_s}\sqrt{\left(\frac{dy}{dx}\right)^2 + 1}$.

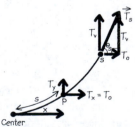

     Center

     ASSESS   The solution to this differential equation, if you're still curious, is $y(x) = \frac{T_0}{\mu}\cosh\left(\frac{\mu}{T_0}x\right) - \frac{T_0}{\mu}$. This is a catenary, and it looks similar to a parabola.

69.   **INTERPRET**   We need to compare the speed of a hammer-throw hammer with that of a "speeding bullet." The hammer is moving in a circle. We use the vector sum of the forces, and the angle at which the hammer is "sagging" as it goes around the circle.

DEVELOP   The sum of the forces on the hammer is equal to the centripetal force on the hammer. We resolve the tension into horizontal and vertical components and equate the vertical component with the force of gravity. The horizontal component must provide the centripetal force.

EVALUATE   We find the components of tension: $T_y = T\sin\theta = mg \rightarrow T = \dfrac{mg}{\sin\theta}$. $T_x = T\cos\theta = mg\dfrac{\cos\theta}{\sin\theta}$. The x component is the centripetal force:

$$m g\frac{\cos\theta}{\sin\theta} = m\frac{v^2}{R} \rightarrow v = \sqrt{\frac{gR}{\tan\theta}} = \sqrt{\frac{(9.8 \text{ m/s}^2)(2.4 \text{ m})}{\tan(10°)}} = 11.5 \text{ m/s}$$

ASSESS   This would barely be speeding in a 25 mph zone.

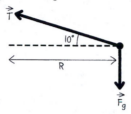

71.   **INTERPRET**   We need to find the minimum radius, at a given speed for a vertical circle, if the acceleration is not to exceed six times that of gravity. We find the centripetal acceleration of the plane, and remember that gravity is also a factor.

DEVELOP   We'll have to convert the speed and the acceleration to SI units. The equation for centripetal acceleration is $a_c = \frac{v^2}{r}$, which we will solve for r.

EVALUATE   The speed is $1.8 \times (340 \text{ m/s}) = 612$ m/s. The centripetal acceleration is 5 g's: 6 total, but gravity provides one no matter how fast he's going since this is at the bottom of a vertical loop: $a_c = 5 \times g = 49$ m/s^2.

$$a_c = \frac{v^2}{r} \rightarrow r = \frac{v^2}{a_c} = 7.64 \text{ km}$$

ASSESS   If he wants to make a smaller loop, he'll have to slow down!

6 WORK, ENERGY, AND POWER

EXERCISES

Section 6.1 Work

13. **INTERPRET** The relevant physical quantity here is the work done by a person sliding a box.

    DEVELOP If you push parallel to a level floor, the applied force equals the frictional force (since the acceleration is zero), $F_a = f_k$. Since the normal force equals the weight of the box: $n = mg$, the applied force is

    $$F_a = \mu_k n = \mu_k mg$$

    Equation 6.1, $W = F\Delta x$, can then be used to find the work done.

    EVALUATE The applied force is constant and parallel to the displacement. So using Equation 6.1, we get

    $$W_a = F_a \Delta r = f_k \Delta r = \mu_k n \Delta r = \mu_k mg \Delta r = (0.21)(50 \text{ kg})(9.8 \text{ m/s}^2)(4.8 \text{ m}) = 494 \text{ J}$$

    ASSESS The units are correct, $1 \text{ J} = 1 \text{ N} \cdot \text{m} = 1 \text{ kg} \cdot \text{m/s}^2 \cdot \text{m}$. If the floor were frictionless ($\mu_k = 0$), then the work done would be $W_a = 0$, as expected.

15. **INTERPRET** This problem is about the work done by gravity on the water falling from a certain height.

    DEVELOP Since the density of water is 1000 kg/m^3, the mass of a cubic meter of water is 1000 kg, and the force of gravity at the Earth's surface on a cubic meter of water is

    $$F_g = mg = (1000 \text{ kg})(9.8 \text{ m/s}^2) = 9800 \text{ N}$$

    vertically downward. We can then use Equation 6.1, $W = F\Delta x$, to find the work done.

    EVALUATE The displacement of the water is parallel to this, so the work done by gravity on the water is

    $$W_g = F_g \Delta y = (9800 \text{ N})(980 \text{ m}) = 9.6 \times 10^6 \text{ J}$$

    ASSESS The units are correct, $1 \text{ J} = 1 \text{ N} \cdot \text{m}$. The greater the distance the water falls, the larger the amount of work done by gravity.

17. **INTERPRET** This problem is about the work done by the elevator cable on the elevator as it accelerates upward.

    DEVELOP To give the elevator a constant upward acceleration a_y, the tension in the cable must satisfy

    $$T - mg = ma_y \quad \rightarrow \quad T = m(g + a_y)$$

    We can then use Equation 6.1, $W = F\Delta x$, to find the work done.

    EVALUATE Acting over a parallel displacement $\Delta y = h$ upward, the amount of work done by the cable on the elevator is

    $$W_T = T\Delta y = m(g + a_y)\Delta y = m(g + 0.1g)h = 1.1mgh$$

    ASSESS The units are correct, $1 \text{ J} = 1 \text{ N} \cdot \text{m}$. The greater the upward acceleration a_y, the larger the amount of work must be done by the cable. Of course, if the elevator undergoes free fall with $a_y = -g$, then the tension in the cable is zero, and no work is done on the elevator.

19. **INTERPRET** This problem is about taking the scalar product of two vectors.

    DEVELOP From Equation 6.3, we see that the scalar product of two vectors $\vec{A}$ and $\vec{B}$ is defined as

    $$\vec{A} \cdot \vec{B} = AB\cos\theta$$

EVALUATE   The scalar products are: **(a)** $\vec{A} \cdot \vec{B} = (15 \text{ u})(6.5 \text{ u}) \cos 27° = 86.9 \text{ u}^2$, and **(b)** $\vec{A} \cdot \vec{B} = (15 \text{ u})(6.5 \text{ u}) \cos 78° = 20.3 \text{ u}^2$.

Note that we used u as the symbol for the unspecified units.

ASSESS   The scalar product $\vec{A} \cdot \vec{B}$ is proportional to $\cos\theta$. It is at a maximum when $\cos\theta = 1$, or $\theta = 0$. This corresponds to the situation where $\vec{A}$ and $\vec{B}$ are parallel to each other. On the other hand, when $\vec{A}$ and $\vec{B}$ are perpendicular to each other, $\theta = 90°(\cos\theta = 0)$, and $\vec{A} \cdot \vec{B} = 0$.

21.   INTERPRET   This problem is about the distance a stalled car can be moved with a given amount of work done by a person.

DEVELOP   Equation 6.2, $W = F\Delta r \cos\theta$, applies here. The displacement is horizontal but the applied force is at an angle $\theta = 17°$ with the horizontal.

EVALUATE   Using Equation 6.2, the distance the car moves is

$$\Delta x = \frac{W}{F_x} = \frac{W}{F\cos\theta} = \frac{860 \text{ J}}{(470 \text{ N})\cos 17°} = 1.91 \text{ m}$$

ASSESS   Only the horizontal component of the force, $F_x = F\cos\theta$, does the work.

Section 6.2 Forces that Vary

23.   INTERPRET   The problem is about work done to stretch a spring.

DEVELOP   The problem can be solved by using Equation 6.10, $W = \frac{1}{2}kx^2$, which represents the work done to stretch the spring from its equilibrium position by an amount x.

EVALUATE   **(a)** The amount of work done in stretching 10 cm (0.1 m) from equilibrium is

$$W = \frac{1}{2}kx^2 = \frac{1}{2}(200 \text{ N/m})(0.1 \text{ m})^2 = 1.0 \text{ J}$$

(b) Similarly, to stretch from 10 cm to 20 cm from equilibrium requires

$$W = \frac{1}{2}k\left(x_2^2 - x_1^2\right) = \frac{1}{2}(200 \text{ N/m})[(0.2 \text{ m})^2 - (0.1 \text{ m})^2] = 3.0 \text{ J}$$

ASSESS   The work done in stretching the spring from 10 cm to 20 cm is greater than that from equilibrium to 10 cm. This is because, according to Hooke's law, $F_{\text{on spring}} = kx$, the force required is much greater.
Another way of solving part **(b)** is to note that the work to stretch the spring from 0 to 20 cm is four times the work in part **(a)**, or 4.0 J, so the work in part **(b)** is 4.0 J − 1.0 J = 3.0 J.

25.   INTERPRET   The problem is about work done to stretch a spring. We want to find out how much the spring can be stretched with a given amount of work.

DEVELOP   The problem can be solved by using Equation 6.10, $W = \frac{1}{2}kx^2$, which represents the work done to stretch the spring from its equilibrium position by an amount x.

EVALUATE   Since the work done on a spring in stretching it a distance x from its unstretched length is $W = \frac{1}{2}kx^2$, the length stretched is

$$x = \sqrt{\frac{2W}{k}} = \sqrt{\frac{2(8.5 \text{ J})}{190 \text{ N/m}}} = 0.299 \text{ m} = 29.9 \text{ cm}$$

ASSESS   With a fixed amount of work W, the extent of stretching is inversely proportional to $\sqrt{k}$. This means that the stiffer the spring (greater k), the less it will be stretched, and vice versa.

Section 6.3 Kinetic Energy

27.   INTERPRET   How much work is done in accelerating a particle from rest to some final speed? We use the work-energy theorem.

DEVELOP   The relationship between work and kinetic energy is $W = \Delta K$. $K \equiv \frac{1}{2}mv^2$, so we can use the mass of a proton $(1.67 \times 10^{-27} \text{ kg})$ and the given velocity $v = 2.1 \times 10^7$ m/s to find the change in K and thus the work.

EVALUATE $W = \Delta K = \frac{1}{2}mv_f^2 - \frac{1}{2}mv_i^2 = \frac{1}{2}mv_f^2 = \frac{1}{2}(1.67 \times 10^{-27} \text{ kg})(2.1 \times 10^7 \text{ m/s})^2 = 3.68 \times 10^{-13}$ J.

ASSESS   In nuclear physics problems such as this one, energies are often given in the more conveniently sized unit of *electron volts*. 1 eV $= 1.6 \times 10^{-19}$ J, so the amount of work done in this case is 2.3 MeV.

29. **INTERPRET**   The object of interest is the skateboarder. We are asked to find the amount of work done on the skateboarder between the top and bottom of the hill.

DEVELOP   The relevant equations here are Equations 6.12 and 6.14 (work-energy theorem), which can be combined as $W_{net} = \Delta K = \frac{1}{2}m(v_2^2 - v_1^2)$. Given the initial velocity v_1 and the final velocity v_2, the work done on the skateboarder can be readily calculated.

EVALUATE   The above equation gives

$$W_{net} = \Delta K = \frac{1}{2}m\left(v_2^2 - v_1^2\right) = \frac{1}{2}(60 \text{ kg})[(10 \text{ m/s})^2 - (5.0 \text{ m/s})^2] = 2.25 \text{ kJ}$$

ASSESS   By work-energy theorem, the change in kinetic energy of an object is equal to the net work done on the object. Therefore, the greater the difference in kinetic energy, ΔK, the more the work required.

31. **INTERPRET**   The object of interest is the baseball. We are interested in the speed of the baseball as it hits the ground and its kinetic energy.

DEVELOP   In this problem we shall solve part (**b**) first for the speed, and then part (**a**) for the kinetic energy. The speed of the baseball (magnitude of the velocity) follows from Equation 2.11, $v^2 = v_0^2 + 2a\Delta y$. Once v is known, its kinetic energy can be calculated using Equation 6.13, $K = \frac{1}{2}mv^2$.

EVALUATE   Using Equation 2.11 with $y = 0$ at ground level and ignoring the air resistance, (**b**) the final speed of the ball is

$$v = \sqrt{-2g(y - y_0)} = \sqrt{-2(9.8 \text{ m/s}^2)(-16 \text{ m})} = 17.7 \text{ m/s}$$

(**a**) The kinetic energy is

$$K = \frac{1}{2}mv^2 = \frac{1}{2}(0.150 \text{ kg})(17.7 \text{ m/s})^2 = 23.5 \text{ J}$$

ASSESS   The kinetic energy can be related to the initial height as

$$K = \frac{1}{2}m(2gy_0) = mgy_0$$

Thus, the greater the initial height the ball is dropped, the greater the final speed when it hits the ground.

Section 6.4 Power

33. **INTERPRET**   We convert power from kcal/day to Watts.

DEVELOP   We can use standard unit-conversion techniques: 1 calorie = 4.19 Joule.

EVALUATE

$$\frac{2000 \text{ kcal}}{1 \text{ d}} \times \frac{1 \text{ d}}{24 \text{ h}} \times \frac{1 \text{ h}}{3600 \text{ s}} \times \frac{1000 \text{ cal}}{1 \text{ kcal}} \times \frac{1 \text{ J}}{4.19 \text{ cal}} = 5.52 \text{ J/s} = 5.52 \text{ W}$$

ASSESS   This is an *average* power. Human power output is higher during exercise.

35. **INTERPRET**   This problem is about the power output of a car battery, or the rate at which energy is drained.

DEVELOP   According to Equation 6.15, if the amount of work ΔW is done in time Δt, then the average power is $\bar{P} = \Delta W/\Delta t$.

EVALUATE   Using Equation 6.15, the power output for each of the three cases is

(**a**) $\bar{P} = \dfrac{\Delta W}{\Delta t} = \dfrac{(1 \text{ kWh})}{(1/60) \text{ h}} = 60 \text{ kW}$

(**b**) $\bar{P} = \dfrac{\Delta W}{\Delta t} = \dfrac{(1 \text{ kWh})}{1 \text{ h}} = 1 \text{ kW}$

(**c**) $\bar{P} = \dfrac{\Delta W}{\Delta t} = \dfrac{(1 \text{ kWh})}{24 \text{ h}} = \dfrac{(1000 \text{ Wh})}{24 \text{ h}} = 41.7 \text{ W}$

ASSESS   From Equation 6.15, we see that when the amount of work done is kept fixed, the average power is inversely proportional to Δt. Thus, the average power output is the greatest in case (**a**) and smallest in case (**c**).

37. **INTERPRET**   This problem is about the total work done, given average power and time.

DEVELOP   According to Equation 6.15, if the average power is $\bar{P}$, then the amount of work done over a period Δt is $\Delta W = \bar{P}\Delta t$. In SI units, 1 hp $= 746$ W.

EVALUATE   Working at constant power output, Equation 6.15 gives the total work (energy output) as

$$\Delta W = \bar{P}\Delta t = (3.5 \text{ hp})(746 \text{ W/hp})(3600 \text{ s}) = 9.40 \times 10^6 \text{ J}$$

ASSESS   Given a constant average power, the work done is proportional to the time interval Δt.

39. **INTERPRET**   In this problem we are asked to estimate the power out, or rate of work, while doing deep knee bends at a given rate.

DEVELOP   The work done against gravity in raising or lowering a weight through a height, h, has magnitude $W_g = F_g h$. The body begins and ends each deep knee bend at rest ($\Delta K = 0$), so the muscles do a total work (down and up) of $W_{tot} = 2W_g = 2F_g h$ for each complete repetition.

Next, we assume that the lower extremities comprise 35% of the body mass, and are not included in the moving mass, then for a 75 kg person, F_g is about

$$F_g = 0.65(75 \text{ kg})(9.8 \text{ m/s}^2) = 480 \text{ N}$$

We estimate that h is somewhat greater than 25% of the body height, or about 45 cm $= 0.45$ m.

EVALUATE   With all the assumptions made above, the muscle power output for one repetition per second is about

$$\bar{P} = \frac{W_{tot}}{\Delta t} = \frac{2F_g h}{\Delta t} = \frac{2(480 \text{ N})(0.45 \text{ m})}{1 \text{ s}} = 430 \text{ W}$$

ASSESS   A power output of 430 W for deep knee bends is reasonable. Walking at 3 mph has approximately the same output.

41. **INTERPRET**   We find the time necessary to provide a required quantity of energy, given the power.

DEVELOP   We use the definition of power: $P = \frac{W}{t}$. The work is $W = 20$ kJ, and the power is $P = 900$ W.

EVALUATE $P = \dfrac{W}{t} \rightarrow t = \dfrac{W}{P} = \dfrac{20 \times 10^3 \text{ J}}{900 \text{ W}} = 22 \text{ s}$

ASSESS   This time is a reasonable amount for the given amount of defrosting.

PROBLEMS

43. **INTERPRET**   The problem is about calculating work, given force and displacement. The object of interest is the box, which is being pushed up a ramp.

DEVELOP   We shall use Equation 6.5, $W = \vec{F} \cdot \Delta\vec{r}$, to calculate the work done in pushing the box up the ramp.

EVALUATE   **(a)** The box has risen $\Delta y = 1$ m vertically. This means that the displacement up the ramp (parallel to the applied force) is $\Delta r = \frac{\Delta y}{\sin\theta} = \frac{1 \text{ m}}{\sin 30°} = 2$ m. Therefore, the work done during this process is

$$W_a = \vec{F}_a \cdot \Delta\vec{r} = (200 \text{ N})(2 \text{ m}) = 400 \text{ J}$$

(b) To find the mass, we first note that the work done by gravity is

$$W_g = \vec{F}_g \cdot \Delta\vec{r} = (-mg\hat{j}) \cdot (\Delta x\hat{i} + \Delta y\hat{j}) = -mg\Delta y = -mg\Delta r \sin\theta$$

On the other hand, the work done by friction is

$$W_f = \vec{f}_k \cdot \Delta\vec{r} = -f_k\Delta r = -\mu_k n\ \Delta r = -\mu_k(mg\cos\theta)\Delta r$$

Since the speed remains unchanged, by work-energy theorem, $W = \Delta K$, the total work must be zero:

$$W_{tot} = W_a + W_g + W_f = 0$$

This implies

$$W_a = -W_g - W_f = mg\Delta r \sin\theta + \mu_k(mg\cos\theta)\Delta r = mg\Delta r(\sin\theta + \mu_k \cos\theta)$$

from which the mass is found to be

$$m = \frac{W_a}{g\Delta r(\sin\theta + \mu_k \cos\theta)} = \frac{F_a}{g(\sin\theta + \mu_k \cos\theta)} = \frac{200 \text{ N}}{(9.8 \text{ m/s}^2)[\sin 30° + (0.18)\cos 30°]}$$

$$= 31.1 \text{ kg}$$

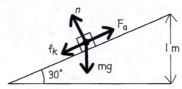

ASSESS   The mass could also be found by solving Newton's second law, with zero acceleration:

$$F_{net} = F_a - mg(\sin\theta + \mu_k\cos\theta) = 0 \quad\Rightarrow\quad m = \frac{F_a}{g(\sin\theta + \mu_k\cos\theta)}$$

45. **INTERPRET**   The problem is about calculating the average force, given work and displacement. The object of interest is the locomotive is pulling a train.

DEVELOP   We shall use Equation 6.5, $W = \vec{F}\cdot\Delta\vec{r}$, to solve for the average force in the coupling between the locomotive and the rest of the train.

EVALUATE   The distance pulled by the locomotive is $\Delta r = 180$ km $= 1.80\times10^5$ m. If we define the average force by $W = F_{av}\,\Delta r$, then

$$F_{av} = \frac{W}{\Delta r} = \frac{7.9\times10^{11}\text{ J}}{1.80\times10^5\text{ m}} = 4.39\times10^6\text{ N}$$

ASSESS   The average force depends only on the total work done and the displacement. The train's mass is not required to answer this question.

47. **INTERPRET**   In this problem, we are asked to evaluate the scalar products between different pairs of unit vectors $\hat{i},\hat{j}$, and $\hat{k}$.

DEVELOP   As shown in Equation 6.3, the scalar product of two vectors $\vec{A}$ and $\vec{B}$ is defined as

$$\vec{A}\cdot\vec{B} = AB\cos\theta$$

where A and B are the magnitudes of the vectors and θ is the angle between them. With this definition, the scalar products between different pairs of unit vectors can be readily computed.

EVALUATE   (a) The dot product of any vector with itself equals its magnitude squared, $\vec{A}\cdot\vec{A} = A^2\cos0° = A^2$. For any unit vector ($|\hat{i}| = |\hat{j}| = |\hat{k}| = 1$), $\hat{n}\cdot\hat{n} = 1$. That is

$$\hat{i}\cdot\hat{i} = \hat{j}\cdot\hat{j} = \hat{k}\cdot\hat{k} = 1$$

(b) If two vectors $\vec{A}$ and $\vec{B}$ are perpendicular, then their dot product is zero, $\vec{A}\cdot\vec{B} = AB\cos90° = 0$. Since the unit vectors $\hat{i},\hat{j}$, and $\hat{k}$ are mutually perpendicular, hence,

$$\hat{i}\cdot\hat{j} = \hat{j}\cdot\hat{k} = \hat{k}\cdot\hat{i} = 0$$

(c) Using the distributive law, we have

$$\vec{A}\cdot\vec{B} = (A_x\hat{i} + A_y\hat{j} + A_z\hat{k})\cdot(B_x\hat{i} + B_y\hat{j} + B_z\hat{k}) = A_xB_x\hat{i}\cdot\hat{i} + A_xB_y\hat{i}\cdot\hat{j} + A_xB_z\hat{i}\cdot\hat{k} + A_yB_x\hat{j}\cdot\hat{i}$$
$$+ A_yB_y\hat{j}\cdot\hat{j} + A_yB_z\hat{j}\cdot\hat{k} + A_zB_x\hat{k}\cdot\hat{i} + A_zB_y\hat{k}\cdot\hat{j} + A_zB_z\hat{k}\cdot\hat{k}$$
$$= A_xB_x + A_yB_y + A_zB_z$$

where we have used the results from (a) and (b). The expression indeed agrees with Equation 6.4.

ASSESS   The quantity $\vec{A}\cdot\vec{B}$ is a scalar formed by two vectors. Scalar product is commutative: $\vec{A}\cdot\vec{B} = \vec{B}\cdot\vec{A}$, and also distributive: $\vec{A}\cdot(\vec{B}+\vec{C}) = \vec{A}\cdot\vec{B} + \vec{A}\cdot\vec{C}$.

49. **INTERPRET**   In this problem we are asked to find the angle between two vectors.

DEVELOP   As shown in Equation 6.3, the scalar product of two vectors $\vec{A}$ and $\vec{B}$ is defined as

$$\vec{A}\cdot\vec{B} = AB\cos\theta$$

where A and B are the magnitudes of the vectors and θ is the angle between them. With this definition, the angle between two vectors is given by

$$\cos\theta = \frac{\vec{A}\cdot\vec{B}}{AB}$$

Consider two vectors of the form $\vec{A} = a_x\hat{i} + a_y\hat{j}$, and $\vec{B} = b_x\hat{i} + b_y\hat{j}$. The scalar product of the two vectors is $\vec{A} \cdot \vec{B} = a_x b_x + a_y b_y$. Since the magnitudes of $\vec{A}$ and $\vec{B}$ are $A = \sqrt{a_x^2 + a_y^2}$ and $B = \sqrt{b_x^2 + b_y^2}$, using Equation 6.3, the scalar product can also be written as

$$\vec{A} \cdot \vec{B} = a_x b_x + a_y b_y = \sqrt{a_x^2 + a_y^2}\,\sqrt{b_x^2 + b_y^2}\,\cos\theta$$

where θ is the angle between $\vec{A}$ and $\vec{B}$. Solving for θ, we have

$$\theta = \cos^{-1}\frac{a_x b_x + a_y b_y}{\sqrt{a_x^2 + a_y^2}\,\sqrt{b_x^2 + b_y^2}}$$

EVALUATE   The expression above gives the angle between any two vectors. We use θ_{AB} to denote the angle between $\vec{A}$ and $\vec{B}$. Therefore:

$$\theta_{AB} = \cos^{-1}\left[\frac{3(-1) + 2(6)}{\sqrt{3^2 + 2^2}\,\sqrt{(-1)^2 + 6^2}}\right] = 65.8°$$

$$\theta_{AC} = \cos^{-1}\left[\frac{3(7) + 2(-2)}{\sqrt{13}\,\sqrt{7^2 + (-2)^2}}\right] = 49.6°$$

$$\theta_{BC} = \cos^{-1}\left[\frac{(-1)7 + 6(-2)}{\sqrt{37}\,\sqrt{53}}\right] = 115°$$

ASSESS   Note that for any three vectors in a plane, one of the above angles is equal to the sum of the other two. In our case, we have $\theta_{BC} = \theta_{AB} + \theta_{AC}$.

51.   **INTERPRET**   In this problem we are asked to find the work done by a non-constant force that varies with position.
DEVELOP   Since we are dealing with a varying force $F(x)$, to evaluate the work done, we need to integrate using Equation 6.8:

$$W = \int_{x_1}^{x_2} F\,dx$$

The integral can be approximated as $W \approx \sum_{i=1}^{N} W_i = \sum_{i=1}^{N} F(x_i)\Delta x$, as given in Equation 6.6.

EVALUATE   (a) The work done on the particle moving from $x = 0$ to $x = 6$ m is

$$W = \int_{x_1}^{x_2} F\,dx = \int_0^6 ax^2\,dx = \frac{1}{3}a\,|x^3\,|_0^6 = \frac{1}{3}(5\text{ N/m}^2)(6\text{ m})^3 = 360\text{ J}$$

(b) Using Equation 6.6 with $\Delta x = 2$ m, the work done may be approximated as $W \approx \sum_{i=1}^{3} F(x_i)\Delta x$, where $x_i = 1$ m, 3 m, 5 m are the midpoints. The result is

$$W \approx (5\text{ N/m}^2)[(1\text{ m})^2 + (3\text{ m})^2 + (5\text{ m})^2](2\text{ m}) = 350\text{ J}$$

The percent error is only $\delta = (100\%)\frac{(360\text{ J}-350\text{ J})}{360\text{ J}} = 2.78\%$.
(c) Now with $\Delta x = 1$ m, and $x_i = 0.5$ m, 1.5 m, 2.5 m, 3.5 m, 4.5 m, and 5.5 m, the work done is approximately equal to

$$W \approx \sum_{i=1}^{6} F(x_i)\Delta x = (5\text{ N/m}^2)(0.5\text{ m})^2[(1\text{ m})^2 + (3\text{ m})^2 + (5\text{ m})^2 + (7\text{ m})^2 + (9\text{ m})^2 + (11\text{ m})^2](1\text{ m}) = 357.5\text{ J}$$

and the percent error is $\delta = (100\%)\frac{(360\text{ J}-357.5\text{ J})}{360\text{ J}} = 0.694\%$.
(d) Similarly, with $\Delta x = 0.5$ m, we have

$$W \approx \sum_{i=1}^{12} F(x_i)\Delta x = (5\text{ N/m}^2)(0.25\text{ m})^2(1^2 + 3^2 + \cdots + 23^2)(0.5\text{ m}) = 359.375\text{ J}$$

with $\delta = 0.174\%$. (Note that the direct calculation of the sum is tedious, but we can use the formula for the sum of the squares of the first n numbers, namely $\sum_1^n k^2 = \frac{1}{6}n(n+1)(n+2)$. The sum in question is $\sum_{k=1}^{12}(2k-1)^2 = \sum_{k=1}^{23} k^2 - \sum_{k=1}^{11}(2k)^2 = 4324 - 2024 = 2300$.)

ASSESS   The problem demonstrates that the smaller the width Δx in $W \approx \sum_{i=1}^{3} F(x_i)\Delta x$, the more accurate the approximation. In fact, the approximation becomes exact in the limit $\Delta x \to 0$, as shown in Equation 6.7:

$$\lim_{\Delta x \to 0} \sum_{i=1}^{N} F(x_i)\Delta x = \int_{x_1}^{x_2} F(x)\,dx$$

53.   **INTERPRET**   In this problem we are asked to find the work done by a non-constant force that varies with position.

DEVELOP   Since we are dealing with a varying force $F(x)$, to evaluate the work done, we need to integrate using Equation 6.8:

$$W = \int_{x_1}^{x_2} F\,dx$$

With $F = a\sqrt{x}$, we obtain $W_{x_1 \to x_2} = \int_{x_1}^{x_2} ax^{1/2}dx = \frac{2}{3}ax^{3/2}\Big|_{x_1}^{x_2} = \left(\frac{2a}{3}\right)\left(x_2^{3/2} - x_1^{3/2}\right).$

EVALUATE   In each case, the work done is

(a) $W_{0 \to 3} = \dfrac{2}{3}(9.5 \text{ N/m}^{1/2})(3 \text{ m})^{3/2} = 32.9 \text{ J}$

(b) $W_{3 \to 6} = \dfrac{2}{3}(9.5 \text{ N/m}^{1/2})[(6 \text{ m})^{3/2} - (3 \text{ m})^{3/2}] = 60.2 \text{ J}$

(c) $W_{6 \to 9} = \dfrac{2}{3}(9.5 \text{ N/m}^{1/2})[(9 \text{ m})^{3/2} - (6 \text{ m})^{3/2}] = 77.9 \text{ J}$

ASSESS   Since the force increases with x (as $\sqrt{x}$), more work is done as the object moves further in the x direction.

55.   **INTERPRET**   As you push the swing, you are doing work against gravity. While gravitational force is constant, the path is a circular arc.

DEVELOP   Since the path of the swing is a circular arc (radius L and differential arc length $|d\vec{r}| = L\,d\theta$), only the tangential (or parallel) components of any forces acting do work on the swing; the radial (or perpendicular) components do no work since the scalar product with the path element is zero. Thus, the tension in the chains and the radial components of gravity or the applied force do no work.

EVALUATE   If you pull slowly, so that the tangential acceleration is zero, then $F_\parallel = mg\sin\theta$, and the work you do is

$$W = \int_0^\phi \vec{F} \cdot d\vec{r} = \int_0^\phi F_\parallel \, |dr| = \int_0^\phi mg\sin\theta \cdot L\,d\theta = mgL|-\cos\theta|_0^\phi = mgL(1 - \cos\phi)$$

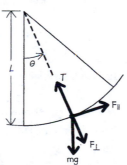

ASSESS   The result can also be written as $W = mgh$, where $h = L(1 - \cos\phi)$ is the vertical distance measured from the bottom of the swing. Thus, the work is the energy required to lift the child on the swing by a vertical distance h.

57.   **INTERPRET**   This problem is about the distance the plane can be towed with a given amount of work done by the tractor.

DEVELOP   Equation 6.2, $W = F\Delta r\cos\theta$, applies here. The displacement is horizontal but the applied force (tension in the link) is at an angle $\theta = 22°$ with the horizontal.

EVALUATE   Using Equation 6.2, the distance the plane moves is

$$\Delta x = \frac{W}{F_x} = \frac{W}{F\cos\theta} = \frac{8.7 \times 10^6 \text{ J}}{(4.1 \times 10^5 \text{ N})\cos 22°} = 22.9 \text{ m}$$

ASSESS   Only the horizontal component of the force, $F_x = F\cos\theta$, does the work.

59.   **INTERPRET**   In this problem we are asked to find the work done by a non-constant force that varies with position.
DEVELOP   Since we are dealing with a varying force $F(x)$, to evaluate the work done, we need to integrate using Equation 6.8:

$$W = \int_{x_1}^{x_2} F\,dx$$

With $F = ax^{3/2}$, we obtain $W_{x_1 \to x_2} = \int_{x_1}^{x_2} ax^{3/2}\,dx = \dfrac{2}{5}ax^{5/2}\Big|_{x_1}^{x_2} = \left(\dfrac{2a}{5}\right)\left(x_2^{5/2} - x_1^{5/2}\right).$

EVALUATE   The work required in moving the object from $x = 0$ to $x = 14$ m is

$$W_{0 \to 14\ \text{m}} = \left(\frac{2}{5}\right)(0.75\ \text{N/m}^{3/2})(14\ \text{m})^{5/2} = 220\ \text{N}\cdot\text{m} = 220\ \text{J}$$

ASSESS   Since the force increases with x (as $x^{3/2}$), more work is done as the object moves further in the x direction.

61.   **INTERPRET**   In this problem the pump (with a given power) is doing work against gravity to deliver water to a tank above the ground. The quantity of interest is the amount of water that the pump can deliver during a given time interval.
DEVELOP   According to Equation 6.15, if the average power is $\bar{P}$, then the amount of work done over a period Δt is $\Delta W = \bar{P}\Delta t$. Since the work required in lifting an object of mass m to a vertical height h is $W = mgh$, the rate at which the mass can be delivered is given by

$$\bar{P} = \frac{\Delta W}{\Delta t} = \left(\frac{\Delta m}{\Delta t}\right)gh \quad \to \quad \frac{\Delta m}{\Delta t} = \frac{\bar{P}}{gh}$$

In SI units, 1 hp = 746 W.
EVALUATE   Using the expression above, we find the rate to be

$$\frac{\Delta m}{\Delta t} = \frac{\bar{P}}{gh} = \frac{(0.5\ \text{hp})(746\ \text{W/hp})}{(9.8\ \text{m/s}^2)(60\ \text{m})} = 0.634\ \text{kg/s}$$

Since the mass of 1 gallon (1 gal $= 3.786 \times 10^{-3}\,\text{m}^3$) of water is $(1000\ \text{kg/m}^3)(3.786 \times 10^{-3}\,\text{m}^3) = 3.786$ kg, the rate can also be rewritten as

$$\frac{\Delta m}{\Delta t} = \left(0.634\ \frac{\text{kg}}{\text{s}}\right)\left(60\ \frac{\text{s}}{\text{min}}\right)\left(3.786\ \frac{\text{kg}}{\text{gal}}\right) = 10.1\ \text{gal/min}$$

ASSESS   Given a constant average power, the rate of delivery $\Delta m/\Delta t$ is inversely proportional to the vertical distance h. The greater the h, the slower is the rate, as we expect.

63.   **INTERPRET**   This problem is about the total work done, given the average power and time. The object of interest is the runner.
DEVELOP   According to Equation 6.15, if the average power is $\bar{P}$, then the amount of work done over a period Δt is $\Delta W = \bar{P}\Delta t$. In this problem, $\bar{P}$ is a function of the speed.
EVALUATE   The runner's average power output is

$$\bar{P} = m(bv - c) = (54\ \text{kg})[(4.27\ \text{J/kg}\cdot\text{m})(5.2\ \text{m/s}) - 1.83\ \text{W/kg}] = 1.10 \times 10^3\ \text{W}$$

Therefore, over the race time, $\Delta t = \frac{\Delta x}{v} = \frac{10000\ \text{m}}{5.2\ \text{m/s}} = 1.92 \times 10^3$ s, the runner's work output is

$$\Delta W = \bar{P}\Delta t = (1.10 \times 10^3\ \text{W})(1.92 \times 10^3\ \text{s}) = 2.12 \times 10^6\ \text{J} = 0.588\ \text{kWh}$$

ASSESS   Long-distance running is an intense physical activity, so we expect a much higher average power output (about 1.10 kW) than other activities such as walking (about 250 W), swimming (about 480 W) or playing basketball (880 W).

65.   **INTERPRET**   In this problem a constant average power is supplied to the car as it climbs a slope against the air resistance. We want to know the angle of the slope if the car is moving at a steady speed.

DEVELOP   At constant velocity, there is no change in kinetic energy, so the net work done on the car is zero. Therefore, the power supplied by the engine equals the power expended against gravity and air resistance. The power can be found from Equation 6.19, $P = \vec{F} \cdot \vec{v}$.

EVALUATE   Since gravity $\vec{F}_g$ makes an angle of $\theta + 90°$ with the velocity $\vec{v}$ (where θ is the slope angle to the horizontal), the power done against gravity is

$$P_g = \vec{F}_g \cdot \vec{v} = mgv\cos(\theta + 90°) = -mgv\sin\theta$$

Similarly, the air resistance makes an angle of $180°$ to the velocity and

$$P_{air} = \vec{F}_{air} \cdot \vec{v} = F_{air}v\cos 180° = -F_{air}v$$

In SI units, $v = 60$ km/h $= 16.7$ m/s. Since the car moves at a constant speed, $P_{tot} = P_{car} + P_g + P_{air} = 0$, or

$$P_{car} = -P_g - P_{air} = mgv\sin\theta + F_{air}v$$

The equation can be solved to give

$$\theta = \sin^{-1}\left(\frac{P_{car} - F_{air}v}{mgv}\right) = \sin^{-1}\left(\frac{38000 \text{ W} - (450 \text{ N})(16.7 \text{ m/s})}{(1400 \text{ kg})(9.8 \text{ m/s}^2)(16.7 \text{ m/s})}\right) = 7.67°$$

ASSESS   To see that the result makes sense, we first note that increasing P_{car}, the power output of the car's engine, will allow the car to climb a steeper slope (with larger angle θ). On the other hand, when $P_{car} < P_{air}$, we get a negative value for θ. This means that the car's power is not large enough to overcome the air resistance, and the car won't be able to climb the slope at all.

67.   **INTERPRET**   The object of interest here is the chest. The physical quantity we are asked to find is the power needed to push the chest against friction.

DEVELOP   If you push parallel to a level floor, the applied force equals the frictional force (since the acceleration is zero), $F_a = f_k$. Since the normal force equals the weight of the box: $n = mg$, the applied force is

$$F_a = \mu_k n = \mu_k mg$$

Equation 6.19, $P = \vec{F} \cdot \vec{v}$, can then be used to find the power needed. To find the work, we use Equation 6.1, $W = F\Delta x$.

EVALUATE   **(a)** The power required is

$$P_a = F_a v = \mu_k mgv = (0.78)(95 \text{ kg})(9.8 \text{ m/s}^2)(0.62 \text{ m/s}) = 450 \text{ W}$$

which is about 0.6 hp.

(b) The work done by the applied force acting over a displacement $\Delta x = 11$ m is

$$W_a = F_a\Delta x = \mu_k mg\Delta x = (0.78)(95 \text{ kg})(9.8 \text{ m/s}^2)(11 \text{ m}) = 7.99 \times 10^3 \text{ J}$$

ASSESS   An alternative way to calculate the power is to note that the time required to push the chest 11 m is $\Delta t = \frac{\Delta x}{v} = \frac{11 \text{ m}}{0.62 \text{ m/s}} = 17.74$ s. Using Equation 6.17, we have

$$W_a = P_a\Delta t = (450 \text{ W})(17.74 \text{ s}) = 7.99 \times 10^3 \text{ J}$$

69.   **INTERPRET**   This problem is about the total work done, given the power and time. The object of interest is the machine.

DEVELOP   The power given in this problem is time-varying. Therefore, to find the work done in a given time interval, we need to integrate Equation 6.18: $W = \int_{t_1}^{t_2} P\,dt$.

EVALUATE   With $P = ct^2$, we obtain

$$W_{t_1 \to t_2} = \int_{t_1}^{t_2} ct^2\,dt = \frac{1}{3}ct^3\Big|_{t_1}^{t_2} = \frac{c}{3}(t_2^3 - t_1^3) = \frac{1}{3}\left(18\frac{\text{W}}{\text{s}^2}\right)[(20 \text{ s})^3 - (10 \text{ s})^3] = 42 \text{ kJ}$$

ASSESS   Since the power increases with t (as t^2), more work is done by the machine as time goes on. For example, the work done between $t = 20$ s and $t = 30$ s would be greater than the work done between $t = 10$ s and $t = 20$ s.

71.   **INTERPRET**   This problem is about the total work done, given the power and time. The object of interest is the machine. We want to show that the total work done could be finite, even though the machine runs forever.

DEVELOP   The power given in this problem is time-varying. Therefore, to find the work done in a given time interval, we need to integrate Equation 6.18: $W = \int_{t_1}^{t_2} P\,dt$.

EVALUATE   With $P = P_0 t_0^2/(t + t_0)^2$, we obtain

$$W = \int_0^\infty \frac{P_0 t_0^2}{(t + t_0)^2}\,dt = P_0 t_0^2 \int_0^\infty \frac{dt}{(t + t_0)^2} = \frac{P_0 t_0^2}{(t + t_0)}\bigg|_0^\infty = P_0 t_0$$

ASSESS   The result shows that even though the machine operates forever, the total amount of work done is finite. This is of no surprise because the power output decreases with time.

73.   **INTERPRET**   In this problem we are asked to find the work done by a non-constant force that varies with position. We want to show that even though the force could become arbitrarily large as x approaches a certain value, the work done could remain finite.

DEVELOP   Since we are dealing with a varying force $F(x)$, to evaluate the work done, we need to integrate using Equation 6.8: $W = \int_{x_1}^{x_2} F\,dx$.

EVALUATE   With $F = b/\sqrt{x}$, we obtain

$$W_{x_1 \to x_2} = \int_{x_1}^{x_2} bx^{-1/2}\,dx = 2bx^{1/2}\bigg|_{x_1}^{x_2} = 2b\left(\sqrt{x_2} - \sqrt{x_1}\right)$$

Thus, we see that $W_{x_1 \to x_2}$ is finite as $x_1 \to 0$. In fact, $W \to 2b\sqrt{x_2}$, for $x_1 \to 0$.

ASSESS   The result demonstrates that even though a function $F(x)$ could diverge at some value $x = x_0$, the integral $\int F(x)\,dx$ can be finite at $x = x_0$.

75.   **INTERPRET**   In this problem we are asked to find the work done by a non-constant force that varies with position. The force is applied to stretch a spring.

DEVELOP   Since we are dealing with a varying force $F(x)$, to evaluate the work done, we need to integrate using Equation 6.8: $W = \int_{x_1}^{x_2} F\,dx$.

EVALUATE   The spring force is a restoring force (opposite to the displacement); the work we must do (against this force) to stretch the spring is $\int F\,dx$.

77.   **INTERPRET**   We need to calculate the work done against a given vector force, along a vector path. We will use the most general integral form of work.

DEVELOP   We calculate the work with $W = \int_{s_1}^{s_2} \vec{F} \cdot d\vec{s}$. The path taken fits the equation $y = ax^2 - bx$, where $a = 2$ m^{-1} and $b = 4$, so $\frac{dy}{dx} = 2ax - b$ and we can use $d\vec{s} = dx\hat{i} + (2ax - b)dx\hat{j}$. The force is $\vec{F} = cxy\hat{i} + d\hat{j}$, where $c = 10$ N/m^2 and $d = 15$ N. The position x goes from zero to 3 m.

EVALUATE

$$W = \int_{x=0}^{x=3\text{m}} (cxy\hat{i} + d\hat{j}) \cdot (\hat{i} + (2ax - b)\hat{j})\,dx = \int_0^3 [cxy + d(2ax - b)]\,dx$$

$$W = \int_0^3 [cx(ax^2 - bx) + d(2ax - b)]\,dx = \int_0^3 [cax^3 - cbx^2 + 2adx - bd]\,dx$$

$$W = \left[\tfrac{1}{4}cax^4 - \tfrac{1}{3}cbx^3 + adx^2 - bdx\right]_0^3 = [405 \text{ J} - 360 \text{ J} + 270 \text{ J} - 180]$$

$$W = 135 \text{ J}$$

ASSESS   We're not sure what to compare this to, but the answers are similar in magnitude to those of the next problem, which uses the same force over different paths. The units work out right.

79.   **INTERPRET**   We need to find the work done by the tractor as it is pulling the jet. The tractor pulls at an angle, but with a constant force and direction, so we can use a simple definition of work, $W = \vec{F} \cdot \vec{s}$.

      DEVELOP   The force exerted by the tractor is $\vec{F} = 4.1 \times 10^5$ N. The distance the jet moves is $s = 22.9$ m. The angle between force and distance is $\theta = 22°$, and we want to do no more than 7 MJ of work. We use $W = \vec{F} \cdot \vec{s} = Fs\cos\theta$.

      EVALUATE $W = \vec{F} \cdot \vec{s} = Fs\cos\theta = (4.1 \times 10^5 \text{ N}) \times (22.9 \text{ m}) \times \cos(22°) = 8.7$ MJ

      ASSESS   In our report, we should suggest that the tractor pull the jet along a line closer to the direction of the jet's motion so that it does not have to pull so hard.

81.   **INTERPRET**   A mass falls a certain distance. We find the force necessary to stop the mass within a second distance. Using the work-energy theorem, we see that the work done by gravity on the way down is equal in magnitude to the work done by the stopping force, since there is no change in kinetic energy between the initial and final state.

      DEVELOP   The height dropped is $h = 0.7$ m, and the stopping distance is $s = 0.02$ m. The mass of the leg is 8 kg. $|W_{down}| = |W_{stop}|$.

      EVALUATE $|W_{down}| = |W_{stop}| \rightarrow mgh = Fs \rightarrow F = (mg)\frac{h}{s}$. The value $\frac{h}{s} = 35$, so the average stopping force is 35 times the weight of the leg.

      ASSESS   The shorter the distance over which something is stopped, the greater the force required. This is why cars are built to "crumple" on impact: The increased distance traveled by the passengers during the crash means a lower average force on their bodies.

      (a) Between $x = 0$ and 1, the work done on the spring is

$$W_{0 \to 1} = \int_0^1 100x^2 dx = \frac{100}{3}x^3 \Big|_0^1 = \frac{100}{3} \text{ J} = 33.3 \text{ J}$$

      (b) Between $x = 1$ and 2, the work done on the spring is

$$W_{1 \to 2} = \int_0^1 100(4x - x^2 - 2)dx = 100\left(2x^2 - \frac{1}{3}x^3 - 2x\right)\Big|_1^2 = 167 \text{ J}$$

      ASSESS   The work done in stretching the spring from $x = 1$ m to $x = 2$ m is about five times that from $x = 0$ to $x = 1$ m. This is due to the fact that the forces acting on the spring depend on x. In the case where the applied force is a constant independent of x, then the work done in stretching the spring from $x = 1$ m to $x = 2$ m and that from $x = 0$ to $x = 1$ m would be the same.

7
CONSERVATION OF ENERGY

EXERCISES

Section 7.1 Conservative and Nonconservative Forces

11. **INTERPRET** In this problem we want to find the work done by the frictional force in moving a block from one point to another over two different paths.

    DEVELOP Figure 7.16 is a plane view of the horizontal surface over which the block is moved, showing the paths (a) and (b). The force of friction is $f = \mu n = \mu mg$ opposite to the displacement. Since f is constant, using Equation 6.1, the work done is

    $$W = \vec{f} \cdot \Delta \vec{r} = -f \Delta r$$

    EVALUATE The work done by friction along path (a) with $\Delta r_a = L + L = 2L$ is

    $$W_a = -f \Delta r_a = -\mu mg(2L) = -2\mu mgL$$

    Similarly, the work done by friction along path (b) with $\Delta r_b = \sqrt{L^2 + L^2} = \sqrt{2}L$ is

    $$W_b = -f \Delta r_b = -\mu mg(\sqrt{2}L) = -\sqrt{2}\mu mgL$$

    ASSESS Since the work done depends on the path chosen, friction is not a conservative force.

Section 7.2 Potential Energy

13. **INTERPRET** The problem is about gravitational potential energy relative to a reference point of zero energy. In Example 7.1, the reference point was taken to be the 33rd floor. In this problem, we take the street level to be our reference point.

    DEVELOP The change in potential energy with a change in the vertical distance Δy is given by Equation 7.3, $\Delta U = mg\Delta y$. Each floor is 3.5 m high.

    EVALUATE (a) The office of the engineer is on the 33rd floor, or is 32 stories above the street level (the first floor) where $U_1 = 0$. Thus, the difference in gravitational potential energy is

    $$\Delta U = U_{33} - U_1 = U_{33} - 0 = mg\ \Delta y = (55 \times 9.8 \text{ N})(32 \times 3.5 \text{ m}) = 60.4 \text{ kJ}$$

    (b) At the 59th floor, $\Delta U = U_{59} - U_1 = (55 \times 9.8 \text{ N})(58 \times 3.5 \text{ m}) = 109 \text{ kJ}$.
    (c) Street level is the zero of potential energy, $U_1 = 0$.

    ASSESS Potential energy depends on the reference point chosen, but potential energy difference between two points does not. What matters physically is the difference in potential energy. The differences in potential energy between any two levels are the same as in Example 7.1, e.g., $U_{59} - U_{33} = (109 \text{ kJ} - 60.4 \text{ kJ}) = 49.0 \text{ kJ}$.

15. **INTERPRET** The problem is about the change in gravitational potential energy as a person comes down from the mountaintop.

    DEVELOP The change in potential energy with a change in the vertical distance Δy is given by Equation 7.3, $\Delta U = mg\Delta y$.

    EVALUATE The change in vertical position is $\Delta y = -3050$ m. Using Equation 7.3, we have

    $$\Delta U = U(\text{sea level}) - U(\text{mt.top}) = mg\Delta y = (75 \text{ kg})(9.8 \text{ m/s}^2)(-3050 \text{ m}) = -2.24 \times 10^6 \text{ J}$$

    ASSESS As you come down from the mountaintop ($\Delta y < 0$), your potential energy decreases. The result does not depend on your choice of reference point (zero potential energy).

17. **INTERPRET**   The problem is about the change in gravitational potential energy as the hiker ascends. Given the position of zero potential energy, we are interested in her altitude.

DEVELOP   The change in potential energy with a change in the vertical distance Δy is given by Equation 7.3, $\Delta U = mg\Delta y = mg(y - y_0)$. Knowing ΔU and y_0 allows us to determine y.

EVALUATE   Equation 7.3 gives

$$\Delta U = U(y) - U(y_0) = mg(y - y_0)$$

From the above expression, we find the altitude of the hiker to be

$$y = y_0 + \frac{\Delta U}{mg} = 1250 \text{ m} + \frac{-2.4 \times 10^5 \text{ J}}{(60 \text{ kg})(9.8 \text{ m/s}^2)} = 842 \text{ m}$$

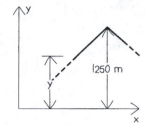

ASSESS   In this problem, the point of zero potential energy is taken to be the top of the mountain with $y_0 = 1250$ m. Since the hiker's potential energy is negative, we expect the hiker's altitude to be lower than y_0.

19. **INTERPRET**   This problem is about the potential energy stored in a spring. We'd like to know how much the spring has to be stretched in order to store a given amount of energy.

DEVELOP   The amount of energy stored in a spring is given by Equation 7.4, $U = \frac{1}{2}kx^2$, where x is the distance stretched (or compressed) from its natural length. This is the equation we shall use to solve for x.

EVALUATE   Assuming one starts stretching from the unstretched position $(x = 0)$, Equation 7.4 gives

$$x = \sqrt{\frac{2U}{k}} = \sqrt{\frac{2(210 \text{ J})}{1.4 \text{ kN/m}}} = 54.8 \text{ cm}$$

ASSESS   The larger the distance x stretched (or compressed), the greater the potential energy U stored in the spring. On the other hand, if U is fixed, then x would decrease with increasing spring constant k.

Section 7.3 Conservation of Mechanical Energy

21. **INTERPRET**   The problem is about conservation of mechanical energy. Both potential energy and kinetic energy are involved. The potential energy stored in the spring has been converted to kinetic energy of the plane. The speed of the plane is the physical quantity of interest.

DEVELOP   Suppose the initial kinetic energy of the plane is $K_0 = \frac{1}{2}mv_0^2$ and the potential energy of the spring is $U_0 = 0$. In the final state, the kinetic energy of the plane becomes $K = 0$ and the potential energy of the spring is $U = \frac{1}{2}kx^2$. These quantities are related by the principle of conservation of mechanical energy given in Equation 7.7:

$$K_0 + U_0 = K + U = \text{constant}$$

EVALUATE   Using Equation 7.7, we have

$$\frac{1}{2}mv_0^2 + 0 = 0 + \frac{1}{2}kx^2$$

which can be solved to give

$$v_0 = \sqrt{\frac{k}{m}}x = \sqrt{\frac{40,000 \text{ N/m}}{10,000 \text{ kg}}}(25 \text{ m}) = 50 \text{ m/s}$$

ASSESS   From the above expression we see that the initial speed of the plane is directly proportional to x. This means that a larger initial speed of the plane will lead to a greater stretching of the spring.

23. **INTERPRET**   We are asked to find the amount a spring compresses as it stops a boxcar. This is a conservation of energy problem involving spring potential energy and kinetic energy. The boxcar has kinetic energy before it hits the spring, and the spring has potential energy when the boxcar is stopped.

    DEVELOP   We use conservation of energy: $U_i + K_i = U_f + K_f$. The initial energy is entirely kinetic, and the final energy is entirely potential. The potential energy of a spring is $U = \frac{1}{2}kx^2$, and the kinetic energy of a moving object is $K = \frac{1}{2}mv^2$. The spring constant of this spring is $k = 2.8$ MN/m. The initial speed of the boxcar is $v = 7.5$ m/s and the mass of the boxcar is $m = 35000$ kg.

    EVALUATE $\cancel{U_i} + K_i = U_f + \cancel{K_f} \rightarrow \frac{1}{\cancel{2}}mv^2 = \frac{1}{\cancel{2}}kx^2 \rightarrow x = v\sqrt{\frac{m}{k}} = 0.84$ m

    ASSESS   Check the units: $\sqrt{\frac{m}{k}}$ has units of seconds, so $[m/s] \times [s] = [m]$ and we're fine. Keep your eyes on this term $\sqrt{\frac{k}{m}}$, though—it becomes very important in later chapters!

Section 7.4 Potential Energy Curves

25. **INTERPRET**   The object of interest is the particle. As it slides along a frictionless track, conversion between gravitational potential energy and kinetic energy takes place, but the overall mechanical energy is conserved. The conservation principle allows us to calculate the speed and position of the particle at various points on the track.

    DEVELOP   Let the kinetic energy of the particle be $K = \frac{1}{2}mv^2$ and the gravitational potential energy be $U = mgy$ (measured above the reference level $y = 0$ in Fig. 7.17). In the absence of friction, by conservation of mechanical energy stated in Equation 7.7, the sum of these is a constant:

$$K + U = \text{constant}$$

    We are given that $v_A = 0$ at $y_A = 3.8$ m, so we can evaluate the constant and express the mechanical energy at any other point in terms of it:

$$\frac{1}{2}mv^2 + mgy = \frac{1}{2}mv_A^2 + mgy_A = mgy_A$$

    EVALUATE   (a) Applying the mechanical energy conservation principle to point B, we have

$$\frac{1}{2}mv_B^2 + mgy_B = mgy_A$$

    Solving for the speed at point B, we find

$$v_B = \sqrt{2g(y_A - y_B)} = \sqrt{2(9.8 \text{ m/s}^2)(3.8 \text{ m} - 2.6 \text{ m})} = 4.85 \text{ m/s}$$

    (b) Similarly, we have $\frac{1}{2}mv_C^2 + mgy_C = mgy_A$, which can be solved to give

$$v_C = \sqrt{2g(y_A - y_C)} = \sqrt{2(9.8 \text{ m/s}^2)(3.8 \text{ m} - 1.3 \text{ m})} = 7.00 \text{ m/s}$$

    (c) The right-hand turning point is the point D where the particle's velocity is instantaneously zero, before changing direction back to the left in Fig. 7.17. Thus, $y_A = y_D$, and from the figure, we estimate that x_D is about 11 m.

    ASSESS   By mechanical energy conservation,

$$K + U = \frac{1}{2}mv^2 + mgy = \text{constant}$$

    we see that the speed of the particle is a maximum at $y = 0$. Similarly, when the speed of the particle is zero, y is at a maximum.

27. **INTERPRET**   This problem is about finding the force acting on a particle, given its potential energy. The motion of the particle is in one dimension.

    DEVELOP   The force on a particle in one-dimensional motion, when the potential energy is a straight line segment, is the negative of the slope of $U(x)$, $F_x = -\frac{dU}{dx}$, as shown in Equation 7.8. This is, of course, just the conservative force represented by the potential energy.

    EVALUATE   Using $F_x = -\frac{\Delta U}{\Delta x}$, we can determine the force (negative of the slopes) as: (a) $F_a = -\frac{3 \text{ J}}{1.5 \text{ m}} = -2$ N; (b) $F_b = 0$ since the slope is 0; (c) $F_c = -\frac{-4 \text{ J}}{0.5 \text{ m}} = 8$ N; (d) $F_d = -\frac{-1 \text{ J}}{1 \text{ m}} = 1$ N; (e) $F_e = -\frac{4 \text{ N}}{1 \text{ m}} = -4$ N; (f) $F_f = 0$ since the slope is 0.

ASSESS   The result indicates that the greater the change of the potential energy $U(x)$ with x, the greater is the force on the particle. Integrating Equation 7.8 leads to Equation 7.2a:

$$\Delta U(x) = -\int F(x)dx$$

But $\int F(x)dx$ is just equal to the work done by the force, W. Therefore, we can say that $W = -\Delta U$.

PROBLEMS

29.   **INTERPRET**   Water is pumped to a higher reservoir to store potential energy. We need to calculate the gravitational potential energy of the reservoir, and the time it would take to drain the reservoir given the power output of the generators. Although it is not stated in the problem, we will assume that the efficiency of the generators is 100%.

    DEVELOP   The mass of the reservoir is $m = 1.6 \times 10^{10}$ kg, and the height is $h = 270$ m. The power output of the station is $P = 1.08$ GW. We use $U = mgh$ to find the potential energy, and $P = \frac{W}{t} = \frac{-\Delta U}{t}$ to find the time.

    EVALUATE

    (a) $U = mgh = (1.6 \times 10^{10}$ kg$) \times (9.8$ m/s$^2) \times (270$ m$) = 4.2 \times 10^{13}$ W $= 42$ TW

    (b) $P = \frac{W}{t} = \frac{-\Delta U}{t} \rightarrow t = \frac{-\Delta U}{P} = \frac{42 \times 10^{12} \text{ J}}{1.08 \times 10^9 \text{ J/s}} = 39000$ s $= 11$ h

    ASSESS   The actual efficiency of the generators will be nowhere near 100%, so the flow necessary to keep the plant running at its rated output will drain the reservoir more rapidly than this.

31.   **INTERPRET**   In this problem we are asked to find the gravitational potential energy of a mass located on an incline. As the mass moves along the incline, its potential energy changes because its vertical distance measured from the bottom of the incline varies.

    DEVELOP   To avoid confusion, let x-y axes refer to the Earth's surface, and x'-y' axes refer to the incline. The gravitational potential energy (g assumed constant, zero at $y = 0$) is

$$U = mgy$$

    EVALUATE   With the coordinate system defined above, for a point on the incline, $y = x' \sin\theta$, and

$$U = mgy = mgx' \sin\theta$$

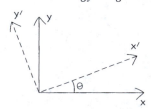

    ASSESS   To see that our result makes sense, we can check the following limits: (i) $\theta = 0$. The situation corresponds to a flat surface, and $U = 0$, since $\sin 0 = 0$, (ii) $\theta = 90°$. This corresponds to a vertical incline with $x' = y$. The potential energy is simply $U = mgy$.

33.   **INTERPRET**   In this problem we place the brick in different orientations and want to find the corresponding gravitational potential energy.

    DEVELOP   The gravitational potential energy of the brick is given by Equation 7.3:

$$\Delta U = mg\Delta y = mg(y - y_0)$$

    where Δy is the vertical distance above the point of zero potential energy.

    In position (a) the center of the brick is a distance

$$\Delta y_a = y_a - y_0 = 10 \text{ cm} - 2.75 \text{ cm} = 7.25 \text{ cm}$$

    In (b), we have $\Delta y_b = \frac{1}{2}\sqrt{(20 \text{ cm})^2 + (5.5 \text{ cm})^2} - 2.75 \text{ cm} = 10.4 \text{ cm} - 2.75 \text{ cm} = 7.62 \text{ cm}$ (see sketch; the center is midway along the diagonal of the face of the brick).

    EVALUATE   From Equation 7.3, the gravitational potential energies at positions (a) and (b) are

$$U_a = mg\Delta y_a = (1.5 \text{ kg})(9.8 \text{ m/s}^2)(7.25 \text{ cm}) = 1.07 \text{ J}$$
$$U_b = (1.5 \text{ kg})(9.8 \text{ m/s}^2)(7.62 \text{ cm}) = 1.12 \text{ J}$$

above the zero energy.

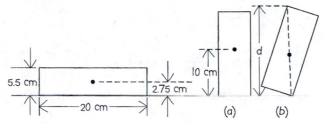

(a)   (b)

Assess Our result indicates that $U_b > U_a$. This makes sense because the center of the brick is higher in (b).

35. **Interpret** This problem is about elastic potential energy stored in a rope. We would like to compare the results obtained here with that of Example 7.3, which used a different force equation.

Develop Since the force varies with x, an integration is needed to find the potential energy. We shall use Equation 7.2a: $\Delta U = -\int_{x_1}^{x_2} F(x)\,dx$.

Evaluate Integrating over $F = -kx + bx^2 - cx^3$, we obtain

$$U = -\int_0^x F(x)\,dx = -\int_0^{2.62}(-kx + bx^2 - cx^3)\,dx = \left(\frac{1}{2}kx^2 - \frac{1}{3}bx^3 + \frac{1}{4}cx^4\right)\Bigg|_0^{2.62}$$

$$= \frac{1}{2}(223\ \text{N/m})(2.62\ \text{m})^2 - \frac{1}{3}(4.1\ \text{N/m}^2)(2.62\ \text{m})^3 + \frac{1}{4}(3.1\ \text{N/m}^3)(2.62\ \text{m})^4$$

$$= 778\ \text{J}$$

In Example 7.3, we find the energy stored to be $U' = 741$ J. Therefore, the percentage difference is

$$(100\%)\frac{U - U'}{U'} = (100\%)\frac{778\ \text{J} - 741\ \text{J}}{741\ \text{J}} = 4.9\%$$

Assess Adding the term $-cx^3$ increases the potential energy of the system. The negative sign increases the restoring force, and thus the work needed to stretch the spring.

37. **Interpret** The problem is about finding the potential energy difference between two points when the force acting on a particle is known.

Develop According to Equation 7.2, when an object moves from x_1 to x_2 under the influence of a force $F(x)$, the change in potential energy is

$$\Delta U = -\int_{x_1}^{x_2} F(x)\,dx$$

Evaluate (a) Integrating over $F(x) = A/x^2$, we obtain

$$U(x_2) - U(x_1) = -\int_{x_1}^{x_2}\frac{A}{x^2}\,dx = A\frac{1}{x}\Bigg|_{x_1}^{x_2} = A\left(\frac{1}{x_2} - \frac{1}{x_1}\right)$$

(b) For $x_1 \to \infty$, $U(x_2) - U(\infty) = \frac{A}{x_2}$. We see that the potential energy remains finite in this limit. In this case, it makes sense to define the zero of potential energy at infinity, $U(\infty) = 0$, so $U(x) = \frac{A}{x}$.

Assess The negative sign in $\Delta U = -\int_{x_1}^{x_2} F(x)\,dx$ means that if the work done by the force is positive, then the potential energy must decrease.

39. **Interpret** The problem is about conservation of mechanical energy. Both potential energy and kinetic energy are involved. The kinetic energy of the truck is converted to gravitational potential energy as it moves uphill.

Develop Initially the kinetic energy of the truck is $K_0 = \frac{1}{2}mv_A^2$ and its gravitational potential energy is $U_0 = 0$ (with respect to the ground). In the final state, all the kinetic energy has been converted to gravitational potential energy: $K = 0$ and $U = mgy$. These quantities are related by the principle of conservation of mechanical energy given in Equation 7.7:

$$K_0 + U_0 = K + U = \text{constant}$$

Evaluate Using Equation 7.7, we have

$$\frac{1}{2}mv_A^2 + 0 = 0 + mgy$$

which can be solved to give $y = \frac{v_A^2}{2g}$. With $L = \frac{y}{\sin\theta}$ and $v_A = 110$ km/h $= 30.6$ m/s, the distance the truck moves up is

$$L = \frac{v_A^2}{2g\sin\theta} = \frac{(30.6 \text{ m/s})^2}{2(9.8 \text{ m/s}^2)\sin 30°} = 95.3 \text{ m}$$

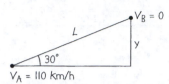

ASSESS   We find the distance traveled up by the truck to be inversely proportional to $\sin\theta$. To see that this makes sense, let's examine some limiting cases. Consider first $\theta = 0$. This corresponds to a flat rod, and without friction, the truck will travel indefinitely ($L \to \infty$). On the other hand, a larger θ means that the hill is steeper, and we expect the distance traveled upward to decrease.

41.   **INTERPRET**   The problem is about conservation of mechanical energy. Both potential energy and kinetic energy are involved. The object of interest is the child on the swing. As the swing moves, conversion between kinetic energy and potential energy takes place, while the total mechanical energy remains constant.

DEVELOP   Let L denote the length of the chain of the swing. Suppose the swing is released from an angle θ_0, measured with the vertical. This position corresponds to a vertical distance of $y_0 = L(1 - \cos\theta_0)$. After the child is released, the swing attains a maximum speed at the lowest point, where all the gravitational potential energy has been converted to kinetic energy. Using conservation of mechanical energy stated in Equation 7.7,

$$K_0 + U_0 = K + U = \text{constant}$$

the maximum speed can be calculated.

EVALUATE   Equation 7.7 implies

$$\frac{1}{2}mv_{\text{max}}^2 + 0 = 0 + mgy_0 = mgL(1 - \cos\theta_0)$$

This gives

$$v_{\text{max}} = \sqrt{2gL(1 - \cos\theta_0)} = \sqrt{2(9.8 \text{ m/s}^2)(3.2 \text{ m})(1 - \cos 50°)} = 4.73 \text{ m/s}$$

ASSESS   The result shows that increasing the initial angle θ_0 gives a greater speed at the bottom of the swing, in agreement with our expectation.

43.   **INTERPRET**   The problem is about conservation of mechanical energy. It involves converting potential energy of elastic and gravitational nature to the kinetic energy of the toboggan. The speed of the toboggan at the bottom of the hill is the physical quantity of interest.

DEVELOP   If friction is neglected everywhere (in the spring, on the snow, through the air, etc.), the mechanical energy of the kids plus toboggan (including potential energy of gravitation and the spring, as well as kinetic energy) is conserved:

$$U_{\text{top}} + K_{\text{top}} = U_{\text{bot}} + K_{\text{bot}}$$

At the top of the hill, $K_{\text{top}} = 0$ (the toboggan starts from rest), and

$$U_{\text{top}} = U_{\text{top,}g} + U_{\text{top,}s} = mgy_{\text{top}} + \frac{1}{2}kx^2$$

while at the bottom, $K_{\text{bot}} = \frac{1}{2}mv^2$ and $U_{\text{bot}} = mgy_{\text{bot}}$ (since $U_{\text{bot,}s} = 0$ when the spring is no longer compressed).

EVALUATE   (a) The kinetic energy at the bottom of the hill is

$$K_{\text{bot}} = mg(y_{\text{top}} - y_{\text{bot}}) + \frac{1}{2}kx^2 = (80 \text{ kg})(9.8 \text{ m/s}^2)(9.5 \text{ m}) + \frac{1}{2}(890 \text{ N/m})(2.6 \text{ m})^2$$
$$= 7.45 \times 10^3 \text{ J} + 3.01 \times 10^3 \text{ J}$$
$$= 1.05 \times 10^4 \text{ J}$$

The speed at the bottom is

$$v_{bot} = \sqrt{\frac{2K_{bot}}{m}} = \sqrt{\frac{2(1.05 \times 10^4 \text{ J})}{80 \text{ kg}}} = 16.2 \text{ m/s}$$

(b) The ratio of the potential energy of the spring to the final kinetic energy is

$$\frac{U_{top,s}}{K_{bot}} = \frac{3.01 \times 10^3 \text{ J}}{1.05 \times 10^4 \text{ J}} = 28.8\%$$

ASSESS   In this problem, the kinetic energy of the toboggan comes from its gravitational potential energy as well as the potential energy of the spring. The speed at the bottom of the hill can be written as

$$v_{bot} = \sqrt{2g(y_{top} - y_{bot}) + \frac{k}{m}x^2}$$

From the expression we see that if the spring constant is reduced (making it less stiff), then the final speed of the toboggan will also decrease.

45.   **INTERPRET**   The problem is about conservation of mechanical energy. The kinetic energy of the car is converted to potential energy stored in the spring, while the total mechanical energy of the system remains conserved. The maximum speed of the car is the physical quantity of interest.

DEVELOP   Suppose in the initial state, the kinetic energy of the car is $K_0 = \frac{1}{2}mv_0^2$ and the potential energy of the spring is $U_0 = 0$. In the final state, the kinetic energy of the car becomes $K = 0$ and the potential energy of the spring is $U = \frac{1}{2}kx^2$. These quantities are related by the principle of conservation of mechanical energy given in Equation 7.7:

$$K_0 + U_0 = K + U = \text{constant}$$

EVALUATE   Using Equation 7.7, we have

$$\frac{1}{2}mv_0^2 + 0 = 0 + \frac{1}{2}kx^2$$

which can be solved to give

$$v_0 = \sqrt{\frac{k}{m}}x = \sqrt{\frac{7.0 \times 10^5 \text{ N/m}}{1400 \text{ kg}}}(0.16 \text{ m}) = 3.58 \text{ m/s} = 12.9 \text{ km/h}$$

ASSESS   From the above expression we see that the initial speed of the car is directly proportional to x, the amount of compression of the spring. If initial speed of the car is such that $v > v_0$, then the compression will exceed the maximum allowed limit of 16 cm and results in a damage to the bumper.

47.   **INTERPRET**   The problem is about conservation of mechanical energy. The object of interest is the pendulum. As it swings, conversion between kinetic energy and gravitational potential energy takes place, while the total mechanical energy remains constant.

DEVELOP   Let L denote the length of the pendulum. Let the pendulum be released from an angle θ_0, measured with the vertical. This position corresponds to a vertical distance of $y_0 = L(1 - \cos\theta_0)$. The pendulum attains a maximum speed at the lowest point of the swing, where all the gravitational potential energy has been converted to kinetic energy. Using conservation of mechanical energy stated in Equation 7.7,

$$K_0 + U_0 = K + U = \text{constant}$$

the length of the pendulum can be calculated.

EVALUATE   Equation 7.7 implies

$$\frac{1}{2}mv_{max}^2 + 0 = 0 + mgy_0 = mgL(1 - \cos\theta_0)$$

This gives

$$L = \frac{v_{max}^2}{2g(1 - \cos\theta_0)} = \frac{(0.55 \text{ m/s})^2}{2(9.8 \text{ m/s}^2)(1 - \cos 8°)} = 1.59 \text{ m}$$

ASSESS   The result shows that L is inversely proportional to $1 - \cos\theta_0$. This means that if the maximum speed at the bottom of the swing is to remain unchanged, then increasing the initial angle θ_0 must be accompanied by a decrease in L.

49.   **INTERPRET**   In this problem we want to find the turning points associated with a potential energy function $U(x)$.

DEVELOP   A particle in a one-dimensional potential well, which conserves mechanical energy, satisfies Equation 7.7 with total energy

$$E = K + U = \frac{1}{2}mv^2 + U(x)$$

At the turning point, $v = 0$ and $K = 0$.

EVALUATE   The turning points in the problem can be located by solving

$$E = U(x) = 3.5 = 7.0 - 8.0x + 1.7x^2 \quad \rightarrow \quad 1.7x^2 - 8.0x + 3.5 = 0$$

where energy is in joules and displacement in meters. The quadratic formula gives

$$x = \frac{8.0 \pm \sqrt{(8.0)^2 - 4(1.7)(3.5)}}{2(1.7)} \text{ m} = 4.22 \text{ m or } 0.488 \text{ m}$$

ASSESS   A turning point is a point where the particle reverses its motion. One may readily verify that $U(4.22 \text{ m}) = U(0.488 \text{ m}) = 3.5$ J.

51.   **INTERPRET**   In this problem we want to find the equilibrium separation between NaCl ions, given the potential energy function $U(x)$.

DEVELOP   At the equilibrium separation, the potential energy is a minimum, or

$$\frac{dU(r)}{dr} = 0$$

By solving the above equation, the equilibrium separation between ions in NaCl can be found.

EVALUATE   Differentiating $U(r)$ with respect to r gives

$$\left.\frac{dU}{dr}\right|_{eq} = 0 = -nbr_{eq}^{-(n+1)} + ar_{eq}^{-2}$$

or

$$r_{eq} = \left(\frac{nb}{a}\right)^{1/(n-1)} = \left(\frac{(8.22)(5.52 \times 10^{-98})}{4.04 \times 10^{-28}}\right)^{1/(8.22-1)} = 2.82 \times 10^{-10} \text{ m} = 2.82 \text{ Å}$$

ASSESS   Our result can be compared with the experimental value of 2.36 Å. Note that angstrom ($1 \text{ Å} = 10^{-10}$ m) is a common non-SI unit of length, used in chemistry and atomic physics.

53.   **INTERPRET**   In this problem we are asked to find the speed of the skier at two different locations, given that the downward slope has a coefficient of friction $\mu_k = 0.11$.

DEVELOP   The work done by friction skiing down a straight slope of length L is

$$W_f = -f_k L = -\mu_k nL = -\mu_k (mg\cos\theta)\left(\frac{h}{\sin\theta}\right) = -\mu_k mgh\cot\theta$$

where $h = L\sin\theta$ is the vertical drop of the slope. The energy principle applied between the start and the first level now gives $\Delta K_{AB} + \Delta U_{AB} = W_{f,AB}$, or

$$\frac{1}{2}mv_B^2 = mg(y_A - y_B) - \mu_k mg(y_A - y_B)\cot\theta_{AB}$$

Similarly, for the motion between the top and the second level, we must include all the work done by friction, so

$$\Delta K_{AC} + \Delta U_{AC} = W_{f,AB} + W_{f,BC}$$

or

$$\frac{1}{2}mv_C^2 = mg(y_A - y_C) - \mu_k mg(y_A - y_B)\cot\theta_{AB} - \mu_k mg(y_B - y_C)\cot\theta_{BC}$$

EVALUATE   Solving the equation for v_B, we obtain

$$v_B = \sqrt{2g(y_A - y_B)(1 - \mu_k \cot\theta_{AB})} = \sqrt{2(9.8 \text{ m/s}^2)(25 \text{ m})(1 - 0.11\cot 32°)} = 20.1 \text{ m/s}$$

Similarly, for v_C, we have

$$v_C = \sqrt{2g[(y_A - y_C) - \mu_k(y_A - y_B)\cot\theta_{AB} - \mu_k(y_B - y_C)\cot\theta_{BC}]}$$
$$= \sqrt{2(9.8 \text{ m/s}^2)[63 \text{ m} - (0.11)(25 \text{ m})\cot 32° - (0.11)(38 \text{ m})\cot 20°]}$$
$$= 30.4 \text{ m/s}$$

ASSESS   Let's consider the case where $\mu_k = 0$. In this limit, the results become

$$v_B = \sqrt{2g(y_A - y_B)} = \sqrt{2(9.8 \text{ m/s}^2)(25 \text{ m})} = 22.1 \text{ m/s}$$
$$v_C = \sqrt{2g(y_A - y_C)} = \sqrt{2(9.8 \text{ m/s}^2)(63 \text{ m})} = 35.1 \text{ m/s}$$

which are the same as the answers obtained in Exercise 20 for the frictionless case.

55.   **INTERPRET**   In this problem we want to find the distance a block slides on a frictional surface after being launched from a compressed spring.

DEVELOP   Suppose the block comes to rest at B, a distance L from its initial position at rest against the compressed spring at A. We shall use Equation 7.5, $\Delta K + \Delta U = W_{nc}$, where W_{nc} is the work done by nonconservative force. This leads to

$$W_{nc} = -\mu_k mgL = \Delta K + \Delta U = -U_s^A = -\frac{1}{2}kx^2$$

since the kinetic energies at A and B and the change in gravitational potential energy are zero.

EVALUATE   Solving the equation above, we obtain

$$L = \frac{kx^2/2}{\mu_k mg} = \frac{(340 \text{ N/m})(0.18 \text{ m})^2/2}{(0.27)(1.5 \text{ kg})(9.8 \text{ m/s}^2)} = 1.39 \text{ m}$$

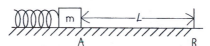

ASSESS   The distance moved by the block is proportional to the spring's potential energy, $kx^2/2$. In addition, it is inversely proportional to the coefficient of friction, μ_k. In the limit $\mu_k = 0$, the surface is frictionless and we expect the mass to travel indefinitely.

57.   **INTERPRET**   In this problem we want to find the final position of a block after being launched from a compressed spring. Its path involves a frictional surface followed by a frictionless curve.

DEVELOP   The energy of the block when it first encounters friction (at point O) is

$$K_0 = \frac{1}{2}kx^2$$

if we take the zero of gravitational potential energy at that level. Crossing the frictional zone, the work done by the friction is

$$W_{nc} = -\mu_k mgL$$

Depending on the ratio of $K_0/|W_{nc}|$, the block will move back and forth several times before losing all its energy and coming to rest.

EVALUATE   Initially the block has an energy

$$K_0 = \frac{1}{2}kx^2 = \frac{1}{2}(200 \text{ N/m})(0.15 \text{ m})^2 = 2.25 \text{ J}$$

The work done by the friction is

$$\Delta E = W_{nc} = -\mu_k mgL = -(0.27)(0.19 \text{ kg})(9.8 \text{ m/s}^2)(0.85 \text{ m}) = -0.427 \text{ J}$$

Since $\frac{K_0}{|W_{nc}|} = 5.27$, five complete crossings are made, leaving the block with energy $K = K_0 - 5|W_{nc}| = 0.113 \text{J}$ on the curved rise side. This remaining energy is sufficient to move the block a distance

$$s = \frac{K}{\mu_k mg} = \frac{0.113 \text{ J}}{(0.27)(0.19 \text{ kg})(9.8 \text{ m/s}^2)} = 0.225 \text{ m} = 22.5 \text{ cm}$$

towards point O, so the block comes to rest $85 \text{ cm} - 22.5 \text{ cm} = 62.5 \text{ cm}$ to the right of point O.

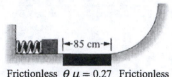

Frictionless θ $\mu = 0.27$ Frictionless

ASSESS   Since $K_0 > |W_{nc}|$, the block does not lose all its energy the first time when it moves across the frictional zone. No energy is lost while it moves along the frictionless curve. The number of times the block moves back and forth across the frictional zone depends on the ratio $K_0/|W_{nc}|$.

59.   **INTERPRET**   The object of interest is the roller coaster, which, after being launched from a compressed spring, moves along a circular loop. The physical quantity we are asked about is the minimum compression length that allows the car to stay on the track.

DEVELOP   If the car stays on the track, the radial component of its acceleration is $a = v^2/R$, and the normal force is greater than zero. Thus,

$$n = \frac{mv^2}{R} + mg\cos\theta \geq 0 \quad \rightarrow \quad v^2 \geq -gR\cos\theta$$

Now $-\cos\theta$ has its maximum value at the top of the loop ($\theta = 180°$), so $v_B^2 \geq gR$ is the condition for the car to stay on the track all the way around. This is the result obtained in Example 5.7.

Once the minimum speed at point B is determined, by applying the mechanical energy conservation principle given in Equation 7.7, the minimum compression length of the spring can be calculated.

EVALUATE   In the absence of friction, the conservation of mechanical energy requires

$$K_A + U_A = K_B + U_B \quad \rightarrow \quad 0 + \frac{1}{2}kx^2 + mgy_A = \frac{1}{2}mv_B^2 + mgy_B$$

Solving for x, we obtain

$$x^2 = \frac{m}{k}\left[v_B^2 + 2g(y_B - y_A)\right] \geq \frac{m}{k}[gR + 2g(2R)] = \frac{5mgR}{k}$$

or

$$x \geq \sqrt{\frac{5mgR}{k}} = \sqrt{\frac{5(840 \text{ kg})(9.8 \text{ m/s}^2)(6.2 \text{ m})}{31,000 \text{ N/m}}} = 2.87 \text{ m}$$

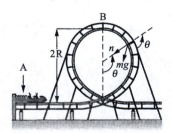

ASSESS   Our result indicates that if the radius of the loop increases, then the amount of spring compression must also go up in order for the car to stay on the track. On the other hand, when a stiffer spring (with larger k) is used, then less compression would be required.

61.   **INTERPRET**   In this problem we want to find the distance a child can move across a frictional surface after sliding down a frictionless incline.

DEVELOP   Suppose the child starts near the hilltop with $K_A = 0$ and stops on rough level ground, $K_B = 0$, after falling through a potential energy difference

$$\Delta U = U_B - U_A = -mg(y_A - y_B)$$

The work done by friction (on level ground, $n = mg$) is

$$W_{nc} = -f_k x = -\mu_k mgx$$

where x is the distance slid across the rough level stretch. The energy principle, Equation 7.5, relates these quantities:

$$\Delta K + \Delta U = W_{nc} \quad \rightarrow \quad 0 - mg(y_A - y_B) = -\mu_k mgx$$

EVALUATE   Solving the above equation for x, we obtain

$$x = \frac{y_A - y_B}{\mu_k} = \frac{7.2 \text{ m}}{0.51} = 14.1 \text{ m}$$

ASSESS   As expected, the distance the child travels is proportional to $\Delta y = y_A - y_B$. This is so because the greater the value of Δy, the more the gravitational potential energy that's been converted to kinetic energy. On the other hand, we expect x to be inversely proportional to the coefficient of friction, μ_k. A smaller μ_k will allow the child to travel a much further distance before losing all its kinetic energy.

63.   **INTERPRET**   The problem is about finding the speed of a particle when the force acting on it is known.

DEVELOP   The force is conservative, so we can apply Equation 7.7, $K_0 + U_0 = K + U$. The subscript 0 refers to the origin, and $K_0 = 0$. According to Equation 7.2, when an object moves from x_1 to x_2 under the influence of a force $F(x)$, the change in potential energy is

$$\Delta U = -\int_{x_1}^{x_2} F(x) dx$$

Once ΔU is known, the speed of the particle can be calculated from

$$K = \frac{1}{2}mv^2 = -\Delta U$$

EVALUATE   **(a)** Integrating $F(x) = a\sqrt{x}$, we obtain

$$\Delta U = U - U_0 = -\int_0^x a\sqrt{x'} \, dx' = -\frac{2}{3}ax^{3/2}$$

Therefore, the speed of the particle as a function of x is

$$v = \sqrt{\frac{2K}{m}} = \sqrt{\frac{-2\Delta U}{m}} = \left(\frac{4a}{3m}x^{3/2}\right)^{1/2} = 2\sqrt{\frac{a}{3m}}x^{3/4}$$

ASSESS   We can check our answer by substituting the result back to the expression for K. This leads to

$$K = \frac{1}{2}mv^2 = \frac{1}{2}m\left(\frac{4a}{3m}x^{3/2}\right) = \frac{2a}{3}x^{3/2}$$

Indeed, we see that $K = -\Delta U$, as required by energy conservation principle.

65.   **INTERPRET**   We find whether a spring-launched block makes it to the top of an incline with friction, and how much kinetic energy it has when it gets there (if it gets there.) We can use energy methods to solve this problem, but friction is a factor so mechanical energy is not conserved.

DEVELOP   The initial energy of the system is spring potential energy. The final energy is gravitational potential, kinetic, and we also count the mechanical energy lost to friction. $U_i = U_f + K_f + W_f$. The weight of the block is $mg = 4.5$ N. The angle of the incline is $\theta = 30°$, and its length is $L = 2.0$ m. The spring constant is $k = 2.0$ kN/m, and the initial spring compression is $x = 0.1$ m. The coefficient of kinetic friction is $\mu = 0.5$. We solve this for K: if K is positive for $s = L$, that is the amount of kinetic energy remaining at the top. If it's negative, we'll go on to find the value of s that makes K zero. Finally, we need to repeat our answer for a block of twice the weight.

EVALUATE

(a) We will define the distance the block travels up the incline as s. The height change for this distance is then $h = s\sin\theta$.

$U_i = U_f + K_f + W_f \rightarrow \frac{1}{2}kx^2 = mgh + \frac{1}{2}mv^2 + F_f s = mgs\sin\theta + \frac{1}{2}mv^2 + \mu mgs\cos\theta$. Solve for K:

$\frac{1}{2}mv^2 = \frac{1}{2}kx^2 - mgs(\sin\theta + \mu\cos\theta) = 1.60$ J. The block reaches the top, with 1.6 J of extra kinetic energy.

(b) We use the same equation for K, but this time with $mg = 9$ N: $\frac{1}{2}mv^2 = \frac{1}{2}kx^2 - mgs(\sin\theta + \mu\cos\theta) = -6.79$ J. It

does *not* reach the top. Now we solve for the distance s, with $v = 0$. $s = \frac{\frac{1}{2}kx^2}{mg(\sin\theta + \mu\cos\theta)} = 1.19$ m.

ASSESS   The mass cannot have negative kinetic energy. We use this to our advantage in the problem.

67.   **INTERPRET**   We use work and energy to find the nature of a repulsive force between two objects, a uranium nucleus and an alpha particle. We know the change in kinetic energy, and we know the distance over which the force acts, so we can use the integral definition of work to find the factor A in the force, which has form $F = \frac{A}{x^2}$.

DEVELOP   We use $W = \Delta K$, where $W = \int_{x_0}^\infty F dx$. The final speed of the α particle is $v = 2.93\times10^5$ m/s, and its mass is $m = 6.7\times10^{-27}$ kg. The initial position is $x_0 = 1.43\times10^{-10}$ m.

EVALUATE

$$\Delta K = W \rightarrow \frac{1}{2}mv^2 = \int_{x_0}^\infty \frac{A}{x^2}dx = A\left[-\frac{1}{x}\right]_{x_0}^\infty = A\left[\frac{1}{x_0}\right] \rightarrow A = \frac{1}{2}mv^2 x_0 = 4.11\times10^{-26}\text{ Jm}$$

ASSESS   We can check our units by noting that this value of A gives us a force in units of $\frac{Jm}{m^2} = \frac{J}{m} = \frac{Nm}{m} = $ N.

69.   **INTERPRET**   We find a "leaping equation" relating the power of an animal to the mass, the push-off distance, and the height reached. We use conservation of energy.

DEVELOP   We will take the *beginning* of the animal's push as zero height. The height when it leaves the ground is d, and the final height is h. Working backward from the top, we see that the change in gravitational potential from h to d must equal the kinetic energy at d. From this we can find the velocity at point d. The work done by the animal as it pushes from zero to d is the kinetic energy at d plus the change in gravitational potential from zero to d. Power is $P = \frac{W}{t}$, and to find the time we use Equation 2.9: $x = x_0 + \frac{1}{2}(v_0 + v)t$.

EVALUATE   The kinetic energy at d is $K_d = \frac{1}{2}mv^2 = mg(h-d)$, so the speed at d is $v = \sqrt{2g(h-d)}$. Solve Equation 2.9 for time:

$$t = 2\frac{x - x_0}{v_0 + v} = 2\frac{d}{v} = 2\frac{d}{\sqrt{2g(h-d)}}$$

Power is

$$P = \frac{W}{t} = \frac{\frac{1}{2}mv^2 + mgd}{2\frac{d}{\sqrt{2g(h-d)}}} = \frac{\frac{1}{2}m(v^2 + 2gd)\sqrt{2g(h-d)}}{2d} = \frac{m(v^2 + 2gd)\sqrt{2g(h-d)}}{4d}$$

ASSESS   We'd better check units on this equation:

$$P = \frac{kg\left(\left(\frac{m}{s}\right)^2 + \left(\frac{m}{s^2}\right)m\right)\sqrt{\left(\frac{m}{s^2}\right)(m)}}{m} = \frac{kg\left(\frac{m^2}{s^2}\right)\frac{m}{s}}{m\,s} = \frac{(kg\,m/s^2)m}{s} = \frac{Nm}{s} = \frac{J}{s}$$

It's good!

GRAVITY

EXERCISES

Section 8.2 Universal Gravitation

13. **INTERPRET** In this problem we want to use astrophysical data to find the Moon's acceleration in its circular orbit about the Earth.

    DEVELOP The gravitational force between two masses m_1 and m_2 is given by Equation 8.1: $F = \frac{Gm_1m_2}{r^2}$, where r is their distance of separation. The acceleration of the Moon in its orbit can be computed by considering the gravitational force between the Moon and the Earth.

    EVALUATE Using Equation 8.1, the gravitational force between the Earth (mass M_E) and the Moon (mass m_m) is

$$F = \frac{GM_E m_m}{r_{mE}^2}$$

    where r_{mE} is the distance between the Moon and the Earth. By Newton's second law, $F = ma$, the acceleration of the Moon is

$$a = \frac{F}{m_m} = \frac{GM_E}{r_{mE}^2} = \frac{(6.67 \times 10^{-11} \text{ N} \cdot \text{m}^2/\text{kg}^2)(5.97 \times 10^{24} \text{ kg})}{(3.85 \times 10^8 \text{ m})^2} = 2.69 \times 10^{-3} \text{ m/s}^2$$

    where we have used the astrophysical data given in Appendix E.

    ASSESS An alternative way to find the acceleration of the Moon as it orbits the Earth is to note that it completes a nearly circular orbit of 385,000 km radius in 27 days. Based on this information, the (centripetal) acceleration of the Moon is

$$a = \frac{v^2}{r_{mE}} = \frac{4\pi^2 r_{mE}}{T^2} = \frac{4\pi^2 (3.85 \times 10^8 \text{ m})}{(27.3 \times 86,400 \text{ s})^2} = 2.73 \times 10^{-3} \text{ m/s}^2$$

    Note that since the Moon's orbit is actually elliptical (with 5.5% eccentricity), the values based on circular orbits are reasonable.

15. **INTERPRET** In this problem we are asked to use astrophysical data to find the gravitational acceleration near the surface of (a) Mercury and (b) Saturn's moon Titan.

    DEVELOP The gravitational force between two masses m_1 and m_2 is given by Equation 8.1: $F = \frac{Gm_1m_2}{r^2}$, where r is their distance of separation. So for an object of mass m near a celestial body of mass M and radius R, the gravitational force between them is $F = \frac{GMm}{R^2}$. By Newton's second law, $F = ma$, the gravitational acceleration near the surface of the gravitating body is

$$a = \frac{GM}{R^2}$$

    EVALUATE With reference to the first two columns in Appendix E, we find

    (a)
$$g_{\text{Mercury}} = \frac{GM_{\text{Mercury}}}{R_{\text{Mercury}}^2} = \frac{(6.67 \times 10^{-11} \text{ N} \cdot \text{m}^2/\text{kg}^2)(0.330 \times 10^{24} \text{ kg})}{(2.44 \times 10^6 \text{ m})^2} = 3.70 \text{ m/s}^2$$

    (b)
$$g_{\text{Titan}} = \frac{GM_{\text{Titan}}}{R_{\text{Titan}}^2} = \frac{(6.67 \times 10^{-11} \text{ N} \cdot \text{m}^2/\text{kg}^2)(0.135 \times 10^{24} \text{ kg})}{(2.58 \times 10^6 \text{ m})^2} = 1.35 \text{ m/s}^2$$

ASSESS   The measured values are $g_{\text{Mercury}} = 3.70$ m/s² and $g_{\text{Titan}} = 1.4$ m/s². So our results are in good agreement with the data.

17.   **INTERPRET**   In this problem we want to find the gravitational force between the astronaut and the space shuttle.

DEVELOP   The gravitational force between two masses m_1 and m_2 separated by a distance r is given by Equation 8.1: $F = \frac{Gm_1 m_2}{r^2}$. This is the equation we shall use to solve for the force between the astronaut and the space shuttle.

EVALUATE   Substituting the values given in the problem statement into Equation 8.1 gives

$$F = \frac{Gm_1 m_2}{r^2} = \frac{(6.67\times10^{-11}\,\text{N}\cdot\text{m}^2/\text{kg}^2)(67\text{ kg})(73{,}000\text{ kg})}{(84\text{ m})^2} = 4.62\times10^{-8}\text{ N}$$

ASSESS   The gravitational force between two masses is always attractive and decreases as $1/r^2$. Thus, when two masses are separated by a very large distance, the force between them becomes hardly noticeable.

Section 8.3 Orbital Motion

19.   **INTERPRET**   The object of interest in this problem is the satellite. This problem explores the connection between the satellite's altitude and the period of its circular orbit.

DEVELOP   The gravitational force between the Earth and the satellite provides the centripetal force to keep the orbit circular. Thus,

$$\frac{GM_E m_s}{r_{sE}^2} = \frac{m_s v^2}{r_{sE}} = \frac{m_s}{r_{sE}}\left(\frac{2\pi r_{sE}}{T}\right)^2 = \frac{4\pi^2 m_s r_{sE}}{T^2}$$

where T is the period and r_{sE} is the distance between the satellite and the Earth, measured from Earth's center. The altitude is $h = r_{sE} - R_E$, where R_E is the radius of the Earth.

EVALUATE   Solving the above equation with $T = 2$ h $= 7200$ s, we obtain

$$r_{sE} = \left(\frac{GM_E T^2}{4\pi^2}\right)^{1/3} = \left(\frac{(6.67\times10^{-11}\,\text{N}\cdot\text{m}^2/\text{kg}^2)(5.97\times10^{24}\text{ kg})(7200\text{ s})^2}{4\pi^2}\right)^{1/3} = 8.06\times10^6\text{ m}$$

The altitude is

$$h = r - R_E = 8.06\times10^6\text{ m} - 6.37\times10^6\text{ m} \approx 1690\text{ km}$$

ASSESS   The radius of the circular orbit is proportional to $T^{2/3}$ (Kepler's third law). This means that if the period T is to be doubled, then the radius has to increase by a factor of $2^{2/3} \approx 1.6$.

21.   **INTERPRET**   In this problem we are asked to find the orbital period of Mars, given its radius.

DEVELOP   Kepler's thirds law shown in Equation 8.4 states that

$$T^2 = \frac{4\pi^2 r^3}{GM} \quad\rightarrow\quad \frac{T^2}{r^3} = \frac{4\pi^2}{GM} = \text{constant}$$

Note that this result is independent of the mass m of the orbiting object. Thus, for two celestial bodies whose semi-major axes are r_1 and r_2, the ratio of their period would be

$$\frac{T_1^2}{T_2^2} = \frac{r_1^3}{r_2^3} \quad\rightarrow\quad T_1 = \left(\frac{r_1}{r_2}\right)^{3/2} T_2$$

EVALUATE   Using the relation derived above, we find the period of Mars to be

$$T_{\text{Mars}} = \left(\frac{r_{\text{Mars}}}{r_E}\right)^{3/2} T_E = (1.52)^{3/2}(1\text{ y}) = 1.87\text{ y}$$

ASSESS   Since Mars has a larger orbit compared to the Earth, we expect it to take longer to complete one revolution. Our result can be compared with the measured value of 1.88 y given in Appendix E.

23.   **INTERPRET**   In this problem we want to find the period of a golf ball orbiting the Moon.

DEVELOP   Kepler's third law shown in Equation 8.4 states that

$$T^2 = \frac{4\pi^2 r^3}{GM} \quad \rightarrow \quad T = 2\pi\sqrt{\frac{r^3}{GM}}$$

Knowing the mass and the radius of the Moon allows us to determine the period of the golf ball.

EVALUATE   Using the equation obtained above, we find the period to be

$$T = 2\pi\sqrt{\frac{R^3}{GM}} = 2\pi\sqrt{\frac{(1.74 \times 10^6 \text{ m})^3}{(6.67 \times 10^{-11} \text{ N} \cdot \text{m}^2/\text{kg}^2)(7.35 \times 10^{22} \text{ kg})}} = 6.51 \times 10^3 \text{ s} = 109 \min$$

ASSESS   This result is independent of the mass m of the golf ball. Any mass thrown into this orbit would have the same period T.

Section 8.4 Gravitational Energy

25.   **INTERPRET**   The problem asks about the change in potential energy as Earth goes from perihelion to aphelion.

DEVELOP   The potential energy difference between two points at distances r_1 and r_2 from the center of a gravitating mass M is, according to Equation 8.5:

$$\Delta U_{12} = -GMm\left(\frac{1}{r_2} - \frac{1}{r_1}\right)$$

Perihelion is the point of closest approach to the Sun, while the aphelion is the most distant point from the Sun. Most planetary orbits are elliptical.

EVALUATE   Using the equation above, the change in potential energy as Earth goes from perihelion to aphelion is

$$\Delta U = -GMm\left(\frac{1}{r_a} - \frac{1}{r_p}\right)$$

$$= -(6.67 \times 10^{-11} \text{ N} \cdot \text{m}^2/\text{kg}^2)(1.99 \times 10^{30} \text{ kg})(5.97 \times 10^{24} \text{ kg})\left(\frac{1}{1.52 \times 10^{11} \text{ m}} - \frac{1}{1.47 \times 10^{11} \text{ m}}\right)$$

$$= 1.77 \times 10^{32} \text{ J}$$

ASSESS   Since $r_a > r_p$, the change in potential energy is positive. In fact, if we choose infinity to be the point of zero potential energy, $U(\infty) = 0$, then the potential energy at a distance r from the gravitating center would be

$$U(r) = -\frac{GMm}{r}$$

With $r_a > r_p$, we readily see that $U(r_a) > U(r_p)$.

27.   **INTERPRET**   The problem asks for the maximum altitude the rocket can reach with an initial launch speed v_0. The problem sounds similar to what we have encountered in Chapter 2, but here the acceleration is not constant,

DEVELOP   If we consider the Earth at rest as approximately an inertial system, then a vertically launched rocket would have zero kinetic energy (instantaneously) at its maximum altitude, and the situation is the same as Example 8.5. Conservation of mechanical energy, $K + U = K_0 + U_0$, can be used to solve for the maximum altitude.

EVALUATE   The conservation equation gives

$$\frac{1}{2}mv_0^2 - \frac{GM_E m}{R_E} = -\frac{GM_E m}{R_E + h}$$

or

$$h = \left(\frac{1}{R_E} - \frac{v_0^2}{2GM_E}\right)^{-1} - R_E$$

$$= \left(\frac{1}{6.37 \times 10^6 \text{ m}} - \frac{(5100 \text{ m/s})^2}{2(6.67 \times 10^{-11} \text{N} \cdot \text{m}^2/\text{kg}^2)(5.97 \times 10^{24} \text{ kg})}\right)^{-1} - 6.37 \times 10^6 \text{ m}$$

$$= 1.67 \times 10^6 \text{ m}$$

ASSESS   If we assume a potential energy change of $\Delta U = mgh$, where $g = 9.8$ m/s^2, then the result would have been

$$h = \frac{v_0^2}{2g} = \frac{(5100 \text{ m/s})^2}{2(9.8 \text{ m/s}^2)} = 1.33 \times 10^6 \text{ m}$$

The decreasing gravitational acceleration $g(r)$ allows the rocket to go higher!

29. **INTERPRET**   The problem asks about the energy needed to put a mass into the Earth's geosynchronous orbit. The energy required is simply equal to the work done by an external agent.

    DEVELOP   The energy of an object at rest ($K_0 = 0$) on the Earth's surface is $E_0 = U_0 = -\frac{GM_E m}{R_E}$ (neglect diurnal rotational energy, etc.), while its total mechanical energy in a circular orbit is

    $$E = U + K = -\frac{GM_E m}{r} + \frac{GM_E m}{2r} = -\frac{GM_E m}{2r}$$

    as given in Equation 8.8b. The difference is the energy required to put the mass into the orbit.

    EVALUATE   The distance, measured from the Earth's center, that corresponds to the geosynchronous orbit is (see Example 8.2) $r = 4.22 \times 10^7$ m. This, the energy necessary to put a mass of $m = 1$ kg into a circular geosynchronous orbit is

    $$\Delta E = E - E_0 = (6.67 \times 10^{-11} \text{ N} \cdot \text{m}^2/\text{kg}^2)(5.97 \times 10^{24} \text{ kg})(1 \text{ kg})\left(\frac{1}{6.37 \times 10^6 \text{ m}} - \frac{1}{2(4.22 \times 10^7 \text{ m})}\right) = 5.78 \times 10^7 \text{ J}$$

    ASSESS   As expected, positive work needs to be done to increase both the gravitational potential energy (making it less negative) and the kinetic energy of the mass.

31. **INTERPRET**   This problem is about finding the minimum speed required for an object to escape forever from a planet of known mass.

    DEVELOP   The escape speed is the speed that makes the total energy zero:

    $$E = U + K = -\frac{GMm}{r} + \frac{1}{2}mv^2 = 0$$

    Solving for v, we find $v_{esc} = \sqrt{\frac{2GM}{R}}$, where $r = R$ is the radius of the planet.

    EVALUATE   From the equation above, the radius of the planet is

    $$R = \frac{2GM}{v_{esc}^2} = \frac{2(6.67 \times 10^{-11} \text{ N} \cdot \text{m}^2/\text{kg}^2)(2.9 \times 10^{24} \text{ kg})}{(7.1 \times 10^3 \text{ m/s})^2} = 7.67 \times 10^6 \text{ m}$$

    ASSESS   The more massive the gravitating body, the greater the speed is required to escape from its gravitational field.

33. **INTERPRET**   This problem is about finding the would-be radius of the Earth if the escape speed is 30 km/s.

    DEVELOP   The escape speed is the speed that makes the total energy zero:

    $$E = U + K = -\frac{GMm}{r} + \frac{1}{2}mv^2 = 0$$

    Solving for v, we find $v_{esc} = \sqrt{\frac{2GM}{R}}$, where $r = R$ is the radius of the gravitating mass.

    EVALUATE   Given that $v_{esc} = 30$ km/s $= 30,000$ m/s, the would-be radius of the Earth is

    $$R = \frac{2GM_E}{v_{esc}^2} = \frac{2(6.67 \times 10^{-11} \text{ N} \cdot \text{m}^2/\text{kg}^2)(5.97 \times 10^{24} \text{ kg})}{(3.0 \times 10^4 \text{ m/s})^2} = 8.85 \times 10^5 \text{ m}$$

    ASSESS   Expressed in terms of the real value $R_E = 6.37 \times 10^6$ m, we have $R \approx 0.139 R_E$. The relationship between R and v_{esc} with M kept fixed is

    $$\frac{v_{esc,1}}{v_{esc,2}} = \sqrt{\frac{R_2}{R_1}}$$

    Thus, if we use the fact that the escape speed from the Earth ($R_E = 6.37 \times 10^6$ m) is $v_{ecs,0} = 11.2$ km/s, we can apply the above equation to find the would-be radius if the escape speed were $v_{esc} = 30$ km/s. The result is

    $$\frac{R}{R_E} = \left(\frac{v_{esc,0}}{v_{esc}}\right)^2 = \left(\frac{11.2 \text{ km/s}}{30 \text{ km/s}}\right)^2 \approx 0.139$$

    which is the same as what was found before.

PROBLEMS

35.   **INTERPRET**   This problem explores the gravitational acceleration of a gravitating body as a function of altitude h.

DEVELOP   Using Equation 8.1, the gravitational force between a mass m and a planet of mass M_p is $F = \frac{GM_p m}{r^2}$, where r is their distance of separation, measured from the center of the planet. From Newton's second law, $F = ma$, the acceleration of gravity at any altitude $h = r - R_p$, above the surface of a spherical planet of radius R_p, is

$$g(h) = \frac{GM_p}{(R_p + h)^2} = \frac{GM_p}{R_p^2}\left(\frac{R_p}{R_p + h}\right)^2 = g(0)\left(\frac{R_p}{R_p + h}\right)^2$$

where $g(0)$ is the value at the surface. Once the ratio $g(h)/g(0)$ is known, we can readily find the altitude h in terms of R_p.

EVALUATE   Solving for h, we find

$$\frac{h}{R_p} = \sqrt{\frac{g(0)}{g(h)}} - 1$$

Therefore, for $\frac{g(h)}{g(0)} = \frac{1}{2}$, we have $\frac{h}{R_p} = \sqrt{2} - 1 = 0.414$.

ASSESS   To see that the result makes sense, we take the limit $h = 0$ where the object rests on the surface of the planet. In this limit, we recover $g(0)$ as the gravitational acceleration. The equation also shows that $g(h)$ decreases as the altitude h is increased, and $g(h)$ approaches zero as $h \to \infty$.

37.   **INTERPRET**   In this problem we want to find the Moon's acceleration in its circular orbit about the Earth. In addition, we want to use the result to confirm the inverse-square law for the gravitational force.

DEVELOP   The centripetal force that keeps the Moon's orbit circular is:

$$F = \frac{m_m v^2}{r_{mE}} = \frac{4\pi^2 m_m r_{mE}}{T^2}$$

The equation allows us to compute the acceleration of the Moon in its circular orbit.

EVALUATE   Substituting the values given in the problem statement, we find

$$a = \frac{v^2}{r_{mE}} = \frac{4\pi^2 r_{mE}}{T^2} = \frac{4\pi^2 (3.84 \times 10^8\,\text{m})}{(27.3 \times 86,400\,\text{s})^2} = 2.73 \times 10^{-3}\,\text{m/s}^2$$

If the gravitational force varies as the inverse square of the distance, the ratio of the accelerations

$$\frac{a}{g_E} = \frac{2.73 \times 10^{-3}\,\text{m/s}^2}{9.81\,\text{m/s}^2} = 2.78 \times 10^{-4}$$

should equal $\left(\frac{R_E}{r_{mE}}\right)^2 = \left(\frac{6.37 \times 10^6}{3.85 \times 10^8}\right)^2 = 2.74 \times 10^{-4}$. This is true, to within the accuracy of the data used.

ASSESS   Using Equation 8.1, the gravitational force between the Earth (mass M_E) and the Moon (mass m_m) is

$$F = \frac{GM_E m_m}{r_{mE}^2}$$

where r_{mE} is the distance between the Moon and the Earth. Using the inverse-square force law, the expression for the acceleration a can be rewritten as

$$a = \frac{F}{m_m} = \frac{GM_E}{r_{mE}^2} = \left(\frac{GM_E}{R_E^2}\right)\left(\frac{R_E}{r_{mE}}\right)^2 = g_E\left(\frac{R_E}{r_{mE}}\right)^2$$

where $g_E = 9.81\,\text{m/s}^2$ is the gravitational acceleration near the surface of the Earth. Thus, we see that gravitational force obeys the inverse-square law. In fact, the same result (within numerical accuracy) is obtained by using

$$a = \frac{F}{m_m} = \frac{GM_E}{r_{mE}^2} = \frac{(6.67 \times 10^{-11}\,\text{N} \cdot \text{m}^2/\text{kg}^2)(5.97 \times 10^{24}\,\text{kg})}{(3.84 \times 10^8\,\text{m})^2} = 2.70 \times 10^{-3}\,\text{m/s}^2$$

39.   **INTERPRET**   In this problem we are asked to find the speed and period of an object orbiting about a gravitating body—the white dwarf.

DEVELOP   The gravitational force between the spaceship and the white dwarf provides the centripetal force for the spaceship to move in a circular path about the white dwarf:

$$F = \frac{GMm}{r^2} = \frac{mv^2}{r}$$

Solving for the orbital speed gives $v = \sqrt{\frac{GMm}{r}}$ (Equation 8.3).

EVALUATE   (a) The radius of a low orbit is approximately the radius of the white dwarf, or R_E, so Equation 8.3 gives

$$v = \sqrt{\frac{GM}{R_E}} = \sqrt{\frac{(6.67 \times 10^{-11}\ \text{N} \cdot \text{m}^2/\text{kg}^2)(1.99 \times 10^{30}\ \text{kg})}{6.37 \times 10^6\ \text{m}}} = 4.56 \times 10^6\ \text{m/s}$$

or about 1.5% of the speed of light.

(b) The orbital period is $T = \frac{2\pi R_E}{v} = \frac{2\pi(6.37 \times 10^6\ \text{m})}{4.56 \times 10^6\ \text{m/s}} = 8.77$ s. It is very short.

ASSESS   According to Kepler's third law, the relationship between T and M is given by

$$T^2 = \frac{4\pi^2 r^3}{GM} \quad \Rightarrow \quad MT^2 = \frac{4\pi^2 R^3}{G} = \text{constant}$$

Thus, we see that if the mass of the gravitating body M is increased while keeping its radius R constant, then its period T must decrease.

41.   **INTERPRET**   In this problem we want to compare the orbital periods of two satellites located at different distances from the center of the Earth.

DEVELOP   Kepler's third law shown in Equation 8.4 states that

$$T^2 = \frac{4\pi^2 r^3}{GM} \quad \rightarrow \quad \frac{T^2}{r^3} = \frac{4\pi^2}{GM} = \text{constant}$$

Thus, for the two satellites A and B, the ratio of their period would be

$$\frac{T_A^2}{T_B^2} = \frac{r_A^3}{r_B^3} \quad \rightarrow \quad \frac{T_A}{T_B} = \left(\frac{r_A}{r_B}\right)^{3/2}$$

EVALUATE   With $r_A = 2r_B$, the ratio of their period is

$$\frac{T_A}{T_B} = \left(\frac{r_A}{r_B}\right)^{3/2} = 2^{3/2} = 2.83$$

ASSESS   By Kepler's third law, the orbital period is proportional to $r^{3/2}$. Therefore, the further away the satellite is from the Earth, the longer is its period.

43.   **INTERPRET**   This problem asks about the speed of the waste canister as it hits the Sun. The quantity can be computed by applying the conservation of energy to the dropped canister.

DEVELOP   The change of potential energy as the waste canister travels from the Earth's orbit into the Sun can be calculated by using Equation 8.5:

$$\Delta U = -GMm\left(\frac{1}{R_s} - \frac{1}{r_{Es}}\right)$$

where R_s is the radius of the Sun and r_{Es} is the Earth's orbital radius. By conservation of mechanical energy, the decrease in potential energy must be equal to the gain in kinetic energy:

$$\Delta U + \Delta K = 0 \quad \rightarrow \quad \frac{1}{2}mv^2 = GMm\left(\frac{1}{R_s} - \frac{1}{r_{Es}}\right)$$

EVALUATE   Solving for v, we obtain

$$v = \sqrt{2GM\left(\frac{1}{R_s} - \frac{1}{r_{Es}}\right)} = \sqrt{2(6.67 \times 10^{-11}\ \text{N} \cdot \text{m}^2/\text{kg}^2)(1.99 \times 10^{30}\ \text{kg})\left(\frac{1}{6.96 \times 10^8\ \text{m}} - \frac{1}{1.50 \times 10^{11}\ \text{m}}\right)} = 616\ \text{km/s}$$

ASSESS   The speed is very high. Therefore, an enormous amount of energy (1.9×10^{11} J for each kg) would be required to achieve this, making this proposal impractical.

45.   **INTERPRET**   In this problem we want to compare the maximum height attained by a rocket with changing gravitational acceleration and that under the assumption of a constant acceleration.

DEVELOP   If the rocket has an initial vertical speed v_0, we can find the height, h, to which it can rise (where its kinetic energy is instantaneously zero) from the conservation of energy:

$$K_0 + U_0 = K + U \quad \rightarrow \quad \frac{1}{2}mv_0^2 - \frac{GM_E m}{R_E} = -\frac{GM_E m}{R_E + h}$$

On the other hand, if we assume a constant acceleration, then the height attained would be

$$h' = \frac{v_0^2}{2g} = \frac{v_0^2}{2(GM_E/R_E^2)} = \frac{v_0^2 R_E^2}{2GM_E}$$

where we have used $g = GM_E/R_E^2$.

EVALUATE   Solving for h, we obtain

$$h = R_E\left(\frac{1}{1 - v_0^2 R_E/2GM_E} - 1\right) = R_E\left(\frac{1}{1 - h'/R_E} - 1\right) = h'\left(\frac{R_E}{R_E - h'}\right)$$

Since the factor multiplying h' is $R_E/(R_E - h') > 1$, $h > h'$, and the equations of constant gravity underestimate the height. For h' to differ from h by 1%, i.e., $(h - h')/h = 0.01$, we require that $h' = 0.99h$. Thus,

$$\frac{h}{h'} = \frac{1}{1 - h'/R_E} \quad \rightarrow \quad \frac{1}{0.99} = \frac{1}{1 - h'/R_E}$$

or $h' = 0.01R_E$. This gives

$$h = \frac{h'}{0.99} = \frac{R_E}{99} = \frac{6.37 \times 10^6 \text{ m}}{99} = 6.43 \times 10^4 \text{ m} = 64.3 \text{ km}$$

ASSESS   We could have anticipated that $h > h'$ because the force of gravity decreases with increasing altitude.

47.   **INTERPRET**   In this problem we are asked to compare the speed of an object in circular orbit and its escape speed.

DEVELOP   The escape speed is the speed that makes the total energy zero:

$$E = U + K = -\frac{GMm}{r} + \frac{1}{2}mv^2 = 0$$

On the other hand, the gravitational force is what provides the centripetal force for the object to move in a circular path:

$$F = \frac{GMm}{r^2} = \frac{mv^2}{r}$$

EVALUATE   From the above equations we find $v_{esc} = \sqrt{\frac{2GM}{r}}$ and $v = \sqrt{\frac{GMm}{r}}$ for the orbital speed. Thus, the ratio of the escape speed to the orbital speed is

$$\frac{v_{esc}}{v} = \sqrt{2}$$

ASSESS   In order to escape from the orbit, the escape speed must be greater than its present orbital speed. Our calculation shows that the speed must increase by a factor of at least $\sqrt{2}$.

49.   **INTERPRET**   In this problem we want to find the speeds of two meteoroids as they approach the Earth in different manners.

DEVELOP   Conservation of energy applied to the meteoroids gives:

$$K_0 + U_0 = K + U \quad \rightarrow \quad \frac{1}{2}mv_0^2 - \frac{GM_E m}{r_0} = \frac{1}{2}mv^2 - \frac{GM_E m}{r}$$

Solving for v, we obtain

$$v = \sqrt{v_0^2 + 2GM_E\left(\frac{1}{r} - \frac{1}{r_0}\right)}$$

EVALUATE   (a) For the first meteoroid, its speed when it strikes the Earth $(r = R_E)$ is

$$v_1 = \sqrt{(2100 \text{ m/s})^2 + 2(6.67 \times 10^{-11} \text{N} \cdot \text{m}^2/\text{kg}^2)(5.97 \times 10^{24} \text{ kg})\left(\frac{1}{6.37 \times 10^6 \text{ m}} - \frac{1}{2.5 \times 10^8 \text{ m}}\right)}$$

$$= 1.12 \times 10^4 \text{ m/s} = 11.2 \text{ km/s}$$

(b) For the second meteoroid, its speed at the distance of closest approach $(r = 8500 \text{ km})$ is

$$v_2 = \sqrt{(2100 \text{ m/s})^2 + 2(6.67 \times 10^{-11} \text{N} \cdot \text{m}^2/\text{kg}^2)(5.97 \times 10^{24} \text{ kg})\left(\frac{1}{8.50 \times 10^6 \text{ m}} - \frac{1}{2.5 \times 10^8 \text{ m}}\right)}$$

$$= 9.74 \times 10^3 \text{ m/s} = 9.74 \text{ km/s}$$

(c) The escape velocity at a distance of $r = 8500$ km from the center of the Earth is

$$v_{esc} = \sqrt{\frac{2GM_E}{r}} = \sqrt{\frac{2(6.67 \times 10^{-11} \text{N} \cdot \text{m}^2/\text{kg}^2)(5.97 \times 10^{24} \text{ kg})}{8.5 \times 10^6 \text{ m}}} = 9.67 \text{ km/s}$$

Therefore, the second meteoroid will probably not return. Alternatively, v_{esc} at a distance of 250,000 km is 1.78 km/s.

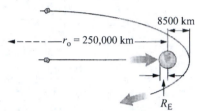

ASSESS   By energy conservation, the change in kinetic energy is equal to the negative of the change of potential energy: $\Delta K = -\Delta U$. Since $\Delta U_1 < \Delta U_2$ (the potential energy at the surface of the Earth is lower than that at $r = 8500$ km), we have $\Delta K_1 > \Delta K_2$, i.e., the first meteoroid has a greater speed.

51.   **INTERPRET**   In this problem we are asked about the speed of the projectile as a function of r, given that its initial launch speed is twice the escape speed.

DEVELOP   Conservation of energy applied to the meteoroids (initially at $r_0 = R$) gives:

$$K_0 + U_0 = K + U \quad \rightarrow \quad \frac{1}{2}mv_0^2 - \frac{GMm}{R} = \frac{1}{2}mv^2 - \frac{GMm}{r} \quad \text{or} \quad v(r) = \sqrt{v_0^2 + 2GM\left(\frac{1}{r} - \frac{1}{R}\right)}$$

On the other hand, the escape speed is the speed that makes the total energy zero:

$$E = U + K = -\frac{GMm}{R} + \frac{1}{2}mv^2 = 0 \quad \text{or} \quad v_{esc} = \sqrt{\frac{2GM}{R}}$$

EVALUATE   If $v_0 = 2v_{esc}$, then the speed as a function of r becomes

$$v(r) = \sqrt{v_0^2 + 2GM\left(\frac{1}{r} - \frac{1}{R}\right)} = \sqrt{\frac{8GM}{R} + 2GM\left(\frac{1}{r} - \frac{1}{R}\right)} = \sqrt{2GM\left(\frac{1}{r} + \frac{3}{R}\right)}$$

ASSESS   When $r = R$, we must recover the initial speed: $v_0 = 2v_{esc}$. This indeed is what we find:

$$v(R) = \sqrt{\frac{8GM}{R}} = 2\sqrt{\frac{2GM}{R}} = 2v_{esc}.$$

53.   **INTERPRET**   In this problem we want to find the impact speeds of two meteoroids as they approach the Earth with different initial speeds.

DEVELOP   Conservation of energy applied to the meteoroids gives:

$$K_0 + U_0 = K + U \quad \rightarrow \quad \frac{1}{2}mv_0^2 - \frac{GM_E m}{r_0} = \frac{1}{2}mv^2 - \frac{GM_E m}{r}$$

Solving for v, we obtain

$$v = \sqrt{v_0^2 + 2GM_E\left(\frac{1}{r} - \frac{1}{r_0}\right)}$$

To find the impact speed, we set $r = R_E$.

EVALUATE   (a) For $v_{10} = 10$ km/s $= 1.0 \times 10^4$ m/s, the impact speed is

$$v_1 = \sqrt{(10{,}000 \text{ m/s})^2 + 2(6.67 \times 10^{-11} \text{N} \cdot \text{m}^2/\text{kg}^2)(5.97 \times 10^{24} \text{ kg})\left(\frac{1}{6.37 \times 10^6 \text{ m}} - \frac{1}{1.6 \times 10^8 \text{ m}}\right)}$$

$$= 1.48 \times 10^4 \text{ m/s} = 14.8 \text{ km/s}$$

(b) For $v_{20} = 20$ km/s $= 2.0 \times 10^4$ m/s, the speed is

$$v_2 = \sqrt{(20{,}000 \text{ m/s})^2 + 2(6.67 \times 10^{-11} \text{N} \cdot \text{m}^2/\text{kg}^2)(5.97 \times 10^{24} \text{ kg})\left(\frac{1}{6.37 \times 10^6 \text{ m}} - \frac{1}{1.6 \times 10^8 \text{ m}}\right)}$$

$$= 2.28 \times 10^4 \text{ m/s} = 22.8 \text{ km/s}$$

ASSESS   The final impact speed depends on the initial speed and the change of gravitational potential energy. Since $v_{20} > v_{10}$, and the change in potential energy is the same for both, therefore, we have $v_2 > v_1$.

55.   **INTERPRET**   In this problem we are asked to find the speed of the satellite at the low point, given its speed at the high point.

DEVELOP   Ignoring effects such as the gravitational influence of other bodies or atmospheric drag, we apply energy conservation ($K_0 + U_0 = K + U$) to the satellite in an elliptical Earth orbit. The speed and distance at perigee (the lowest point) may be related to the same quantities at apogee (the highest point) as:

$$\frac{1}{2}mv_a^2 - \frac{GM_E m}{r_a} = \frac{1}{2}mv_p^2 - \frac{GM_E m}{r_p}$$

Solving for v_p, we obtain

$$v_p = \sqrt{v_a^2 + GM_E\left(\frac{1}{r_p} - \frac{1}{r_a}\right)}$$

EVALUATE   The distances to the two points, as measured from the center of the Earth are

$$r_p = h_p + R_E = 230 \text{ km} + 6370 \text{ km} = 6600 \text{ km} = 6.60 \times 10^6 \text{ m}$$

$$r_a = h_a + R_E = 890 \text{ km} + 6370 \text{ km} = 7260 \text{ km} = 7.27 \times 10^6 \text{ m}$$

Substituting the values given in the problem statement, the speed of the satellite at the perigee is

$$v_p = \sqrt{(7230 \text{ m/s})^2 + 2(6.67 \times 10^{-11} \text{N} \cdot \text{m}^2/\text{kg}^2)(5.97 \times 10^{24} \text{ kg})\left(\frac{1}{6.60 \times 10^6 \text{ m}} - \frac{1}{7.27 \times 10^6 \text{ m}}\right)} = 7.96 \text{ km/s}$$

ASSESS   In the limit where the orbit is circular, $r_p = r_a$, and we recover the expected result $v_p = v_a$. As we shall see in Chapter 11, the same result also follows from the conservation of angular momentum or Kepler's second law, which implies that $v_a r_a = v_p r_p$. Using this relation, we have

$$v_p = \left(\frac{r_a}{r_p}\right)v_a = \left(\frac{7.27 \times 10^6 \text{ m}}{6.60 \times 10^6 \text{ m}}\right)(7.23 \text{ km/s}) = 7.96 \text{ km/s}$$

The speed of the satellite at the low point (perigee) is greater than the speed at the high point (apogee).

57.   **INTERPRET**   In this problem we are asked about the orbital radius, and the kinetic energy and speed of a spacecraft, given its mass and total energy.

DEVELOP   The potential energy, kinetic energy and the total energy of an object in a circular orbit are given by Equations 8.6, 8.8a and 8.8b:

$$U = -\frac{GMm}{r}$$

$$K = \frac{1}{2}mv^2 = \frac{GMm}{2r}$$

$$E = U + K = -\frac{GMm}{2r}$$

These equations can be used to solve for the physical quantities we are interested.

EVALUATE   (a) The orbital radius of the spacecraft is

$$r = -\frac{GMm}{2E} = -\frac{(6.67 \times 10^{-11} \text{N} \cdot \text{m}^2/\text{kg}^2)(1.99 \times 10^{30} \text{ kg})(720 \text{ kg})}{2(-5.3 \times 10^{11} \text{ J})} = 9.02 \times 10^{10} \text{ m}$$

(b) The kinetic energy of the spacecraft is

$$K = \frac{1}{2}mv^2 = \frac{GMm}{2r} = -E = 5.3 \times 10^{11} \text{ J}$$

(c) The speed of the spacecraft is

$$v = \sqrt{\frac{2K}{m}} = \sqrt{\frac{-2E}{m}} = \sqrt{\frac{-2(-5.3 \times 10^{11} \text{ J})}{720 \text{ kg}}} = 3.84 \times 10^4 \text{ m/s}$$

ASSESS From **(a)** we see that the radius of the circular orbit is inversely proportional to $-E$, the negative of the total mechanical energy. The more negative E is, the smaller the orbital radius, and the closer the spacecraft is to the Sun. Also, since $v = \sqrt{GM/r}$, the closer the spacecraft is to the Sun, the greater is its speed.

59.   **INTERPRET**   In this problem we are asked to show Equation 8.5 reduces $\Delta U = mg\Delta r$ when $r_1 \approx r_2$.

DEVELOP   Following the hint, we write $r_2 = r_1 + \Delta r$, and Equation 8.5 becomes

$$\Delta U = GMm\left(\frac{1}{r_1} - \frac{1}{r_2}\right) = GMm\left(\frac{r_2 - r_2}{r_1 r_2}\right) = \frac{GMm}{r_1(r_1 + \Delta r)}\Delta r = \frac{GMm}{r_1^2(1 + \Delta r/r_1)}\Delta r$$

If we neglect Δr compared to r_1 in the denominator of the last step, we obtain

$$\Delta U \approx \frac{GMm}{r_1^2}\Delta r$$

EVALUATE   Since the gravitational acceleration at a distance r_1 from the center of the gravitating body of mass M is $g(r_1) = \frac{GM}{r_1^2}$, the above expression can be rewritten as

$$\Delta U \approx mg(r_1)\Delta r$$

Near the Earth's surface $r_1 \approx R_E$, $g(R_E) = GM/R_E^2 = g = 9.8 \text{ m/s}^2$ is constant, so $\Delta U \approx mg \Delta r$.

ASSESS   When $r_1 \approx r_2$, the gravitational accelerations at these two distances are very close, $g(r)_1 \approx g(r_2)$. In this limit, the change in gravitational potential energy only depends on $\Delta r = r_2 - r_1$. The equation $\Delta U = mg\Delta y$ is precisely what we have used in Chapters 2 and 3 where the kinematics setting has always been taken to be close to the Earth's surface, with g assuming a constant value of $\approx 9.8 \text{ m/s}^2$.

61.   **INTERPRET**   In this problem we are asked to derive the period of a "binary system" that consists of two objects of equal mass M orbiting each other.

DEVELOP   The gravitational force between two masses separated by a distance $2r$ is given by Equation 8.1:

$$F = \frac{GM^2}{(2r)^2} = \frac{GM^2}{4r^2}$$

The gravitational force is also the centripetal force that keeps the two masses orbiting about a common center.

EVALUATE   The centripetal force acting on one of the masses is

$$F = \frac{Mv^2}{r} = \frac{M}{r}\left(\frac{2\pi r}{T}\right)^2 = \frac{4\pi^2 Mr}{T^2}$$

Equating the two expressions gives

$$\frac{GM^2}{4r^2} = \frac{4\pi^2 Mr}{T^2}$$

which leads to

$$T^2 = \frac{16\pi^2 r^3}{GM}$$

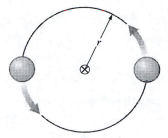

ASSESS   Our result is in complete agreement with Kepler's law which states that

$$\frac{T^2}{r^3} = \text{constant}$$

63.   **INTERPRET**   We estimate the ratio of tidal forces due to the Sun and the Moon, and compare that ratio with the ratio of gravitational forces due to the Sun and the Moon. Tidal forces are proportional to the change of force of gravity with distance.

DEVELOP   We'll differentiate $F = G\frac{m_1 m_2}{r^2}$, and find the ratio of the derivative for the Moon

$(m_2 = m_m = 7.35 \times 10^{22}$ kg, $r_m = 3.84 \times 10^8$ m) to the derivative for the Sun

$(m_2 = m_s = 1.99 \times 10^{30}$ kg, $r_s = 1.5 \times 10^{11}$ m). We will also compare the ratios of the forces directly.

EVALUATE $\frac{dF}{dr} = -2G\frac{m_1 m_2}{r^3}$. The ratio of tidal forces is then

$$\frac{\left.\frac{dF}{dr}\right|_{Sun}}{\left.\frac{dF}{dr}\right|_{Moon}} = \frac{-2G\frac{m_1 m_s}{r_s^3}}{-2G\frac{m_1 m_m}{r_m^3}} = \frac{r_m^3 m_s}{r_s^3 m_m} = 0.45$$

The ratio of gravitational forces is $\dfrac{F_s}{F_m} = \dfrac{G\frac{m_1 m_s}{r_s^2}}{G\frac{m_1 m_m}{r_m^2}} = \dfrac{r_m^2 m_s}{r_s^2 m_m} = 177$

ASSESS   We see that the gravitational force due to the Sun is much higher than that due to the Moon, but the tidal force due to the Moon is higher than that due to the Sun.

65.   **INTERPRET**   From the height jumped by Olympic high-jumpers, calculate whether a jumper can reach escape velocity on an asteroid. We will use conservation of energy here on Earth to calculate the speed with which they jump, and compare it with the equation for escape velocity.

DEVELOP   To find the speed at which the Olympians jump, use conservation of energy: $U_1 + K_1 = U_2 + K_2$, and solve for v in the initial kinetic energy. The h in the potential energy terms is the height of their center of mass, which we will assume starts from about $h_1 = 1.0$ m above the ground. We will also need an expression for escape speed from an asteroid of a given radius, assuming the asteroid is the same density as the Earth. The height to which they can jump on Earth is $h_2 = 7.67$ ft $= 2.34$ m.

EVALUATE   First we find their launch speed on Earth.

$$U_1 + K_1 = U_2 + \cancel{K_2} = mgh_1 + \frac{1}{2}mv^2 = mgh_2 \rightarrow v = \sqrt{2g(h_2 - h_1)} = 5.12 \text{ m/s}$$

We will also need the density of Earth, since we're assuming an asteroid is the same density:

$$\rho = \frac{m_E}{\frac{4}{3}\pi r_E^3} = \frac{3m_E}{4\pi r_E^3} = 5.49 \times 10^3 \text{ kg/m}^3$$

Now we must express the escape speed from a spherical asteroid in terms of the density and radius:

$$v = \sqrt{\frac{2GM}{r}} = \sqrt{\frac{2G(\frac{4}{3}\pi r^3 \rho)}{r}} = r\sqrt{\frac{8G\pi\rho}{3}}. \text{ Solve this for radius: } r = v\sqrt{\frac{3}{8\pi\rho G}} = 2.9 \text{ km}$$

ASSESS   So we would expect that athletes could leap off asteroids of 2.9 km or smaller radius. This solution assumes that the asteroid is spherical, and that the athlete is not encumbered by a spacesuit.

67.   **INTERPRET**   Where, along a line between the Earth and the Moon, is the gravitational force between the two equal? We use the equation for gravitational force.

DEVELOP   We use $F = G\frac{m_1 m_2}{r^2}$. There are two distances: the distance to Earth, which we'll call r_E, and the distance to the Moon, r_M. These two distances should add to the radius of the Moon's orbit, $r = 3.84 \times 10^8$ m. The mass of the Moon is $m_M = 7.35 \times 10^{22}$ kg, and the mass of the Earth is $m_E = 5.97 \times 10^{24}$ kg.

EVALUATE   The force on the spacecraft of mass m from the Earth is $F_E = G\frac{m_E m}{r_E^2}$. The force on the spacecraft due to the Moon is $F_M = G\frac{m_M m}{r_M^2} = G\frac{m_M m}{(r - r_E)^2}$. We set the two equal to each other:

$$\cancel{G}\frac{m_M \cancel{m}}{(r - r_E)^2} = \cancel{G}\frac{m_E \cancel{m}}{r_E^2} \rightarrow m_E(r - r_E)^2 = m_M r_E^2 \rightarrow m_E(r^2 - 2rr_E + r_E^2) = m_M r_E^2$$

$$\rightarrow (m_E - m_M)r_E^2 + (-2rm_E)r_E + (m_E r^2) = 0 \rightarrow r_E = \frac{2rm_E \pm \sqrt{4r^2 m_E^2 - 4(m_E - m_M)m_E r^2}}{2(m_E - m_M)}$$

$$\rightarrow r_E = (3.46 \times 10^8 \text{ m}) \text{ or } (4.32 \times 10^8 \text{ m})$$

ASSESS   One of these two answers is on the *other* side of the Moon—that's not the one we want. The crossover point is at a distance of 3.46×10^5 km from the Earth, or 90% of the way there.

SYSTEMS OF PARTICLES

9

EXERCISES

Section 9.1 Center of Mass

13. **INTERPRET** This problem is about center of mass. Our system consists of three masses located at the vertices of an equilateral triangle. If two masses are known and the location of the center of mass is also known, then the third mass can be calculated.

DEVELOP The center of mass of a system of particles is given by Equation 9.2:

$$\vec{r}_{cm} = \frac{\sum_i m_i \vec{r}_i}{\sum_i m_i} = \frac{\sum_i m_i \vec{r}_i}{M}$$

We shall choose x-y coordinates with origin (0,0) at the midpoint of the base. With this arrangement, the center of the mass is located at $x_{cm} = 0$ and $y_{cm} = y_3/2$, where y_3 is the position of the third mass (and of course, $y_1 = y_2 = 0$ for the equal masses $m_1 = m_2 = m$ on the base).

EVALUATE Using Equation 9.2, the y coordinate of the center of mass is

$$y_{cm} = \frac{m_1 y_1 + m_2 y_2 + m_3 y_3}{m_1 + m_2 + m_3} = \frac{m(0) + m(0) + m_3 y_3}{m + m + m_3} = \frac{m_3 (2 y_{cm})}{2m + m_3}$$

Solving for m_3, we have $2m + m_3 = 2m_3$, or $m_3 = 2m$.

ASSESS From symmetry consideration, it is apparent that $x_{cm} = 0$. On the other hand, we have $m + m = 2m$ at the bottom two vertices of the triangle. Since $y_{cm} = y_3/2$, i.e., y_{cm} is halfway to the top vertex, we expect the mass there to be $2m$ (See Example 9.2).

15. **INTERPRET** This problem is about locating the center of mass. Our system consists of three equal masses located at the vertices of an equilateral triangle of side L.

DEVELOP We take x-y coordinates with origin at the center of one side as shown. The center of mass of a system of particles is given by Equation 9.2:

$$\vec{r}_{cm} = \frac{\sum_i m_i \vec{r}_i}{\sum_i m_i} = \frac{\sum_i m_i \vec{r}_i}{M}$$

EVALUATE From the symmetry (for every mass at x, there is an equal mass at $-x$), we have $x_{cm} = 0$. As for y_{cm}, since $y = 0$ for the two masses on the x-axis, and $y_3 = L\sin 60° = \frac{L\sqrt{3}}{2}$ for the third mass, Equation 9.2 gives

$$y_{cm} = \frac{m_1 y_1 + m_2 y_2 + m_3 y_3}{m_1 + m_2 + m_3} = \frac{m(0) + m(0) + mL\sqrt{3}/2}{m + m + m} = \frac{\sqrt{3}}{6} L = 0.289L$$

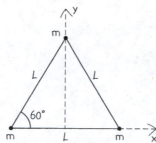

ASSESS   From symmetry consideration, it is apparent that $x_{cm} = 0$. On the other hand, we have $m + m = 2m$ at the bottom two vertices of the triangle, and m at the top of the vertex. Therefore, we should expect y_{cm} to be one third of y_3. This indeed is the case, as y_{cm} can be rewritten as $y_{cm} = y_3/3$.

17.   **INTERPRET**   We are asked to find the center of mass of a given distribution of objects. The objects are in a straight line, so we can do the problem one-dimensionally.

DEVELOP   Distances are given from the back of the canoe: The paddler ($m_1 = 76$ kg) is at $x_1 = 0.85$ m, and the passenger ($m_2 = 52$ kg) is at $x_2 = 3.9$ m. We use $x_{cm} = \frac{1}{M}\sum m_i x_x$ to find the distance from the back of the canoe to the center of mass.

EVALUATE
$$x_{cm} = \frac{1}{M}\sum m_i x_x = \frac{1}{m_1 + m_2}[m_1 x_1 + m_2 x_2] = 2.1 \text{ m}$$

ASSESS   This is near the center of the two people, but closer to the heavier one as we might expect.

Section 9.2 Momentum

19.   **INTERPRET**   The object of interest is the skater. We want to find her velocity after she tosses a snowball in a certain direction.

DEVELOP   On frictionless ice, momentum would be conserved in the process. Since the initial momentum of the skater-snowball system is zero, their final total momentum must also be zero:
$$0 = m_1 \vec{v}_1 + m_2 \vec{v}_2$$
where subscripts 1 and 2 refer to the snowball and skater, respectively.

EVALUATE   By momentum conservation, the final velocity of the skater is
$$\vec{v}_2 = -\frac{m_1}{m_2}\vec{v}_1 = -\frac{12 \text{ kg}}{60 \text{ kg}}[(3.0 \text{ m/s})\hat{i} + (4.0 \text{ m/s})\hat{j}] = -(0.6 \text{ m/s})\hat{i} - (0.8 \text{ m/s})\hat{j}$$

ASSESS   As expected, the skater moves in the opposite direction of the snowball. This is a consequence of momentum conservation.

21.   **INTERPRET**   In this problem we want to find the final speed of a moving toboggan as some snow drops onto it.

DEVELOP   Since there is no net external horizontal force, the total momentum of the snow-toboggan system is conserved. The initial momentum of the system is $P_i = m_t v_{ti}$. Since the snow and the toboggan move together with the same speed v_f, the final momentum is $P_f = (m_t + m_s)v_f$.

EVALUATE   By momentum conservation, $P_i = P_f$, the final speed of the toboggan-snow system is
$$v_f = \frac{m_t}{m_t + m_s}v_{ti} = \frac{8.6 \text{ kg}}{8.6 \text{ kg} + 15 \text{ kg}}(23 \text{ km/h}) = 8.38 \text{ km/h}$$

ASSESS   To see that our result makes sense, let's consider the following limiting cases: **(i)** $m_s = 0$. In this situation, we have $v_f = v_{ti}$, indicating that the toboggan continues with the same speed. **(ii)** $m_s \to \infty$. In the situation where a large quantity of snow is dumped onto the toboggan, we expect the system to slow down considerably. This indeed is what our equation indicates.

Section 9.3 Kinetic Energy of a System

23.   **INTERPRET**   In this problem we are asked about the energy gained by the baseball pieces after explosion.

DEVELOP   The velocities of the baseball and the two pieces into which it explodes are all along the same line, so the conservation of momentum (before and after) takes the form
$$\vec{P}_i = \vec{P}_f \quad \to \quad (m_1 + m_2)\vec{v}_0 = m_1\vec{v}_1 + m_2\vec{v}_2$$

The initial kinetic energy of the system is $K_i = \frac{1}{2}(m_1 + m_2)v_0^2$, and the total final kinetic energy is $K_f = \frac{1}{2}m_1v_1^2 + \frac{1}{2}m_2v_2^2$. Therefore, the change in kinetic energy is

$$\Delta K = K_f - K_i = \frac{1}{2}m_1v_1^2 + \frac{1}{2}m_2v_2^2 - \frac{1}{2}(m_1 + m_2)v_0^2 = \frac{1}{2}m_1\left(v_1^2 - v_0^2\right) + \frac{1}{2}m_2\left(v_2^2 - v_0^2\right)$$

EVALUATE   Let the forward direction be positive. By momentum conservation, the velocity of the second piece, with mass $m_2 = m - m_1 = 150\text{ g} - 38\text{ g} = 112\text{ g}$, is

$$v_2 = \frac{(m_1 + m_2)v_0 - m_1v_1}{m_2} = \frac{(150\text{ g})(60\text{ km/h}) - (38\text{ g})(85\text{ km/h})}{112\text{ g}} = 51.5\text{ km/h} = 14.3\text{ m/s}$$

In SI units $v_0 = 16.67\text{ m/s}$ and $v_1 = 23.6\text{ m/s}$, the difference in kinetic energy is

$$\Delta K = \Delta K_1 + \Delta K_2 = \frac{1}{2}m_1\left(v_1^2 - v_0^2\right) + \frac{1}{2}m_2\left(v_2^2 - v_0^2\right)$$

$$= \frac{1}{2}(38 \times 10^{-3}\text{ kg})[(23.6\text{ m/s})^2 - (16.7\text{ m/s})^2] + \frac{1}{2}(112 \times 10^{-3}\text{ kg})[(14.3\text{ m/s})^2 - (16.7\text{ m/s})0^2]$$

$$= 5.31\text{ J} - 4.10\text{ J}$$

$$= 1.21\text{ J}$$

ASSESS   The change in kinetic energy for the first piece (ΔK_1) is positive because $v_1 > v_0$, but negative for the second ($\Delta K_2 < 0$ since $v_2 < v_0$). Writing the kinetic energy as $K = K_{cm} + K_{int}$, we can say that the explosion has increased the internal kinetic energy K_{int}.

25.   **INTERPRET**   Before an explosion, an object has kinetic energy K. After the explosion, it has two pieces with twice the original speed. We are asked to find and compare the internal and center-of-mass energies after the explosion.
      DEVELOP   There are no external forces, so the total momentum and the value of K_{cm} will not change. This tells us that v_{cm} in the figure below is equal to v_i. We can use the Pythagorean theorem on the right triangle shown to find v_{rel}^2, and thus the value of K_{int}.
      EVALUATE $v_{rel}^2 = v_f^2 - v_{cm}^2 = (2v_i)^2 - v_i^2 = 3v_i^2$. The initial kinetic energy is $K_i = \frac{1}{2}mv_i^2$, the center-of-mass kinetic energy after the explosion is the same as the initial, $K_{cm} = \frac{1}{2}mv_i^2$, and the internal energy after the explosion is $K_{int} = \frac{1}{2}mv_{rel}^2 = \frac{3}{2}mv_i^2$.

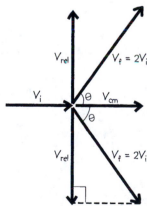

      ASSESS   After this explosion, the internal energy becomes three times larger than the initial energy, but the center-of-mass kinetic energy is unchanged.

Section 9.5 Totally Inelastic Collisions

27.   **INTERPRET**   In this problem, we are asked to show that half of the initial kinetic energy of the system is lost in a totally inelastic collision between two equal masses.
      DEVELOP   Suppose we have two masses m_1 and m_2 moving with velocities $\vec{v}_1$ and $\vec{v}_2$, respectively. After undergoing totally inelastic collision, the two masses stick together and move with final velocity $\vec{v}_f$. Even though the collision is totally inelastic, momentum conservation still applies, and we have (Equation 9.11):

$$m_1\vec{v}_1 + m_2\vec{v}_2 = (m_1 + m_2)\vec{v}_f \quad \rightarrow \quad \vec{v}_f = \frac{m_1\vec{v}_1 + m_2\vec{v}_2}{m_1 + m_2}$$

The initial total kinetic energy of the two-particle system is $K_i = \frac{1}{2}m_1 v_1^2 + \frac{1}{2}m_2 v_2^2$, while the final kinetic energy of the system after collision is

$$K_f = \frac{1}{2}(m_1 + m_2)v_f^2$$

Therefore, the change in kinetic energy is given by

$$\Delta K = K_f - K_i = \frac{1}{2}(m_1 + m_2)v_f^2 - \frac{1}{2}m_1 v_1^2 - \frac{1}{2}m_2 v_2^2 = \frac{1}{2}m_1\left(v_f^2 - v_1^2\right) + \frac{1}{2}m_2\left(v_f^2 - v_2^2\right)$$

EVALUATE   In our case, we have $m_1 = m_2 = m$, $v_1 = v$, and $v_2 = 0$. The initial kinetic energy of the system is $K_i = \frac{1}{2}mv^2$. The final speed is

$$v_f = \frac{m_1 v_1 + m_2 v_2}{m_1 + m_2} = \frac{mv}{m + m} = \frac{1}{2}v$$

Therefore, the change in total kinetic energy is

$$\Delta K = \frac{1}{2}m_1\left(v_f^2 - v_1^2\right) + \frac{1}{2}m_2\left(v_f^2 - v_2^2\right) = \frac{1}{2}m\left(\frac{v^2}{4} - v^2\right) + \frac{1}{2}m\left(\frac{v^2}{4} - 0\right) = -\frac{1}{4}mv^2$$

Thus, we see that half of the total initial kinetic energy is lost in the collision process.

ASSESS   For a totally inelastic collision, one may show that the general expression for ΔK is

$$\Delta K = \frac{1}{2}\frac{(m_1 v_1 + m_2 v_2)^2}{m_1 + m_2} - \frac{1}{2}m_1 v_1^2 - \frac{1}{2}m_2 v_2^2 = -\frac{m_1 m_2}{2(m_1 + m_2)}(v_1 - v_2)^2$$

Clearly, ΔK is always negative, and it depends on the relative speed between m_1 and m_2.

29.   **INTERPRET**   In this problem we want to find the load of a truck that undergoes totally inelastic collision with another load-carrying truck.

DEVELOP   Let's consider the general situation where two objects with masses m_1 and m_2 moving with velocities $\vec{v}_1$ and $\vec{v}_2$, respectively, undergo totally inelastic collision. The two masses stick together and move with final velocity $\vec{v}_f$. Even though the collision is totally inelastic, momentum conservation still applies, and we have (Equation 9.11):

$$m_1 \vec{v}_1 + m_2 \vec{v}_2 = (m_1 + m_2)\vec{v}_f \quad \rightarrow \quad \vec{v}_f = \frac{m_1 \vec{v}_1 + m_2 \vec{v}_2}{m_1 + m_2}$$

In our case, masses m_1 and m_2 represent the total mass of the truck plus the load.

EVALUATE   Let the second truck be initially at rest with $\vec{v}_2 = 0$. The component of Equation 9.11 in the direction of motion of the moving truck is $m_1 v_1 = (m_1 + m_2)v_f$, which can be solved to give

$$m_2 = m_1\left(\frac{v_1}{v_f} - 1\right) = (15{,}000 \text{ kg})\left(\frac{65 \text{ km/h}}{40 \text{ km/h}} - 1\right) = 9375 \text{ kg}$$

where we have used $m_1 = 5500 \text{ kg} + 9500 \text{ kg} = 15{,}000 \text{ kg}$ as the mass of the truck plus the first load. Thus, the load of the second truck is

$$m_{2,\text{load}} = m_2 - m_{\text{truck}} = 9375 \text{ kg} - 5500 \text{ kg} = 3875 \text{ kg}$$

ASSESS   From the above, we find the mass of the second object to be proportional to $v_1 - v_f$. This makes sense because in the limit where m_2 is very small, the final v_f should be very close to v_1. Had the two loads been equal, $m_2 = m_1$, we would find the final speed of $m_1 + m_2$ to be half the initial speed of m_1, $v_f = v_1/2$.

Section 9.6 Elastic Collisions

31.   **INTERPRET**   This problem is about head-on elastic collision. We want to find the speed of the ball after it rebounds elastically from a moving car. The situation is one-dimensional.

DEVELOP   Momentum is conserved in this process. In this one-dimensional case, we may write

$$m_1 v_{1i} + m_2 v_{2i} = m_1 v_{1f} + m_2 v_{2f}$$

Since the collision is completely elastic, energy is conserved:

$$\frac{1}{2}m_1 v_{1i}^2 + \frac{1}{2}m_2 v_{2i}^2 = \frac{1}{2}m_1 v_{1f}^2 + \frac{1}{2}m_2 v_{2f}^2$$

Using the two conservation equations, the final speeds of m_1 and m_2 are (see Equations 9.15a and 9.15b):

$$v_{1f} = \frac{m_1 - m_2}{m_1 + m_2}v_{1i} + \frac{2m_2}{m_1 + m_2}v_{2i} \qquad v_{2f} = \frac{2m_1}{m_1 + m_2}v_{1i} + \frac{m_2 - m_1}{m_1 + m_2}v_{2i}$$

EVALUATE   Let the subscripts 1 and 2 be for the car and the ball, respectively. We choose positive velocities in the direction of the car. Since m_2, m_1 the speed of the ball after it rebounds is

$$v_{2f} = \frac{2m_1}{m_1 + m_2}v_{1i} + \frac{m_2 - m_1}{m_1 + m_2}v_{2i} \approx 2v_{1i} - v_{2i} = 2(14 \text{ m/s}) - (-18 \text{ m/s}) = 46 \text{ m/s}$$

Similarly, the final speed of the car is

$$v_{1f} = \frac{m_1 - m_2}{m_1 + m_2}v_{1i} + \frac{2m_2}{m_1 + m_2}v_{2i} \approx v_{1i} = 14 \text{ m/s}$$

ASSESS   We do not expect the speed of the car to change much after colliding with a ball. However, the ball rebounds with a much greater speed than before. If the car were stationary with $v_{1i} = 0$, then we would find $v_{2f} = -v_{2i} = 18 \text{ m/s}$.

33.   **INTERPRET**   In this problem we are asked to find the speeds of the protons after they collide elastically head-on.
DEVELOP   Consider the general situation where two masses m_1 and m_2 moving with velocities $\vec{v}_1$ and $\vec{v}_2$, undergo elastic collision. Both momentum and energy are conserved in this process. Using the conservation equations, the final speeds of m_1 and m_2 are given by (see Equations 9.15a and 9.15b):

$$v_{1f} = \frac{m_1 - m_2}{m_1 + m_2}v_{1i} + \frac{2m_2}{m_1 + m_2}v_{2i} \qquad v_{2f} = \frac{2m_1}{m_1 + m_2}v_{1i} + \frac{m_2 - m_1}{m_1 + m_2}v_{2i}$$

EVALUATE   We choose positive velocities in the direction of $\vec{v}_1$. With $m_1 = m_2 = m$, $v_1 = v = 11$ Mm/s and $v_2 = -v_1 = -v = -11$ Mm/s, the final speeds are

$$v_{1f} = \frac{m_1 - m_2}{m_1 + m_2}v_{1i} + \frac{2m_2}{m_1 + m_2}v_{2i} = v_{2i} = -v = -11 \text{ Mm/s}$$

$$v_{2f} = \frac{2m_1}{m_1 + m_2}v_{1i} + \frac{m_2 - m_1}{m_1 + m_2}v_{2i} = v_{1i} = v = 11 \text{ Mm/s}$$

ASSESS   In this case, the protons simply exchange places—the final speed of the first proton is equal to the initial speed of the second proton, while the final speed of the second proton is equal to the initial speed of the first proton.

PROBLEMS

35.   **INTERPRET**   In this problem we want to find the center of mass of a pentagon of side a with one triangle missing.
DEVELOP   We choose coordinates as shown. If the fifth isosceles triangle (with the same assumed uniform density) were present, the center of mass of the whole pentagon would be at the origin, so

$$0 = \frac{my_5 + 4my_{cm}}{5m} = \frac{y_5 + 4y_{cm}}{5}$$

where y_{cm} gives the position of the center of mass of the figure we want to find, and y_5 is the position of the center of mass of the fifth triangle. Of course, the mass of the figure is four times the mass of the triangle.
EVALUATE   From symmetry, the x coordinate of the center of mass is $x_{cm} = 0$. Now, to calculate y_{cm}, we make use of the result obtained in Example 9.3 where the center of mass of an isosceles triangle is calculated. This gives $y_5 = -2L/3$. In addition, from the geometry of a pentagon, we have $\tan 36° = \frac{a}{2L}$. Therefore, the y coordinate of the center of mass is

$$y_{cm} = -\frac{1}{4}y_5 = \frac{L}{6} = \frac{a}{12}\cot 36° = 0.115\,a$$

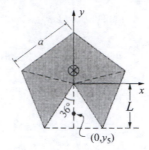

(0,y_5)

ASSESS   From symmetry argument, the center of mass must lie along the line that bisects the figure. With the missing triangle, we expect it to be located above $y = 0$, which would have been the center of mass for a complete pentagon.

37.   **INTERPRET**   We are asked about the motion of the boat, but the problem is fundamentally related to the center of mass of the system.

DEVELOP   This problem is similar to Example 9.4. Take the x axis horizontal from bow to stern, with origin at the center of mass (CM) of the boat and people, which stays fixed, in the absence of external horizontal forces like friction. Then

$$0 = m_p x_{pi} + m_B x_{Bi} = m_p x_{pf} + m_B x_{Bf}$$

where x_{Bi} is the position of the CM of the boat initially, x_{Bf} is its final position, and x_{pi} and x_{pf} are the before and after positions of the people. Note that $x_{pi} < 0$, $x_{Bi} > 0$, $x_{pf} > 0$, and $x_{Bf} < 0$. This equation can be written as

$$m_B(x_{Bi} - x_{Bf}) = m_p(x_{pf} - x_{pi})$$

since $x_{Bi} - x_{Bf}$ is the distance the boat moves relative to the fixed total CM. The distances are related to the dimensions of the boat, since the length of the boat is equal to

$$|x_{pi}| + x_{Bi} + |x_{Bf}| + x_{pf} = x_{pf} - x_{pi} + x_{Bi} - x_{Bf} = 6.5 \text{ m}$$

EVALUATE   Substituting into the first equation, one finds

$$m_B(x_{Bi} - x_{Bf}) = m_p[6.5 \text{ m} - (x_{Bi} - x_{Bf})]$$

Thus, we find

$$x_{Bi} - x_{Bf} = (6.5 \text{ m})\frac{m_p}{m_p + m_B} = (6.5 \text{ m})\frac{1500 \text{ kg}}{1500 \text{ kg} + 12,000 \text{ kg}} = 72.2 \text{ cm}$$

Note that we did not have to assume that the CM of the boat was at the center of the boat.

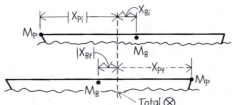

ASSESS   The boat's displacement of 72.2 cm is less than the distance people have walked. This makes sense because the boat is much more massive than the people.

39.   **INTERPRET**   In this problem we are asked about the center of mass of a uniform solid cone. Integration is needed in this problem.

DEVELOP   Choose the y axis along the axis of the cone, and the origin at the center of the base. For mass elements, take disks at height y, of radius $r = R(1 - y/h)$, parallel to the base. Then $dm = \rho \pi r^2 dy$, where ρ is the density of the cone, and $M = \frac{1}{3}\rho \pi R^2 h$ is its mass.

EVALUATE   The CM is along the axis (from symmetry) at a height

$$y_{cm} = \frac{1}{M}\int_0^h y \, dm = \frac{3}{\rho \pi R^2 h}\int_0^h y\rho \pi R^2(1 - y/h)^2 \, dy$$

$$= \frac{3}{h}\int_0^h \left(y - \frac{2y^2}{h} + \frac{y^3}{h^2}\right) dy = \frac{3}{h}\left(\frac{h^2}{2} - \frac{2h^2}{3} + \frac{h^2}{4}\right) = \frac{1}{4}h$$

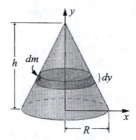

ASSESS   The result makes sense because we expect y_{cm} to be closer to the bottom of the cone because more mass is distributed in this region. Note that y_{cm} is a mass-weighted average of the positions.

41.   **INTERPRET**   We are asked about the compression of the spring due to a totally inelastic collision.

DEVELOP   Since the total momentum of the system is conserved in the process, we have

$$P_i = P_f \quad \rightarrow \quad m_1 v_1 = (m_1 + m_2) v_f$$

The potential energy of the spring at maximum compression equals the kinetic energy of the two-car system prior to contact with the spring: $\frac{1}{2} k x_{max}^2 = \frac{1}{2}(m_1 + m_2) v_f^2$

For **(b)**, we note that when the cars rebound, they are coupled together and both have the same velocity. Since the spring is ideal (by assumption), its maximum potential energy, $\frac{1}{2} k x_{max}^2$, is transformed back into kinetic energy of the cars.

EVALUATE   **(a)** The second car is initially at rest so $v_2 = 0$. By momentum conservation, the speed of the cars after collision is

$$v_f = \frac{m_1 v_1}{m_1 + m_2} = \frac{(9400\ \text{kg})(8.5\ \text{m/s})}{11000\ \text{kg} + 9{,}400\ \text{kg}} = 3.92\ \text{m/s}$$

which leads to

$$x_{max} = v_f \sqrt{\frac{m_1 + m_2}{k}} = (3.92\ \text{m/s}) \sqrt{\frac{11{,}000\ \text{kg} + 9400\ \text{kg}}{3.2 \times 10^5\ \text{N/m}}} = 0.989\ \text{m}$$

(b) With $\frac{1}{2} k x_{max}^2 = \frac{1}{2}(m_1 + m_2) v_f^2$, the rebound speed equals the initial $v_f = 3.92\ \text{m/s}$.

ASSESS   During the first part of the motion-the-collision process, momentum is conserved but energy is not. However, during the second part of the motion-compression and stretching of the spring, energy is conserved. Therefore, the cars rebound with the same speed as that before coming to contact with the spring.

43.   **INTERPRET**   The problem asks about the initial acceleration of the car after being struck by a jet of water. The water jet is the external force.

DEVELOP   The force exerted on the car by the water is the negative of the rate of change of momentum of the water:

$$\vec{F}_c = -\left(\frac{d\vec{p}}{dt} \right)_w$$

If $\vec{v}_i$ and $\vec{v}_f$ are the initial and final velocities of the water for the interval dt, then

$$\left(\frac{d\vec{p}}{dt} \right)_w = \frac{dm}{dt}(\vec{v}_f - \vec{v}_i)$$

where $\vec{v}_i = v_0 \hat{i}$. Next we note that since the water is deflected 90° without loss of relative speed, we should evaluate the change in its velocity in the rest frame of the car,

$$\vec{v}_f - \vec{v}_i = (\vec{v}_f - \vec{v}_c) - (\vec{v}_i - \vec{v}_c) = \vec{v}_f' - \vec{v}_i'$$

where $\vec{v}_c = v_c \hat{i}$ is the velocity of the car,

$$\vec{v}_i' = (v_0 - v_c)\hat{i} = v_r \hat{i}, \qquad \vec{v}_f' = v_r \hat{j}$$

and v_r is the relative velocity. Thus, $\vec{v}_f - \vec{v}_i = v_r(\hat{j} - \hat{i})$. Using this expression, the force exerted on the car by the water can be rewritten as

$$\vec{F}_c = -\frac{dm}{dt}(\vec{v}_f - \vec{v}_i) = -\frac{dm}{dt} v_r(\hat{j} - \hat{i})$$

EVALUATE   If $\vec{F_c}$ is the only force with a horizontal component acting on the car, then

$$a_x = \frac{dv_c}{dt} = \frac{1}{M}\vec{F_c}\cdot\hat{i} = \frac{1}{M}\left[-\left(\frac{dm}{dt}\right)v_r(\hat{i}-\hat{i})\right]\cdot\hat{i} = \frac{v_r}{M}\left(\frac{dm}{dt}\right)$$

Initially, the relative speed is just v_0 ($v_r = v_0 - v_c$ and $v_c = 0$), so the initial acceleration is

$$a_{x0} = \frac{v_0}{M}\left(\frac{dm}{dt}\right)$$

(b) The water ceases to exert a force on the car when $v_r = 0$, or $v_c = v_0$ (which is the car's final speed).

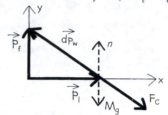

ASSESS   As expected, the acceleration of the car increases with the jet speed v_0 and the water mass rate dm/dt, but decreases with M, the mass of the car.

45.   **INTERPRET**   This is a problem about momentum conservation. The total momentum is not zero. The quantity of interest is the recoil velocity of the thorium nucleus, produced from decay of the ^{238}U nucleus.

   DEVELOP   Momentum is conserved, as in Example 9.6. This gives

$$m_U\vec{v}_U = m_{He}\vec{v}_{He} + m_{Th}\vec{v}_{Th}$$

In terms of the x and y components, we have

$$m_U v_U = m_{He}v_{He,x} + m_{Th}v_{Th,x} = m_{He}v_{He}\cos\phi + m_{Th}v_{Th}\cos\theta \quad (x\ \text{component})$$
$$0 = m_{He}v_{He,y} + m_{Th}v_{Th,y} = m_{He}v_{He}\sin\phi + m_{Th}v_{Th}\sin\theta \quad (y\ \text{component})$$

The equations can be used to solve for the magnitude and direction of $\vec{v}_{TH}$.

   EVALUATE   Solving the two equations, we obtain

$$v_{Th,x} = v_{Th}\cos\theta = \frac{m_U v_U - m_{He}v_{He}\cos\theta}{m_{Th}} = \frac{(238\ \text{u})(5\times10^5\ \text{m/s})-(4\ \text{u})(1.4\times10^7\ \text{m/s})\cos22°}{234\ \text{u}} = 2.87\times10^5\ \text{m/s}$$

and

$$v_{Th}\sin\theta = -\frac{m_{He}v_{He}\sin\phi}{m_{Th}} = -\frac{(4\ \text{u})(1.4\times10^7\ \text{m/s})\sin22°}{234\ \text{u}} = -8.96\times10^4\ \text{m/s}$$

Thus, $v_{Th} = \sqrt{v_{Th,x}^2 + v_{Th,y}^2} = \sqrt{(2.87\times10^5\ \text{m/s})^2 + (-8.96\times10^4\ \text{m/s})^2} = 3.00\times10^5\ \text{m/s}$ and the direction is $\theta = \tan^{-1}(v_{Th,y}/v_{Th,x}) = -17.4°$.

   ASSESS   The fact that θ is negative tells us that the velocity of Thorium is downward, as we expected. Since the velocity of the alpha particle has an upward component, by momentum conservation, the velocity of the Thorium must have downward component to make the y component of the total momentum zero.

47.   **INTERPRET**   This problem is about momentum conservation. The object of interest is the firecracker that has exploded into three pieces. Once the mass and velocity of two of them are known, using momentum conservation, the same quantities can be computed for the third piece.

   DEVELOP   The instant after the explosion (before any external forces have had any time to act appreciably) the total momentum of the pieces of firecracker is still zero:

$$\vec{P}_{tot} = 0 = m_1\vec{v}_1 + m_2\vec{v}_2 + m_3\vec{v}_3$$

The physical quantity of interest here is $\vec{v}_3$.

   EVALUATE   The mass of the third piece is $m_3 = m - m_1 - m_2 = 42\ \text{g} - 12\ \text{g} - 21\ \text{g} = 9\ \text{g}$. Its velocity is

$$\vec{v}_3 = -\frac{m_1\vec{v}_1 + m_2\vec{v}_2}{m_3} = -\frac{(12\ \text{g})(35\ \text{m/s})\hat{i} + (21\ \text{g})(29\ \text{m/s})\hat{j}}{9\ \text{g}} = -(46.7\ \text{m/s})\hat{i} + (67.7\ \text{m/s})\hat{j}$$

ASSESS Since the initial momentum of the firecracker is zero, we expect the momentum of the third piece to completely cancel the momentum of the first two pieces. Thus, $\vec{v}_3$ has components that are opposite of $\vec{v}_1$ and $\vec{v}_2$. Since m_3 is smaller than m_1 and m_2, we expect the magnitude of $\vec{v}_3$ to be greater than the magnitudes of $\vec{v}_1$ and $\vec{v}_2$.

49. **INTERPRET** We are asked to find the angles between the velocities of the two pieces in Exercise 9.25. We have already found, in our solution to Exercise 9.25, the various components of velocity, so we can use trigonometry to find the angle.

DEVELOP We see from the figure in the solution to Exercise 9.25 that the cosine of angle θ is $\frac{v_{cm}}{2v_i}$.

EVALUATE $\cos\theta = \frac{v_{cm}}{2v_i} = \frac{v_i}{2v_i} = \frac{1}{2} \rightarrow \theta = \cos^{-1}\left(\frac{1}{2}\right) = 60°$.

ASSESS This answer is required by conservation of momentum: the initial velocity must equal the velocity of the center of mass after the collision, since there are no external forces.

51. **INTERPRET** In this problem we are asked about the force exerted by the conveyor belt on the cookie sheets. The sheets have mounds of dough falling vertically onto them.

DEVELOP We shall assume that the conveyor belt is horizontal and moving with speed $v = 50$ cm/s, and that the mounds of dough fall vertically, then the change in the horizontal momentum of each mound of mass Δm is $\Delta p = (\Delta m)v$. The average horizontal force needed is equal to the rate at which mounds are dropped (a number N in time Δt, or $N/\Delta t$) times the change in momentum of a mound:

$$\vec{F}_{av} = \left(\frac{N}{\Delta t}\right)\Delta\vec{p} = \left(\frac{N}{\Delta t}\right)(\Delta m)\vec{v}$$

EVALUATE Substituting the values given in the problem statement, we find the average force the conveyor belt exerts on a cookie sheet to be

$$F_{av} = \left(\frac{N}{\Delta t}\right)(\Delta m)v = \left(\frac{1}{2\text{ s}}\right)(0.012\text{ kg})(0.50\text{ m/s}) = 3.0\times10^{-3}\text{ N}$$

ASSESS The average force is just the total change in momentum, $\Delta\vec{P} = N\Delta\vec{p} = N(\Delta m)\vec{v}$, divided by the time, Δt: $\vec{F}_{av} = \frac{\Delta\vec{P}}{\Delta t}$. The greater the change in momentum in a given time interval, the greater is the average force.

53. **INTERPRET** The problem asks about the speed of the drunk driver right before the totally inelastic collision. Energy is not conserved in this process, but momentum is.

DEVELOP If the wreckage skidded on a horizontal road, the work-energy theorem requires that the work done by friction be equal to the change of the kinetic energy of both cars. $W_{nc} = \Delta K$ (see Equation 7.5). Since $W_{nc} = -f_k x = -\mu_k nx = -\mu_k(m_1 + m_2)gx$, and $\Delta K = -\frac{1}{2}(m_1 + m_2)v^2$, where v is the speed of the wreckage immediately after collision, we are lead to

$$\mu_k(m_1 + m_2)gx = \frac{1}{2}(m_1 + m_2)v^2$$

The equation can be used to solve for v. Once v is known, using momentum conservation, the initial speed of the driver can be determined.

EVALUATE From the above, we find the speed of the cars (wreckage) just after the collision to be

$$v = \sqrt{2\mu_k gx}$$

During the instant of time in which the collision occurred, momentum was conserved, so if v_1 was the speed of the drunk driver's car just before the collision (and $v_2 = 0$ for the parked car), then $m_1 v_1 = (m_1 + m_2)v$, or

$$v_1 = \frac{m_1 + m_2}{m_1}v = \frac{m_1 + m_2}{m_1}\sqrt{2\mu_k gx} = \frac{1600\text{ kg} + 1300\text{ kg}}{1600\text{ kg}}\sqrt{2(.77)(9.8\text{ m/s}^2)(25\text{ m})} = 35.2\text{ m/s}$$

This is about 79 mi/h, so the driver was speeding as well as intoxicated.

ASSESS The answer makes sense because increasing v_1 will result in a greater wreckage speed v, and hence a longer skidding distance x.

55.   **INTERPRET**   In this problem we are asked to find the relative speed between the satellite and the booster after an explosion. The impulse of the explosion is given.

DEVELOP   During an explosion, impulses of the same magnitude, but opposite directions, act on the satellite and booster (from Newton's third law, and by momentum conservation). As shown in Equation 9.10b, the impulse $\vec{J}$ is equal to the change of momentum: $\vec{J} = \Delta \vec{p} = m\Delta \vec{v}$. This allows us to find the relative speed between the satellite and the booster.

EVALUATE   Initially both the satellite and the booster are at rest. After explosion, their speeds are

$$v_s = \frac{\Delta p_s}{m_s} = \frac{J}{m_s} \qquad v_b = \frac{\Delta p_b}{m_b} = -\frac{J}{m_b}$$

Thus, the relative speed of separation is

$$|v_s - v_b| = J\left(\frac{1}{m_s} + \frac{1}{m_b}\right) = (350 \text{ N} \cdot \text{s})\left(\frac{1}{950 \text{ kg}} + \frac{1}{640 \text{ kg}}\right) = 0.915 \text{ m/s}$$

ASSESS   The relative speed is shown to depend on J, the magnitude of the impulse. The greater the impulse, the faster the satellite and the booster fly away from each other.

57.   **INTERPRET**   The problem asks about the speed of one of the cars right before its totally inelastic collision with another. Given the road condition, we want to show that one of the cars has a speed of at least 25 km/h. Energy is not conserved in this process, but momentum is.

DEVELOP   If the wreckage skidded on a horizontal road, the work-energy theorem requires that the work done by friction be equal to the change of kinetic energy of both cars. $W_{nc} = \Delta K$ (see Equation 7.5). Since $W_{nc} = -f_k x = -\mu_k n x = -\mu_k (m_1 + m_2)gx$, and $\Delta K = -\frac{1}{2}(m_1 + m_2)v^2$, where v is the speed of the wreckage immediately after collision, we are lead to

$$\mu_k (m_1 + m_2)gx = \frac{1}{2}(m_1 + m_2)v^2$$

Therefore, the speed of the wreckage just after the collision is $v = \sqrt{2\mu_k gx}$. Next, momentum conservation requires that

$$m_1 \vec{v}_1 + m_2 \vec{v}_2 = (m_1 + m_2)\vec{v} \quad \rightarrow \quad \vec{v} = \frac{m_1 \vec{v}_1 + m_2 \vec{v}_2}{m_1 + m_2}$$

Since $\vec{v}_1$ and $\vec{v}_2$ are perpendicular to each other, the scalar product $\vec{v} \cdot \vec{v}$ becomes

$$v^2 = |\vec{v}|^2 = \left(\frac{m_1 \vec{v}_1 + m_2 \vec{v}_2}{m_1 + m_2}\right) \cdot \left(\frac{m_1 \vec{v}_1 + m_2 \vec{v}_2}{m_1 + m_2}\right) = \frac{m_1^2 v_1^2 + m_2^2 v_2^2}{(m_1 + m_2)^2}$$

In the next step, we substitute the maximum speed for one car to get the minimum speed for the other.

EVALUATE   Substituting $v = \sqrt{2\mu_k gx}$ into the above expression leads to

$$v^2 = \frac{m_1^2 v_1^2 + m_2^2 v_2^2}{(m_1 + m_2)^2} = 2\mu_k gx$$

Solving for v_1 gives

$$v_1 = \sqrt{\frac{2\mu_k gx(m_1 + m_2)^2 - m_2^2 v_2^2}{m_1^2}}$$

Consider now the following situations:

(1) Let v_1 and m_1 be the speed and mass of the Toyota and v_2 and m_2 be the speed and mass of the Buick. If the speed of the Buick is $v_2 = 25$ km/h $= 6.94$ m/s, then the speed of the Toyota would be

$$v_1 = \sqrt{\frac{2\mu_k gx(m_1 + m_2)^2 - m_2^2 v_2^2}{m_1^2}}$$

$$= \sqrt{\frac{2(0.91)(9.8 \text{ m/s}^2)(22 \text{ m})(1200 \text{ kg} + 2200 \text{ kg})^2 - (2200 \text{ kg})^2(6.94 \text{ m/s})^2}{(1200 \text{ kg})^2}}$$

$$= 54.66 \text{ m/s} = 197 \text{ km/h}$$

Thus, we conclude that the speed of the Toyota exceeds 25 km/h.

(2) Here let v_1 and m_1 be the speed and mass of the Buick and v_2 and m_2 be the speed and mass of the Toyota. If the speed of the Toyota is $v_2 = 25$ km/h $= 6.94$ m/s, then the speed of the Buick would be

$$v_1 = \sqrt{\frac{2\mu_k gx(m_1 + m_2)^2 - m_2^2 v_2^2}{m_1^2}}$$

$$= \sqrt{\frac{2(0.91)(9.8 \text{ m/s}^2)(22 \text{ m})(2200 \text{ kg} + 1200 \text{ kg})^2 - (1200 \text{ kg})^2(6.94 \text{ m/s})^2}{(2200 \text{ kg})^2}}$$

$$= 30.4 \text{ m/s} = 109 \text{ km/h}$$

Thus, we conclude that the speed of the Buick exceeds 25 km/h.

ASSESS From the analysis above, we conclude that if one car is going at 25 km/h, then the other one must have been speeding, excessively.

59. **INTERPRET** The problem asks about the angle between the initial velocities before a totally inelastic collision takes place.

DEVELOP The collision between the two masses is totally inelastic. By momentum conservation we have

$$m_1 \vec{v}_1 + m_2 \vec{v}_2 = (m_1 + m_2)\vec{v}_f \quad \rightarrow \quad \vec{v}_f = \frac{m_1 \vec{v}_1 + m_2 \vec{v}_2}{m_1 + m_2}$$

which gives

$$v_f^2 = |\vec{v}_f|^2 = \left(\frac{m_1 \vec{v}_1 + m_2 \vec{v}_2}{m_1 + m_2}\right) \cdot \left(\frac{m_1 \vec{v}_1 + m_2 \vec{v}_2}{m_1 + m_2}\right) = \frac{m_1^2 v_1^2 + m_2^2 v_2^2 + 2m_1 m_2 \vec{v}_1 \cdot \vec{v}_2}{(m_1 + m_2)^2}$$

Using the definition of a scalar product, $\vec{v}_1 \cdot \vec{v}_2 = v_1 v_2 \cos\theta$, the angle between $\vec{v}_1$ and $\vec{v}_2$ can be found.

EVALUATE With $m_1 = m_2 = m$ and $v_1 = v_2 = v = 2v_f$, the above equation can be simplified:

$$\frac{v^2}{4} = \frac{m^2 v^2 + m^2 v^2 + 2m^2 v^2 \cos\theta}{4m^2} = \frac{1}{2}v^2(1 + \cos\theta)$$

Therefore, the angle between the two initial velocities is

$$\theta = \cos^{-1}\left(-\frac{1}{2}\right) = 120°$$

ASSESS To see that the result makes sense, suppose $\vec{v}_1$ makes an angle $-60°$ with $+x$ and $\vec{v}_2$ makes an angle $+60°$ with $+x$. The y component of the total momentum cancels. But for the x component, we have

$$mv\cos(-60°) + mv\cos(60°) = (m + m)v_f$$

Solving for v_f, we get $v_f = v/2$. This confirms the result obtained above.

61. **INTERPRET** The collision in this problem is elastic, so both momentum and energy are conserved. We are asked to find the ratio of the two masses if the one that is initially at rest acquires half of the kinetic energy the other object has after collision.

DEVELOP Momentum is conserved in this process. In this one-dimensional case, we may write

$$m_1 v_{1i} + m_2 v_{2i} = m_1 v_{1f} + m_2 v_{2f}$$

Since the collision is completely elastic, energy is conserved:

$$\frac{1}{2}m_1 v_{1i}^2 + \frac{1}{2}m_2 v_{2i}^2 = \frac{1}{2}m_1 v_{1f}^2 + \frac{1}{2}m_2 v_{2f}^2$$

Using the two conservation equations, the final speeds of m_1 and m_2 are (see Equations 9.15a and 9.15b):

$$v_{1f} = \frac{m_1 - m_2}{m_1 + m_2}v_{1i} + \frac{2m_2}{m_1 + m_2}v_{2i} \qquad v_{2f} = \frac{2m_1}{m_1 + m_2}v_{1i} + \frac{m_2 - m_1}{m_1 + m_2}v_{2i}$$

Given that $v_{2i} = 0$, the above expressions may be simplified to

$$v_{1f} = \frac{m_1 - m_2}{m_1 + m_2}v_{1i} \qquad v_{2f} = \frac{2m_1}{m_1 + m_2}v_{1i}$$

Now, if half the kinetic energy of the first object is transferred to the second, then

$$K_{2f} = \frac{1}{2}K_{1i} \quad \rightarrow \quad \frac{1}{2}m_2\left(\frac{2m_1v_{1i}}{m_1+m_2}\right)^2 = \frac{1}{4}m_1v_{1i}^2$$

EVALUATE The above equation can be further simplified to

$$8m_1m_2 = (m_1+m_2)^2 \quad \rightarrow \quad 8\left(\frac{m_1}{m_2}\right) = \left(\frac{m_1}{m_2}+1\right)^2$$

The resulting quadratic equation, $m_1^2 - 6m_1m_2 + m_2^2 = 0$ has two solutions:

$$m_1 = (3 \pm \sqrt{8})m_2 = 5.83m_2 \quad \text{or} \quad (5.83)^{-1}m_2$$

Since the quadratic equation is symmetric in m_1 and m_2, one solution equals the other with m_1 and m_2 interchanged. Thus, one object is 5.83 times more massive than the other.

ASSESS To check that our answer is correct, let's proceed to calculate the kinetic energy of the particles after collision. Using $m_1 = 5.83m_2$, we find

$$K_{2f} = \frac{1}{2}m_2v_{2f}^2 = \frac{1}{2}m_2\left(\frac{2m_1v_{1i}}{m_1+m_2}\right)^2 = \frac{1}{2}\frac{m_1}{5.83}\left(\frac{2m_1}{m_1+m_1/5.83}\right)^2 v_{1i}^2 = \frac{1}{2}\frac{m_1}{5.83}\left(\frac{2}{1+1/5.83}\right)^2 v_{1i}^2 = \frac{1}{4}m_1v_{1i}^2$$

63. **INTERPRET** We are asked to derive Equation 9.15b, $v_{2f} = \frac{2m_1}{m_1+m_2}v_{1i} + \frac{m_2-m_1}{m_1+m_2}v_{2i}$. We will use conservation of momentum, and since this is an elastic collision, conservation of kinetic energy also applies.

DEVELOP We use conservation of momentum, $m_1\vec{v}_{1i} + m_2\vec{v}_{2i} = m_1\vec{v}_{1f} + m_2\vec{v}_{2f}$. This is an elastic collision, so kinetic energy is also conserved, $\frac{1}{2}m_1v_{1i}^2 + \frac{1}{2}m_2v_{2i}^2 = \frac{1}{2}m_1v_{1f}^2 + \frac{1}{2}m_2v_{2f}^2$. We use these two equations to solve for v_{2f}. Much of this problem is done for us already in Equations 9.12a through 9.14.

EVALUATE We begin by solving 9.14 for v_{1f} : $v_{1f} = v_{2f} + v_{2i} - v_{1i}$. When we substitute this value into 9.12, using the sign of v to denote the direction, we obtain $m_1v_{1i} + m_2v_{2i} = m_1(v_{2f} + v_{2i} - v_{1i}) + m_2v_{2f}$. We solve this for v_{2f}:

$$m_1v_{1i} + m_2v_{2i} = m_1v_{2i} - m_1v_{1i} + (m_1+m_2)v_{2f}$$

$$\rightarrow v_{2f} = v_{1i}\frac{m_1+m_1}{m_1+m_2} + v_{2i}\frac{m_2-m_1}{m_1+m_2}$$

$$\rightarrow v_{2f} = \frac{2m_1}{m_1+m_2}v_{1i} + \frac{m_2-m_1}{m_1+m_2}v_{2i}$$

ASSESS We have derived the required equation.

65. **INTERPRET** The problem is about a moving proton colliding elastically with a stationary deuteron. We want to find the fraction of kinetic energy transferred during the process.

DEVELOP With momenta as shown in the sketch (the deuteron's recoil angle θ_{2f} is negative), the components of the conservation of momentum equations for the elastic collision become

$$m_pv_{1i} = m_pv_{1f}\cos\theta_{1f} + m_dv_{2f}\cos\theta_{2f}$$
$$0 = m_pv_{1f}\sin\theta_{1f} + m_dv_{2f}\sin\theta_{2f}$$

In addition, for energy conservation we have

$$\frac{1}{2}m_pv_{1i}^2 = \frac{1}{2}m_pv_{1f}^2 + \frac{1}{2}m_dv_{2f}^2$$

the fraction of initial kinetic energy transferred to the deuteron is

$$\frac{K_{2f}}{K_{1i}} = 1 - \frac{K_{1f}}{K_{1i}} \quad \rightarrow \quad \frac{m_d}{m_p}\left(\frac{v_{2f}}{v_{1i}}\right)^2 = 1 - \left(\frac{v_{1f}}{v_{1i}}\right)^2$$

EVALUATE With $m_d = 2m_p$, the conservation equations become

$$v_{1i} = v_{1f}\cos\theta_{1f} + 2v_{2f}\cos\theta_{2f}$$
$$0 = v_{1f}\sin\theta_{1f} + 2v_{2f}\sin\theta_{2f}$$

and $v_{1i}^2 = v_{1f}^2 + 2v_{2f}^2$. To find the final velocities, eliminate θ_{2f} from the first and second equations and v_{2f} from the third:

$$v_{1i}^2 - 2v_{1i}v_{1f}\theta_{1f} + v_{1f}^2 = 4v_{2f}^2(\sin^2\theta_{2f} + \cos^2\theta_{2f}) = 4v_{2f}^2 = 2v_{1i}^2 - 2v_{1f}^2$$

This results in a quadratic equation for v_{1f}: $3v_{1f}^2 - 2v_{1f}v_{1i}\cos\theta_{1f} - v_{1i}^2 = 0$, with positive solution

$$v_{1f} = \frac{1}{3}v_{1i}(\cos\theta_{1f} + \sqrt{\cos^2\theta_{1f} + 3}) = 0.902v_{1i}$$

where we have used $\theta_{1f} = 37°$. Now, from the kinetic energy equation, $v_{2f} = \sqrt{\frac{1}{2}(v_{1i}^2 - v_{1f}^2)} = 0.305v_{1i}$, and from the transverse momentum equation,

$$\theta_{2f} = \sin^{-1}\left(\frac{-v_{1f}\sin 37°}{2v_{2f}}\right) = \sin^{-1}\left(\frac{-(0.902v_{1i})\sin 37°}{2(0.305v_{1i})}\right) = -62.7°$$

From either v_{1f} or v_{2f}, the fraction of transferred kinetic energy is found to be

$$\frac{K_{2f}}{K_{1i}} = 1 - \frac{K_{1f}}{K_{1i}} = 1 - \left(\frac{v_{1f}}{v_{1i}}\right)^2 = 1 - (0.902)^2 = 18.6\%$$

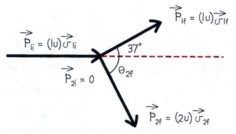

ASSESS   The fraction of energy transfer can also be obtained as

$$\frac{K_{2f}}{K_{1i}} = \frac{m_d}{m_p}\left(\frac{v_{2f}}{v_{1i}}\right)^2 = 2(0.305)^2 = 0.186 = 18.6\%$$

Here one does not need both final velocities to answer this question, but a more complete analysis of this collision, including the deuteron recoil angle, is instructive.

67. **INTERPRET**   In this problem we have an oxygen molecule colliding inelastically with an oxygen atom to form an ozone molecule, and we'd like to find the velocity of the ozone.
     DEVELOP   For this totally inelastic collision, momentum conservation gives

$$m_1\vec{v}_1 + m_2\vec{v}_2 = (m_1 + m_2)\vec{v}_f \quad \rightarrow \quad \vec{v}_f = \frac{m_1\vec{v}_1 + m_2\vec{v}_2}{m_1 + m_2}$$

The velocities of the oxygen molecule (denoted with subscript 1) and oxygen atom (subscript 2) are

$$\vec{v}_1 = (580 \text{ m/s})\hat{i}$$
$$\vec{v}_2 = (870 \text{ m/s})[\cos 27°\hat{i} + \sin 27°\hat{j}] = (775 \text{ m/s})\hat{i} + (395 \text{ m/s})\hat{j}$$

EVALUATE   Substituting the above expressions for $\vec{v}_1$ and $\vec{v}_2$ into the first equation, we obtain

$$\vec{v}_f = \frac{m_1\vec{v}_1 + m_2\vec{v}_2}{m_1 + m_2} = \frac{(32 \text{ u})\vec{v}_1 + (16 \text{ u})\vec{v}_2}{32 \text{ u} + 16 \text{ u}} = \frac{2}{3}\vec{v}_1 + \frac{1}{3}\vec{v}_2$$

$$= \frac{2}{3}(580 \text{ m/s})\hat{i} + \frac{1}{3}[(775 \text{ m/s})\hat{i} + (395 \text{ m/s})\hat{j}]$$

$$= (645 \text{ m/s})\hat{i} + (132 \text{ m/s})\hat{j}$$

ASSESS   The magnitude and direction of $\vec{v}_f$ are $v = 658$ m/s and $\theta_x = 11.5°$. The result is reasonable since we expect the angle to be between 0 and 27°.

69.   **INTERPRET**   This problem is about the height the small ball rebounds after being dropped together with a larger
ball and colliding elastically with it.

DEVELOP   The balls reach the ground, after a vertical fall through a height h, with speed $v_0 = \sqrt{2gh}$. Assume that
they undergo an elastic head-on collision, with the large ball M rebounding from the ground with initial velocity
$v_{2i} = v_0$ (positive upward), and the small ball still falling downward with initial velocity $v_{1i} = -v_0$. Equation 9.15a
gives the final velocity of the small ball as

$$v_f = \left(\frac{m-M}{m+M}\right)(-v_0) + \left(\frac{2M}{m+M}\right)v_0 = \left(\frac{3M-m}{m+M}\right)v_0 \approx 3v_0$$

since $M \gg m$. Once v_f is known, the height it rebounds can be readily calculated by using energy conservation.

EVALUATE   Using $\frac{1}{2}mv_f^2 = mgh_f$, we have

$$h_f = \frac{v_f^2}{2g} = \frac{(3v_0)^2}{2g} = 9\frac{v_0^2}{2g} = 9h$$

or about nine times the original height.

ASSESS   This demonstration, sometimes called a Minski cannon, is striking. Try it with a new tennis ball and
properly inflated basketball.

71.   **INTERPRET**   The block slides down a frictionless incline, colliding with another block and rebounds. We want to
calculate the time elapsed before the two blocks collide again.

DEVELOP   Assuming conservation of energy and a smooth transition from incline to horizontal surface, when the
blocks first collide, the smaller one has speed $v_{1i} = \sqrt{2gh} \equiv v_0$, and the larger one is at rest, $v_{2i} = 0$. Since $m_2 = 4m_1$,
the velocities (positive to the right) of the blocks after their head-on, elastic collision (at $t = 0$) are

$$v_{1f} = \left(\frac{m_1 - m_2}{m_1 + m_2}\right)v_{1i} = \left(\frac{1-4}{1+4}\right)v_0 = -\frac{3}{5}v_0$$

$$v_{2f} = \left(\frac{2m_1}{m_1 + m_2}\right)v_{1i} = \frac{2}{5}v_0$$

The larger block moves with constant speed to the right; its position, relative to the bottom of the incline, is

$$x_2(t) = x_{20} + v_{2f}t = 1.4 \text{ m} + \frac{2}{5}v_0 t$$

The smaller block takes time $t_1 = \frac{1.4 \text{ m}}{3v_0/5}$ to get back to the incline, and $t_2 = 2(3v_0/5)/a$ to go up and down the incline,
where $a = g\sin 30° = g/2$. (Use Equation 2.7, with initial speed $-3v_0/5$ up the incline and final speed $+3v_0/5$ down
the incline, to calculate t_2.) It then proceeds with constant speed in pursuit of the larger block, its position being

$$x_1(t) = \frac{3}{3}v_0(t - t_1 - t_2), \quad \text{for} \quad t \geq t_1 + t_2$$

The blocks collide for the second time when $x_1(t) = x_2(t)$.

EVALUATE   The condition $x_1(t) = x_2(t)$ implies

$$\frac{3}{5}v_0(t - t_1 - t_2) = 1.4 \text{ m} + \frac{2}{5}v_0 t$$

Solving for t, we find

$$t = \frac{5[1.4 \text{ m} + (3v_0/5)(t_1 + t_2)]}{v_0} = (7 \text{ m}/v_0) + 3(t_1 + t_2)$$

Numerically, we have $v_0 = \sqrt{2(9.8 \text{ m/s}^2)(0.25 \text{ m})} = 2.21 \text{ m/s}$. This gives

$$t_1 = \frac{1.4 \text{ m}}{3(2.21 \text{ m/s})/5} = 1.05 \text{ s}$$

$$t_2 = \frac{6v_0}{5a} = \frac{12v_0}{5g} = \frac{12(2.21 \text{ m/s})}{5(9.8 \text{ m/s}^2)} = 0.542 \text{ s}$$

Thus, $t = 7.95$ s.

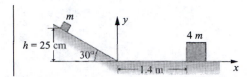

ASSESS   This problem is rather involved. However, the validity of our result can be checked by substituting the numerical values for t_1 and t_2 back to the equations in the intermediate steps.

73.   **INTERPRET**   The problem is about finding the fraction of the initial kinetic energy transferred from one block to the second block in the course of a collision. The fraction is related to the mass ratio.

DEVELOP   With $v_{2i} = 0$, Equations 9.15a and 9.15b become

$$v_{1f} = \left(\frac{m_1 - m_2}{m_1 + m_2}\right)v_{1i} \qquad v_{2f} = \left(\frac{2m_1}{m_1 + m_2}\right)v_{1i}$$

The fraction of the initial kinetic energy transferred to m_2 is

$$\frac{K_{2f}}{K_{1i}} = \frac{\frac{1}{2}m_2 v_{2f}^2}{\frac{1}{2}m_1 v_{1i}^2}$$

EVALUATE   Substituting the expressions for v_{1f} and v_{2f} into the equation above, we obtain

$$\frac{K_{2f}}{K_{1i}} = \frac{\frac{1}{2}m_2 v_{2f}^2}{\frac{1}{2}m_1 v_{1i}^2} = \frac{m_2}{m_1}\left(\frac{2m_1}{m_1 + m_2}\right)^2 = \frac{4m_1 m_2}{(m_1 + m_2)^2} = \frac{4(m_1/m_2)}{(1 + m_1/m_2)^2}$$

Since the result is symmetric in m_1 and m_2, i.e., one can also write $\frac{K_{2f}}{K_{1i}} = \frac{4(m_2/m_1)}{(1+m_2/m_1)^2}$, the same fraction of initial energy is transferred when m_1 is initially at rest.

ASSESS   The fraction of energy transfer reaches a maximum of unity when the mass ratio equals one. This corresponds to the situation where, after a head-on collision, m_1 stops and m_2 moves on with the initial speed of m_1.

75.   **INTERPRET**   We find the center of mass of a slice of pizza, with central angle θ and radius R. We assume constant mass density of the slice. We would expect that it's along the center of the slice, and closer to the crust than to the tip.

DEVELOP   The equation for center of mass is $\vec{r}_{cm} = \frac{1}{M}\int \vec{r}\,dm$. This is used in Example 9.3 to show that the center of mass of a triangle of height L is located at a distance $\frac{2}{3}L$ from the apex. We will use this result to break our slice into triangular sections of height R and width $R\,d\theta$ as shown in the figure below.

The mass density of the slice is $\mu = \frac{M}{A} = \frac{M}{\frac{\theta}{2\pi}(\pi R^2)} = \frac{2M}{\theta R^2}$, and the area of one triangular section is $dA = \frac{1}{2}R(R\,d\phi) =$

$\frac{1}{2}R^2 d\phi$. The distance to the center of mass of one section is $\frac{2}{3}R$, so the distance along horizontal axis is $x = \frac{2}{3}R\cos\phi$. We will integrate these triangular sections from $\phi = -\frac{\theta}{2}$ to $\phi = \frac{\theta}{2}$.

EVALUATE

$$\vec{r}_{cm} = \frac{1}{M}\int \vec{r}\,dm \to x_{cm} = \frac{1}{M}\int x\,dm = \frac{1}{M}\int_{-\frac{\theta}{2}}^{\frac{\theta}{2}}\left(\frac{2}{3}R\cos\phi\right)(\mu\,dA)$$

$$x_{cm} = \frac{1}{M}\int_{-\frac{\theta}{2}}^{\frac{\theta}{2}}\left(\frac{2}{3}R\cos\phi\right)\left(\frac{2M}{\theta R^2}\frac{1}{2}R^2\,d\phi\right) = \frac{2R}{3\theta}\int_{-\frac{\theta}{2}}^{\frac{\theta}{2}}\cos\phi\,d\phi$$

$$x_{cm} = \frac{2R}{3\theta}[\sin\phi]_{-\theta/2}^{\theta/2} = \frac{2R}{3\theta}\left[2\sin\left(\frac{\theta}{2}\right)\right] = \frac{4R}{3\theta}\sin\left(\frac{\theta}{2}\right)$$

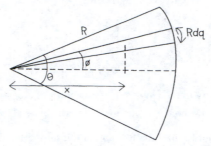

ASSESS   We can check this answer by letting $\theta = 2\pi$: In that case, the center of mass is at $x_{cm} = 0$ as we would expect. If $\theta = \frac{\pi}{4}$ (1/8 of the pizza) then the center of mass is at $x_{cm} = 0.65R$, which matches our original prediction.

77. **INTERPRET**   We use conservation of momentum to find the change in orbital speed of a planet when it's hit by another planet.

DEVELOP   One of the two planets is about the size of Mars: $m_2 = 6.42 \times 10^{23}$ kg. The other has the rest of the mass that currently makes up the Earth-Moon system. The incoming planet had speed $v_2 = 10$ km/s. We do not know the initial speed of the first planet, so we will use it as the reference frame and call its initial speed zero. We will find the mass m_1, then use conservation of momentum $m_1 v_{1i} + m_2 v_{2i} = (m_1 + m_2)v_f$. The final velocity v_f is the change in orbital velocity.

EVALUATE $m_{\text{Earth}} + m_{\text{Moon}} = (5.97 \times 10^{24}$ kg$) + (7.35 \times 10^{22}$ kg$) = 6.04 \times 10^{24}$ kg, so the initial mass of the first planet is $m_1 = (6.04 \times 10^{24}$ kg$) - m_{\text{Mars}} = 5.40 \times 10^{24}$ kg.

$$m_1 \cancel{v_{1i}} + m_2 v_{2i} = (m_1 + m_2)v_f \rightarrow v_f = v_{2i}\left(\frac{m_2}{m_1 + m_2}\right) = 1.0 \text{ km/s}$$

ASSESS   The incoming planet is a little more than 1/10 the mass of the first planet, so the final speed is about 1/10 the initial speed of the incoming planet. This is reasonable.

79. **INTERPRET**   We work as accident reconstructionists in this problem: From the length and direction of skid-marks, we find the initial speed of both vehicles and see if either of them was speeding. We use work and energy to find the speed of the combined vehicles after the collision, and then use conservation of momentum to find the individual speeds of the vehicles before the collision.

DEVELOP   We use work and energy to find the speed of the combined vehicles after the collision: The work done by friction equals the change in kinetic energy. $W = \Delta K$. With this speed and the direction $\theta = 24°$ east of north, we can find the north and east components of momentum. Since momentum is conserved, we then know the momentum of each car, and with the cars' masses, we can find their initial speeds.

The eastbound car has mass $m_1 = 900$ kg, and the northbound pickup has mass $m_2 = 1500$ kg. The coefficient of friction is $\mu = 0.8$, and they slide for a distance $s = 16.33$ m.

EVALUATE   First, we find the speed of the cars after the collision.

$$W = \Delta K \rightarrow -\mu\,\cancel{(m_1 + m_2)}gs = -\frac{1}{2}\cancel{(m_1 + m_2)}v_f^2 \rightarrow v_f = \sqrt{2\mu gs} = 16 \text{ m/s}$$

The components of momentum are then $p_e = p_f \sin\theta = (m_1 + m_2)v_f \sin\theta = 15{,}600$ kg m/s, and $p_n = p_f \cos\theta = (m_1 + m_2)v_f \cos\theta = 35{,}000$ kg m/s.

Momentum is conserved. The initial eastward momentum is that of the car:

$$m_1 v_1 = p_e \rightarrow v_1 = \frac{p_e}{m_1} = 17.4 \text{ m/s} \times \frac{2.24 \text{ mph}}{1 \text{ m/s}} = 38.9 \text{ mph}$$

The initial northward momentum is that of the pickup:

$$m_2 v_2 = p_n \rightarrow v_2 = \frac{p_n}{m_2} = 23.4 \text{ m/s} \times \frac{2.24 \text{ mph}}{1 \text{ m/s}} = 52.4 \text{ mph}$$

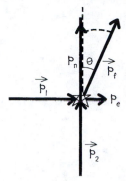

ASSESS   To the correct number of significant figures, the car was moving at 40 mph, and the truck was moving at 50 mph. They were both speeding, which is a reasonable guess in most traffic accidents.

81.   **INTERPRET**   If everyone jumped at once, would there be an earthquake? We find the Earth's recoil motion from this event using center of mass, and compare the energy with that of a magnitude 7.0 earthquake.

DEVELOP   First, we should note that unless we got everyone to the same side of the Earth, there would be no recoil at all since the forces involved would be (roughly) symmetric. Let's go ahead and put everyone in India, though, just for the sake of argument. If all the people are initially at a height $h = 4$ m, and then jump, the center of mass of the system remains in the same place so we can calculate how much the Earth moves. We can also calculate the initial gravitational potential energy of that many people that high up. There are 6.5 billion people, with mass 70 kg each. The radius of the Earth is $r = 6.38 \times 10^6$ m, and its mass is $m = 5.97 \times 10^{24}$ kg.

EVALUATE   The mass of the people is $m = (6.5 \times 10^9) \times (70 \text{ kg}) = 4.55 \times 10^{11}$ kg. The center of mass of the system of Earth and people (all people in the same spot, 4 meters above the surface) is $r_{cm} = \frac{1}{m_E + m}[m(r_E + 4 \text{ m})] = 4.862482 \times 10^{-7}$ m. This is the distance from the center of the Earth and the center of mass *before* the jump. After the jump, $r_{cm} = \frac{1}{m_E + m}[m(r_E)] = 4.862479 \times 10^{-7}$ m. The center of mass moves $x = 3 \times 10^{-13}$ m, which is a fraction of the diameter of a hydrogen atom. The energy of the jump is just the potential energy of the people before they jump: $U = mgh = 17.8 \times 10^{12}$ J. A magnitude 7.0 earthquake has energy 2×10^{15} J, so the jump energy is about 100 times less than this.

ASSESS   The energy of everyone jumping at once would correspond (roughly) to a magnitude 5 earthquake, which would be noticeable but probably not damaging. The Earth would not move any measurable amount.

10 ROTATIONAL MOTION

EXERCISES

Section 10.1 Angular Velocity and Acceleration

15. **INTERPRET** We are asked to compute the linear speed at some location on Earth. The problem involves the rotational motion of the Earth.

DEVELOP We first calculate the angular speed of the Earth using Equation 10.1:

$$\omega_E = \frac{\Delta\theta}{\Delta t} = \frac{1 \text{ rev}}{1 \text{ d}} = \frac{2\pi \text{ rad}}{86,400 \text{ s}} = 7.27 \times 10^{-5} \text{ s}^{-1}$$

The linear speed can then be computed using Equation 10.3: $v = \omega r$.

EVALUATE (a) On the equator,

$$v = \omega_E R_E = (7.27 \times 10^{-5} \text{ s}^{-1})(6.37 \times 10^6 \text{ m}) = 463 \text{ m/s}$$

(b) At latitude θ, $r = R_E \cos\theta$ so $v = \omega_E r = (463 \text{ m/s})\cos\theta$.

ASSESS The angle $\theta = 0$ corresponds to the Equator. So the result found in (b) agrees with (a). In addition, if we take $\theta = 90°$, then we are at the poles, and the linear speed is zero there.

17. **INTERPRET** The problem asks about the linear speed of the straight wood saw, but it's equivalent to finding the linear speed of the circular saw.

DEVELOP We first convert the angular speed to rad/s:

$$\omega = \frac{\Delta\theta}{\Delta t} = \frac{3500 \text{ rev}}{1 \text{ min}} = \frac{2\pi(3500) \text{ rad}}{60 \text{ s}} = 367 \text{ rad/s}$$

The linear speed can then be computed using Equation 10.3: $v = \omega r$.

EVALUATE The radius of the circular saw is $r = 12.5$ cm $= 0.125$ m. Therefore, its linear speed is
$$v = \omega r = (367 \text{ rad/s})(0.125 \text{ m}) = 45.8 \text{ m/s}$$

ASSESS The linear speed of the saw is more than 100 mi/h!

19. **INTERPRET** In this problem we are asked to find the angular speed and number of revolutions of a turbine, given its angular acceleration. The key to this type of rotational problem is to identify the analogous situation for linear motion. The analogies are summarized in Table 10.1.

DEVELOP Given a constant angular acceleration α, the angular velocity and angular position at a later time t can be found using Equations 10.7 and 10.8:

$$\omega = \omega_0 + \alpha t$$

$$\theta = \theta_0 + \omega_0 t + \frac{1}{2}\alpha t^2$$

EVALUATE (a) The initial and final angular velocities are $\omega_0 = 0$ and

$$\omega = 3600 \text{ rpm} = \frac{3600 \text{ rev}}{1 \text{ min}} = \frac{2\pi(3600) \text{ rad}}{60 \text{ s}} = 377 \text{ rad/s}$$

Therefore, the amount of time it takes to reach this angular speed is

$$t = \frac{\omega - \omega_0}{\alpha} = \frac{377 \text{ rad/s} - 0}{0.52 \text{ rad/s}^2} = 725 \text{ s} = 12.1 \text{ min}$$

(b) Using Equation 10.8, we find the number of turns made during this time interval to be

$$\theta = \theta_0 + \omega_0 t + \frac{1}{2}\alpha t^2 = \frac{1}{2}\alpha t^2 = \frac{1}{2}(0.52 \text{ rad/s}^2)(725 \text{ s})^2 = 1.37 \times 10^5 \text{ rad} = 2.17 \times 10^4 \text{ rev}$$

ASSESS   The turbine turns very fast. After 12.1 min, it has reached an angular speed of 377 rad/s, or 60 rev/s!

Section 10.2 Torque

21.   **INTERPRET**   In this problem we are asked to find the torque produced by the frictional force about the wheel's axis.

   DEVELOP   The torque produced by a force is given by Equation 10.10:

$$\tau = rF\sin\theta$$

   where $r_\perp = r\sin\theta$ is the perpendicular distance between the rotation axis and the line of action of the force F. The frictional force acts tangent to the circumference of the wheel and hence perpendicular ($\theta = 90°$) to the radius at the point of contact.

   EVALUATE   Using Equation 10.10, we find the torque to be

$$\tau = rf = (0.5 \text{ m})(320 \text{ N}) = 160 \text{ N} \cdot \text{m}$$

   opposite to the direction of rotation.

   ASSESS   Since frictional force opposes motion, the torque it produces tends to slow down the rotation.

23.   **INTERPRET**   In this problem we want to find the force required to produce a specific torque.

   DEVELOP   The torque produced by a force is given by Equation 10.10:

$$\tau = rF\sin\theta = r_\perp F = rF_\perp$$

   where $r_\perp = r\sin\theta$ is the perpendicular distance (lever arm) between the rotation axis and the line of action of the force F. Alternatively, one can think of $F_\perp = F\sin\theta$ as the effective force.

   EVALUATE   Using Equation 10.10, the magnitude of the applied force may be obtained as $F = \frac{\tau}{r\sin\theta}$. The forces are

   (a)

$$F = \frac{\tau}{r\sin\theta} = \frac{35.0 \text{ N} \cdot \text{m}}{(0.24 \text{ m}) \sin 90°} = 146 \text{ N}$$

   and **(b)** $F = \frac{\tau}{r\sin\theta} = \frac{35.0 \text{ N·m}}{(0.24 \text{ m}) \sin 110°} = 155 \text{ N}$.

   ASSESS   To produce a specific torque most effectively, the applied force should be at the right angle to $\vec{r}$, the position vector from the axis of rotation to the point where the force is applied. This would yield an effective force $F_\perp = F\sin\theta = F$.

25.   **INTERPRET**   We find the torque on a wheel, given that it has a slight off-center mass at a given radius and angle to the horizontal. We use the definition of torque, $\tau = rF\sin\theta$.

   DEVELOP   We start with a sketch of the system, shown in the figure below. From this, we see that the angle between the force and the position vector $\vec{R}$ is $\theta = 90° - \phi$. The valve stem is located a distance $R = 0.32$ m from the center, and has mass $m = 0.025$ kg.

   EVALUATE $\tau = rF\sin\theta = R(mg)\cos\phi = 0.072 \text{ Nm}$

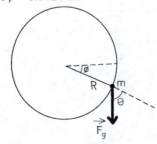

   ASSESS   We see from our equation that the torque would be zero if the angle ϕ were 90°.

Section 10.3 Rotional Inertia and the Analog of Newton's Law

27.   **INTERPRET**   We want to find the moment of inertia of a shaft that has a shape of a solid cylinder.

   DEVELOP   The rotational inertia of a solid cylinder or disk about its axis is $I = \frac{1}{2}MR^2$ (see Table 10.2).

EVALUATE   The rotational inertia of the shaft is

$$I = \frac{1}{2}MR^2 = \frac{1}{2}(6.8\times10^3 \text{ kg})(0.85 \text{ m/2})^2 = 6.14\times10^2 \text{ kg}\cdot\text{m}^2$$

ASSESS   The numerical value is reasonable, given its mass and radius, and the units $(\text{kg}\cdot\text{m}^2)$ are correct.

29. **INTERPRET**   In this problem we are asked to find the minimum mass of a wheel, given its diameter and rotational inertia.

    DEVELOP   Any part of the wheel has a distance from the center less than or equal to the maximum radius. Therefore, using Equation 10.12, we obtain the following inequality:

    $$I = \sum m_i r_i^2 \le \left(\sum m_i\right) r_{max}^2$$

    EVALUATE   (a) The above equation implies that

    $$M = \sum m_i \ge \frac{I}{r_{max}^2}$$

    Therefore, the minimum mass is

    $$M_{min} = \frac{I}{r_{max}^2} = \frac{7.8 \text{ kg}\cdot\text{m}^2}{(0.92 \text{ m/2})^2} = 36.9 \text{ kg}$$

    (b) If not all the mass of the wheel is concentrated at the rim, the total mass is greater than this minimum.

    ASSESS   To have the same rotational inertia I, we can have some of the mass of wheel concentrated near the axis of rotation. Their contribution to I would be small since $I = \sum m_i r_i^2$.

31. **INTERPRET**   By assuming Earth to be a solid sphere with uniform mass distribution, we want to estimate its rotational inertia, and the torque needed to change the length of the day by one second every century.

    DEVELOP   From Table 10.2, the rotational inertia of a solid sphere of radius R and mass M is

    $$I = \frac{2}{5}MR^2$$

    Once I is known, the torque needed to slow down the rotation can be found by using Equation 10.11: $\tau = I\alpha$.

    EVALUATE   (a) For a uniform solid sphere, and an axis through the center,

    $$I_E = \frac{2}{5}M_E R_E^2 = \frac{2}{5}(5.97\times10^{24} \text{ kg})(6.37\times10^6 \text{ m})^2 = 9.69\times10^{37} \text{ kg}\cdot\text{m}^2$$

    (b) The angular speed of rotation of Earth is $\omega = \frac{2\pi}{T}$, where the period is $T = 1 \text{ d} = 86,400$ s. If the period were to change by 1s per century, i.e.,

    $$\frac{dT}{dt} = \frac{1 \text{ s}}{100\times3.16\times10^7 \text{ s}} = 3.16\times10^{-10}$$

    this would correspond to an angular acceleration of

    $$\alpha = \frac{d\omega}{dt} = \frac{d}{dt}\left(\frac{2\pi}{T}\right) = -\frac{2\pi}{T^2}\frac{dT}{dt}$$

    Therefore, a torque of magnitude

    $$|\tau| = I|\alpha| = -\frac{2\pi I}{T^2}\frac{dT}{dt} = \frac{2\pi(9.69\times10^{37} \text{ kg}\cdot\text{m}^2)}{(86,400 \text{ s})^2}(3.16\times10^{-10}) = 2.58\times10^{19} \text{ N}\cdot\text{m}$$

    would be required.

    ASSESS   The torque in (b) is actually generated by tidal friction between the Moon and the Earth. Note that the Earth has a core of denser material, so its actual rotational inertia is less than that obtained in (a).

33. **INTERPRET**   We are asked about the rotational inertia of a Frisbee, given its mass distribution, and the torque required to generate the rotation.

DEVELOP   The rotational inertia (about a perpendicular axis through the center) of the Frisbee is the sum of that for a disk $(I_d = \frac{1}{2} M_d R^2)$ and a ring $(I_r = M_r R^2)$:

$$I = I_d + I_r = \frac{1}{2} M_d R^2 + M_r R^2$$

Next, using Equation 10.11, $\tau = I\alpha$, the torque needed to rotate the Frisbee can be calculated.

EVALUATE   Since half of the mass is spread uniformly in a disk, and the other half concentrated in the rim, we have $M_d = M_r = \frac{1}{2}(108 \text{ g}) = 54$ g. Thus, the rotational inertia is

$$I = \frac{1}{2} M_d R^2 + M_r R^2 = \frac{1}{2}(M_d + 2M_r)R^2 = \frac{1}{2}[0.054 \text{ kg} + 2(0.054 \text{ kg})](0.12 \text{ m})^2$$
$$= 1.17 \times 10^{-3} \text{ kg} \cdot \text{m}^2$$

(b) Using Equation 10.9, the angular acceleration is found to be

$$\alpha = \frac{\omega^2 - \omega_0^2}{2(\theta - \theta_0)} = \frac{(550 \text{ rpm})^2 - 0}{2(0.25 \text{ rev})} = \frac{(57.6 \text{ rad/s})^2}{\pi \text{ rad}} = 1056 \text{ rad/s}^2$$

Thus, the torque required is

$$\tau = I\alpha = (1.17 \times 10^{-3} \text{ kg} \cdot \text{m}^2)(1056 \text{ rad/s}^2) = 1.24 \text{ N} \cdot \text{m}$$

ASSESS   The numerical values all look reasonable, and the units are correct: α in rad/s^2, and torque τ in N $\cdot$ m.

Section 10.4 Rotional Energy

35.   **INTERPRET**   The problem asks about the rotational kinetic energy of a rotating circular saw. In addition, we also want to find the power required for the saw to start from rest and reach a given angular speed.

DEVELOP   The rotational kinetic energy of the saw can be found by using Equation 10.18:

$$K = \frac{1}{2} I \omega^2$$

where $I = \frac{1}{2} MR^2$ for a disk. The required power is simply equal to the rate of the work done on the saw.

EVALUATE   **(a)** With $\omega = 3500 \text{ rpm} = \frac{2\pi(3500) \text{ rad}}{60 \text{ s}} = 367$ rad/s, the rotational kinetic energy is

$$K = \frac{1}{2} I \omega^2 = \frac{1}{2}\left(\frac{1}{2} MR^2\right)\omega^2 = \frac{1}{4}(0.85 \text{ kg})(0.125 \text{ m})^2(367 \text{ rad/s})^2 = 446 \text{ J}$$

(b) The power required is $P_{av} = \frac{W}{\Delta t} = \frac{\Delta K}{\Delta t} = \frac{446 \text{ J}}{3.2 \text{ s}} = 139$ W.

ASSESS   To check the answer, we present an alternative approach to computing K. In this problem, the angular acceleration is

$$\alpha = \frac{\omega - \omega_0}{t} = \frac{367 \text{ rad/s}}{3.2} = 115 \text{ rad/s}^2$$

and the angular displacement is (Using Equation 10.9)

$$\theta = \theta_0 + \frac{\omega^2 - \omega_0^2}{2\alpha} = \frac{(367 \text{ rad/s})^2}{2(115 \text{ rad/s}^2)} = 586 \text{ rad}$$

With these quantities, the rotational energy may be calculated as

$$K = W = \tau\theta = I\alpha\theta = \frac{1}{2} MR^2 \alpha\theta = \frac{1}{2}(0.85 \text{ kg})(0.125 \text{ m})^2(115 \text{ rad/s}^2)(586 \text{ rad}) = 446 \text{ J}$$

which is same as before.

37.   **INTERPRET**   The kinetic energy of the baseball consists of two parts: the kinetic energy of the center of mass, K_{cm}, and the rotational kinetic energy, K_{rot}. We want to find the fraction of total kinetic energy that's K_{rot}.

DEVELOP   The total kinetic energy has center-of-mass energy and internal rotational energy associated with spin about the center of mass (see Equation 10.20):

$$K_{tot} = K_{cm} + K_{rot} = \frac{1}{2} Mv^2 + \frac{1}{2} I_{cm} \omega^2$$

EVALUATE   For a solid sphere, $I_{cm} = 2MR^2/5$. Therefore, the fraction of the total kinetic energy that's rotational is

$$\frac{K_{rot}}{K_{tot}} = \frac{I_{cm}\omega^2}{Mv^2 + I_{cm}\omega^2} = \frac{(2MR^2/5)\omega^2}{Mv^2 + (2MR^2/5)\omega^2} = \frac{2R^2\omega^2}{5v^2 + 2R^2\omega^2}$$

$$= \frac{2(0.037 \text{ m/s})^2(42 \text{ rad/s})^2}{5(33 \text{ m/s})^2 + 2(0.037 \text{ m/s})^2(42 \text{ rad/s})^2}$$

$$= 8.86 \times 10^{-4} = 0.0886\%$$

ASSESS   Rotational kinetic energy constitutes a very small fraction of the total kinetic energy. This is reasonable because the linear speed at a point on the surface of the baseball due to rotation is only $\omega R = (42 \text{ rad/s})(0.037 \text{ m}) = 1.55 \text{ m/s}$.

39.   **INTERPRET**   We need to find the energy stored in a massive flywheel. We are given the size and mass of the flywheel, so we can calculate the rotational inertia. Given the speed, we can calculate the kinetic energy. We also need to find the power output of a generator if the speed of the flywheel changes a given amount in a certain time.
DEVELOP   The energy stored in the flywheel is $K = \frac{1}{2}I\omega^2$. We will need to convert the angular speed in rpm to rad/s, and calculate the rotational inertia of the flywheel disk using $I = \frac{1}{2}mR^2$. The power output in the second part is $P = \frac{\Delta K}{t}$.
The mass of the flywheel is $m = 7.7 \times 10^4$ kg, the radius is $R = 2.4$ m, and the initial rotation rate is 360 rpm. In the second part, the rotation rate goes to 300 rpm in $t = 3$ s.
EVALUATE
(a) $I = \frac{1}{2}mR^2 = 222 \times 10^3$ kg m².

$$\omega = 360\frac{\text{rotation}}{\text{min}} \times \frac{1 \text{ min}}{60 \text{ s}} \times \frac{2\pi \text{ radians}}{1 \text{ rotation}} = 37.7 \text{ rad/s}$$

$$K = \frac{1}{2}I\omega^2 = 1.58 \times 10^8 \text{ J}$$

(b) $\omega_2 = 300$ rpm $= 31.4$ rad/s $\to K_2 = 1.09 \times 10^8$ J

$$\Delta K = 4.82 \times 10^7 \text{ J} \to P = \frac{\Delta K}{t} = 16.1 \text{ MW}$$

ASSESS   This is a good way of generating enormous power pulses.

Section 10.5 Rolling Motion

41.   **INTERPRET**   What fraction of a disk's kinetic energy is rotational if it's rolling? We use the relationship between ω and v.
DEVELOP   We use the total kinetic energy $K = \frac{1}{2}Mv^2 + \frac{1}{2}I\omega^2$, with $v = \omega R$ and $I = \frac{1}{2}MR^2$. The rotational fraction is $f = \frac{K_{rot}}{K}$.
EVALUATE $$f = \frac{\frac{1}{2}I\omega^2}{\frac{1}{2}Mv^2 + \frac{1}{2}I\omega^2} = \frac{(\frac{1}{2}MR^2)\omega^2}{M(\omega R)^2 + (\frac{1}{2}MR^2)\omega^2} = \frac{\frac{1}{2}}{1 + \frac{1}{2}} = \frac{1}{3}$$

ASSESS   This is consistent with what we noted in the previous problem: the rotational inertia is $(\frac{1}{2})MR^2$, so the rotational kinetic energy is $(\frac{1}{2})$ the translational kinetic energy when it rolls without slipping.

PROBLEMS

43.   **INTERPRET**   The problem is about the rotational motion of the wheel. The key to this type of rotational problem is to identify the analogous situation for linear motion. The analogies are summarized in Table 10.1.
DEVELOP   For constant angular acceleration, the final angular speed is given by Equation 10.9:

$$\omega^2 = \omega_0^2 + 2\alpha(\theta - \theta_0)$$

Similarly, to find the time it takes for the wheel to make 2 turns, we solve Equation 10.8:

$$\theta = \theta_0 + \omega_0 t + \frac{1}{2}at^2$$

EVALUATE   (a) The angular acceleration is $\alpha = 18$ rpm/s $= 1080$ rev/min². Using Equation 10.7, we find the final angular speed to be

$$\omega = \sqrt{\omega_0^2 + 2\alpha(\theta - \theta_0)} = \sqrt{0 + 2(1080 \text{ rev/min}^2)(2 \text{ rev})} = 65.7 \text{ rpm}$$

(b) Equation 10.8 gives $\theta - \theta_0 = 0 + \frac{1}{2}\alpha t^2$ or

$$t = \sqrt{\frac{2\theta}{\alpha}} = \sqrt{\frac{2(2\text{ rev})}{1080\text{ rev/min}^2}} = 0.061\text{ min} = 3.65\text{ s}$$

ASSESS   Another way to answer **(b)** is to use Equation 10.8:

$$\omega = \omega_0 + \alpha t \quad \rightarrow \quad t = \frac{\omega - \omega_0}{\alpha} = \frac{65.7\text{ rpm}}{1080\text{ rev/min}^2} = 0.061\text{ min} = 3.65\text{ s}$$

45.   **INTERPRET**   The problem is about the angular velocity (rotation rate) of the CD, with the linear speed specified at some point on the CD.

DEVELOP   Equation 10.3 gives the relation between linear speed and angular speed,

$$\omega = \frac{v}{r}$$

where r is the distance from the center of rotation. With $v = 130$ cm/s being a constant, the angular speed can be found once r is specified.

EVALUATE   **(a)** With $r = 6.0$ cm, we have

$$\omega = \frac{v}{r} = \frac{130\text{ cm/s}}{6.0\text{ cm}} = 21.7\ \text{rad/s} = 207\text{ rpm}$$

for a point on the CD's outer edge.

(b) With $r = 3.75$ cm, we have

$$\omega = \frac{v}{r} = \frac{130\text{ cm/s}}{3.75\text{ cm}} = 34.7\ \text{rad/s} = 331\text{ rpm}$$

ASSESS   Note that radians are a dimensionless angular measure, i.e., pure numbers; therefore angular speed can be expressed in units of inverse seconds.

47.   **INTERPRET**   We are asked to find the angular deceleration of the machinery, given its initial and final angular speeds and the angular displacement.

DEVELOP   The three quantities, the initial and final angular speeds and the angular displacement, are related by Equation 10.9:

$$\omega^2 = \omega_0^2 + 2\alpha(\theta - \theta_0)$$

This is the equation we shall use to solve for a.

EVALUATE   Using Equation 10.9, we find the angular deceleration to be

$$\alpha = \frac{\omega^2 - \omega_0^2}{2(\theta - \theta_0)} = \frac{(440\text{ rpm})^2 - (680\text{ rpm})^2}{2(180\text{ rev})} = -747\text{ rev/min}^2 = -1.30\text{ rad/s}^2$$

ASSESS   A negative acceleration is a positive deceleration. The amount of time it takes for this deceleration is

$$\omega = \omega_0 + \alpha t \quad \rightarrow \quad t = \frac{\omega - \omega_0}{\alpha} = \frac{440\text{ rpm} - 680\text{ rpm}}{-747\text{ rev/min}^2} = 0.32\text{ min} = 19\text{ s}$$

During this time interval, the angular displacement is

$$\theta = \theta_0 + \omega_0 t + \frac{1}{2}at^2 = 0 + (680\text{ rpm})(0.32\text{ min}) + \frac{1}{2}(-747\text{rpm/min}^2)(0.32\text{ min})^2$$
$$= 180\text{ rev}$$

This is indeed the value given in the problem statement.

49.   **INTERPRET**   This problem is about the angular motion of the pulley. We want to find the applied torque that keeps the pulley from rotating.

DEVELOP   If the pulley and string are not moving (no rotation and no slipping) the net torque on the pulley is zero, and the tensions in the string on either side are equal to the weights tied on either end. The tensions are perpendicular to the radii to each point of application. Therefore, the torques due to the tensions have a magnitude

$$TR = mgR,$$

but are in opposite directions. Thus, the condition for no rotation is

$$0 = \tau_{app} - m_1 gR + m_2 gR$$

where we chose positive torques in the direction of the applied torque, which is opposite to the torque produced by the greater mass.

EVALUATE   Using the values given in the problem statement, we find

$$\tau_{app} = (m_1 - m_2)gR = (0.47 \text{ kg} - 0.22 \text{ kg})(9.8 \text{ m/s}^2)(0.06 \text{ m}) = 0.147 \text{ N} \cdot \text{m}$$

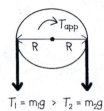

$$T_1 = m_1 g \ > \ T_2 = m_2 g$$

ASSESS   The torque τ_{app} equals the torque which would be produced by balancing the pulley with 250 g added to the side with lesser mass.

51.   **INTERPRET**   We are asked to find the rotational inertia of a thick ring with inner and outer radii R_1 and R_2. The mass distribution is continuous, so we need to do an integral.

DEVELOP   For a thick ring, the ring-shaped mass elements used in Example 10.7 have mass

$$dm = \sigma dA = \frac{M}{\pi(R_2^2 - R_1^2)} 2\pi r dr$$

where $\sigma = M/A$ is the mass density (units: kg/m^2). Note that the ring only extends in radius from R_1 to R_2. The rotational inertia can then be obtained by integrating over

$$I = \int_{R_1}^{R_2} r^2 dm$$

EVALUATE   Upon carrying out the integration, the rotational inertia about an axis perpendicular to the ring and through its center is

$$I = \int_{R_1}^{R_2} r^2 dm = M \int_{R_1}^{R_2} \frac{2\pi r^3 dr}{\pi(R_2^2 - R_1^2)} = \frac{M(R_2^4 - R_1^4)}{2(R_2^2 - R_1^2)} = \frac{M}{2}(R_1^2 + R_2^2)$$

ASSESS   To see that the result makes sense, let's consider the following limits: **(i)** $R_1 \to 0$. In this case, we have a disk with radius R_2 and $I = \frac{1}{2}MR_2^2$. **(ii)** $R_1 \to R_2$. In this limit, we have a thin ring with $I = MR_2^2$.

53.   **INTERPRET**   We are asked about the rotational inertia of a propeller, treating it as a uniform thin rod. In addition, we want to find the time it takes to change its angular speed, given the torque.

DEVELOP   The rotational inertia of a thin rod of length L is (from Table 10.2)

$$I = \frac{1}{3}ML^2$$

where the axis of the rotation passes through one of the endpoints of the rod. For part **(b)**, we note that the average torque is related to the average rate of change of angular speed (angular acceleration) as

$$\bar{\tau} = I\bar{\alpha} = I\frac{\Delta\omega}{\Delta t}$$

EVALUATE   **(a)** The rotational inertia of one blade is $\frac{1}{3}ML^2$ (see Table 10.2). The propeller has three such blades, so

$$I = 3\left(\frac{1}{3}ML^2\right) = ML^2 = (10 \text{ kg})(1.25 \text{ m})^2 = 15.6 \text{ kg} \cdot \text{m}^2$$

(b) The engine torque is the only one considered. Therefore, with $\omega_0 = 1400 \text{ rpm} = 147 \text{ rad/s}$ and $\omega = 1900 \text{ rpm} = 199 \text{ rad/s}$, we have

$$\Delta t = \frac{I\Delta\omega}{\bar{\tau}} = \frac{(15.6 \text{ kg} \cdot \text{m}^2)(199 \text{ rad/s} - 147 \text{ rad/s})}{2700 \text{ N} \cdot \text{m}} = 0.303 \text{ s}$$

ASSESS   Our result shows that Δt is inversely proportional to the applied torque. Increasing the torque reduces the time required to speed up the propeller.

55.   **INTERPRET**   We are asked to find the time it takes for the space station to start from rest and reach a certain angular speed, with a given thrust.

DEVELOP   Suppose that any difference in the distance to the axis from the center to the rockets, or any part of the ring, is negligible compared to the radius of the ring, $R = 11$ m. The angular acceleration (about the axis perpendicular to the ring and through its center) of the ring is

$$\alpha = \frac{\tau}{I} = \frac{2FR}{MR^2} = \frac{2F}{MR}$$

Starting from rest ($\omega_0 = 0$), it would take time $t = \frac{\omega}{\alpha}$ for the ring to reach a final angular speed ω. The angular displacement during this time interval may be obtained by using Equation 10.8:

$$\theta = \theta_0 + \omega_0 t + \frac{1}{2}at^2$$

EVALUATE   Since $g = \omega^2 R$ specifies the desired rate of rotation, we find

$$t = \frac{\omega}{\alpha} = \sqrt{\frac{g}{R}}\left(\frac{MR}{2F}\right) = \sqrt{\frac{9.8 \text{ m/s}^2}{11 \text{ m}}}\frac{(5\times10^5 \text{ kg})(11 \text{ m})}{2(100 \text{ N})} = 2.60\times10^4 \text{ s} = 7.21 \text{ h}$$

(b) From either Equation 10.8, we find the number of revolutions to be

$$\theta - \theta_0 = \frac{1}{2}\alpha t^2 = \frac{1}{2}\omega t = \frac{1}{2}\sqrt{\frac{g}{R}}\,t = \frac{1}{2}\sqrt{\frac{9.8 \text{ m/s}^2}{11 \text{ m}}}(2.60\times10^4 \text{ s}) = 1.23\times10^4 \text{ rad} = 1953 \text{ rev}$$

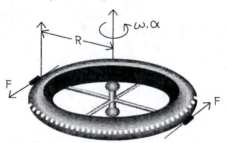

ASSESS   Since $t \sim 1/F$, a larger thrust will enable the space station to reach the desirable spin rate more rapidly. On the other hand, larger mass M and larger R will give a greater rotational inertia, and hence more time would be required to attain the desired spin rate.

57.   **INTERPRET**   We are asked to find the coefficient of friction between block and slope, given the acceleration of the block. The rotational motion of the solid drum that is connected to the block by a string must be considered.

DEVELOP   The equations of motion for the block and drum are

$$mg\sin\theta - \mu_k n - T = ma$$
$$n - mg\cos\theta = 0$$
$$TR = I\alpha$$

where $I = \frac{1}{2}MR^2$, and $a = \alpha R$ (non-slip string). These equations allow us to determine μ_k.

EVALUATE   The drum's equation gives the tension,

$$T = \frac{I\alpha}{R} = \frac{(MR^2/2)(a/R)}{R} = \frac{1}{2}Ma$$

so that μ_k can be found from the block's equations,

$$\mu_k = \frac{mg\sin\theta - ma - Ma/2}{mg\cos\theta} = \frac{(2.4 \text{ kg})(9.8 \text{ m/s}^2)\sin30° - (2.4 \text{ kg} + 0.425 \text{ kg})(1.6 \text{ m/s}^2)}{(2.4 \text{ kg})(9.8 \text{ m/s}^2)\cos30°} = 0.36$$

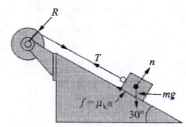

ASSESS   To see that our expression for μ_k makes sense, let's check some limits: (**i**) If $a = 0$, then $\mu_k = \tan\theta$. This is precisely the equation we obtained in Chapter 5 (see Example 5.10). (**ii**) $M = 0$ and $\mu_k = 0$. The situation corresponds to a block of mass m sliding down a frictionless slope with acceleration $a = g\sin\theta$.

59.   **INTERPRET**   In this problem we want to find the angular speed of the potter's wheel after some force has been exerted. The force produces a torque that causes the wheel to rotate.

DEVELOP   The work-energy theorem for constant torque (Equation 10.19) gives

$$W = \tau\,\Delta\theta = \Delta K = \frac{1}{2}I\omega^2$$

since the wheel starts from rest. The equation allows us to determine the angular velocity ω.

EVALUATE   Since the force acting on the wheel is circumferential, the resulting torque is $\tau = FR$. In addition, the rotational inertia of a disk is $I = \frac{1}{2}MR^2$, and $\Delta\theta = \frac{1}{8}$ rev $= \frac{\pi}{4}$ rad. Thus, we have

$$\omega^2 = \frac{2\tau\Delta\theta}{I} = \frac{2FR\Delta\theta}{MR^2/2} = \frac{4F\Delta\theta}{MR}$$

or

$$\omega = \sqrt{\frac{4F\Delta\theta}{MR}} = \sqrt{\frac{4(75\text{ N})(\pi\text{ rad}/4)}{(120\text{ kg})(0.45\text{ m})}} = 2.09\text{ rad/s}$$

ASSESS   The greater the force exerted on the wheel, the larger the angular speed. On the other hand, larger M and R result in a larger rotational inertia, and smaller ω if the same force is applied.

61.   **INTERPRET**   The problem is about the rotational motion of a hollow basketball rolling down an incline. We want to know its speed after it has traveled a certain distance.

DEVELOP   The ball rolls through a vertical height $h = L\sin\theta$, where L is the distance along the incline, which makes an angle θ with the horizontal. For a hollow ball, the rotational inertia is $I_{cm} = \frac{2}{3}MR^2$ (see Table 10.2). Next, conservation of mechanical energy, applied to the hollow basketball starting from rest and rolling without slipping down an incline, yields

$$Mgh = \frac{1}{2}I_{cm}\omega^2 + \frac{1}{2}Mv_{cm}^2$$

where $v_{cm} = \omega R$. The equation allows us to determine v_{cm}.

EVALUATE   Upon substituting $I_{cm} = \frac{2}{3}MR^2$, we obtain

$$Mgh = \frac{1}{2}\left(\frac{2}{3}MR^2\right)\left(\frac{v_{cm}}{R}\right)^2 + \frac{1}{2}Mv_{cm}^2 = \frac{5}{6}Mv_{cm}^2$$

or

$$v_{cm} = \sqrt{\frac{6gh}{5}} = \sqrt{\frac{6gL\sin\theta}{5}} = \sqrt{\frac{6(9.8\text{ m/s})(8.4\text{ m})\sin 30°}{5}} = 7.03\text{ m/s}$$

ASSESS   In general, for an object having a rotational inertia of $I_{cm} = cMR^2$, one can go through a similar derivation and show that the center-of-mass velocity is given by

$$v_{cm} = \sqrt{\frac{2gh}{1+c}}$$

If we ignore the rolling motion of the basketball, the velocity then becomes $v_{cm} = \sqrt{2gh}$, which is what we expect from considering only the linear motion.

63.   INTERPRET   The kinetic energy of the wheel consists of two parts: the kinetic energy of the center of mass, K_{cm}, and the rotational kinetic energy, K_{rot}. We want to find out how changing the moment of inertia and mass of the wheel affects the total kinetic energy.

DEVELOP   The total kinetic energy of the wheel has center-of-mass energy and internal rotational energy associated with spin about the center of mass (see Equation 10.20):

$$K_{tot} = K_{cm} + K_{rot} = \frac{1}{2}Mv_{cm}^2 + \frac{1}{2}I_{cm}\omega^2$$

With the condition for rolling without slipping, $v = \omega R$, the total kinetic energy can be rewritten as

$$K_{tot} = K_{cm} + K_{rot} = \frac{1}{2}Mv_{cm}^2 + \frac{1}{2}I_{cm}\left(\frac{v_{cm}}{R}\right)^2 = \frac{1}{2}Mv_{cm}^2\left(1 + \frac{I_{cm}}{MR^2}\right)$$

The initial condition is $\frac{I_{cm}}{MR^2} = 0.40 = 40\%$. After the redesign,

$$\frac{I'_{cm}}{M'R^2} = \frac{0.9I_{cm}}{(0.8M)R^2} = 1.125\frac{I_{cm}}{MR^2} = (1.125)(0.40) = 0.45$$

EVALUATE   The fractional decrease in kinetic energy is

$$\frac{K - K'}{K} = 1 - \frac{K'}{K} = 1 - \frac{M'(1 + I'_{cm}/M'R^2)}{M(1 + I_{cm}/MR^2)} = 1 - \frac{0.8M}{M}\frac{(1 + 0.45)}{(1 + 0.40)} = 0.171 = 17.1\%$$

ASSESS   Initially, K_{cm} accounts for $\frac{1}{1.4} = 71.4\%$ of the total kinetic energy, while K_{rot} account for the remaining $\frac{0.4}{1.4} = 28.6\%$. After the redesign, $M \rightarrow M' = 0.8M$, the translational kinetic energy decreases by 20%, while the rotational kinetic energy goes down by 10% ($I_{cm} \rightarrow I'_{cm} = 0.9I_{cm}$). Therefore, the total kinetic energy is now

$$(0.8)\frac{1}{1.4} + (0.9)\frac{0.4}{1.4} = 0.829 = 82.9\%$$

of the original. This is a 17.1% decrease.

65.   INTERPRET   In this problem we want to find the new rotational inertia of a circular disk after a hole has been drilled off its center.

DEVELOP   Equation 10.17 shows that the rotational inertia of an object is the sum of the rotational inertias of the pieces, so

$$I_{disk} = I_{hole} + I_{remainder}$$

The hint expresses this fact as $I_{remainder} = I_{disk} - I_{hole}$. Here I_{disk} is the rotational inertia of the whole disk about an axis perpendicular to the disk and through its center, which is $\frac{1}{2}MR^2$ (see Example 10.7). On the other hand, I_{hole} is the rotational inertia of the material removed to form the hole, which the parallel axis theorem gives as

$$I_{hole} = M_{hole}h^2 + I_{cm}$$

Here, $h = R/4$ is the distance of the hole's CM from the axis of the disk, and $I_{cm} = \frac{1}{2}M_{hole}(R/4)^2$ is the rotational inertia of the hole material about a parallel axis through its CM. With these equations, we can readily determine $I_{remainder}$.

EVALUATE   Since the planar mass density of the disk (assumed to be uniform) is $\sigma = M/\pi R^2$, the mass of the hole material is

$$M_{hole} = \sigma A_{hole} = \frac{M}{\pi R^2}\pi(R/4)^2 = \frac{M}{16}$$

Therefore, the rotational inertia of the hole is

$$I_{hole} = M_{hole}h^2 + I_{cm} = \frac{M}{16}\left(\frac{R}{4}\right)^2 + \frac{1}{2}\frac{M}{16}\left(\frac{R}{4}\right)^2 = \frac{3}{512}MR^2$$

and

$$I_{remainder} = I_{disk} - I_{hole} = \frac{1}{2}MR^2 - \frac{3}{512}MR^2 = \frac{253}{512}MR^2 = 0.494MR^2$$

ASSESS   If the hole drilled were concentric with the disk, we would have ($h = 0$)

$$I'_{\text{hole}} = I_{\text{cm}} = \frac{1}{2}\frac{M}{16}\left(\frac{R}{4}\right)^2 = \frac{1}{512}MR^2$$

and

$$I'_{\text{remainder}} = I_{\text{disk}} - I'_{\text{hole}} = \frac{1}{2}MR^2 - \frac{1}{512}MR^2 = \frac{255}{512}MR^2 = 0.498MR^2$$

The same result is obtained if we use the formula $\frac{1}{2}M'(R_1^2 + R_2^2)$ derived in Problem 51, with
$M' = \pi R^2 - \pi(R/4)^2 = (15/16)\pi R^2 = (15/16)M$, $R_1 = R$ and $R_2 = R/4$.

67.   **INTERPRET**   In this problem we want to find out how high up the hill the motorist can go with the total mechanical energy the system has while traveling on the flat road.

DEVELOP   If possible losses are neglected, the mechanical energy of the motorcycle and rider is conserved as it coasts uphill, so the total kinetic energy at the bottom equals the gain in potential energy at the highest point,

$$K_{\text{trans}} + K_{\text{rot}} = M_{\text{tot}}gh$$

The translation kinetic energy of the cycle and rider (including the wheels) and the rotational kinetic energy of the wheels (about their CM) are, assuming rolling without slipping,

$$K_{\text{trans}} = \frac{1}{2}M_{\text{tot}}v^2, \qquad K_{\text{rot}} = 2\left(\frac{1}{2}I\omega^2\right) = I\left(\frac{v}{R}\right)^2$$

These expressions can be combined to solve for h.

EVALUATE   Substituting the second equation into the first, and using $v = 85$ km/h $= 23.6$ m/s, we find the maximum vertical height reached to be

$$h = \frac{v^2}{2g}\left(1 + \frac{2I}{M_{\text{tot}}R^2}\right) = \frac{(23.6 \text{ m/s})^2}{2(9.8 \text{ m/s}^2)}\left(1 + \frac{2(2.1 \text{ kg}\cdot\text{m}^2)}{(395 \text{ kg})(0.26 \text{ m})^2}\right) = 32.9 \text{ m}$$

ASSESS   If the rolling motion is ignored, the result would be $h = \frac{v^2}{2g}$, which is what we expect from considering only the linear motion.

69.   **INTERPRET**   In this problem we are given a disk with non-uniform mass density, and asked to find its total mass and rotational inertia.

DEVELOP   As mass elements, choose thin rings of width dr and radius r (as in Example 10.7) so that

$$dm = \rho(r)dV = \left(\frac{\rho_0 r}{R}\right)2\pi rw\, dr = \frac{2\pi\rho_0 w}{R}r^2 dr$$

The total mass is $M = \int_0^R dm$ and the rotational inertia about the disk axis is $I = \int_0^R r^2 dm$.

EVALUATE   **(a)** The disk's total mass is

$$M = \int_0^R dm = \frac{2\pi\rho_0 w}{R}\int_0^R r^2 dr = \frac{2\pi\rho_0 wR^2}{3}$$

(b) The disk's rotational inertia about a perpendicular axis through its center is

$$I = \int_0^R r^2 dm = \frac{2\pi\rho_0 w}{R}\int_0^R r^4 dr = \frac{2\pi\rho_0 wR^4}{5} = \frac{3}{5}\left(\frac{2\pi\rho_0 wR^2}{3}\right)R^2 = \frac{3}{5}MR^2$$

ASSESS   Our result for I is intermediary between a disk of uniform density and a ring, i.e., $\frac{1}{2}MR^2 < I < MR^2$, if expressed in terms of the total mass M, but is less than a disk of uniform density ρ_0, i.e., $I < \frac{1}{2}\rho_0\pi R^4 w$, since ρ_0 is the maximum density.

71.   **INTERPRET**   We are asked to show that the rotational inertia of a planar object around an axis perpendicular to the plane of the object is equal to the sum of the rotational inertias around two perpendicular axes within the plane of the object. We will use the integral form of rotational inertia, since this is a continuous distribution.

DEVELOP   The rotational inertia is $I = \int r^2 dm$. We will set up our coordinate system such that the two rotational axes within the plane of the object are the x and y coordinate axes.

EVALUATE   The rotational inertia around the x axis is $I_x = \int r^2 dm = \int y^2 dm$. The rotational inertia around the y axis is $I_y = \int r^2 dm = \int x^2 dm$. The sum of the two is $I_x + I_y = \int x^2 dm + \int y^2 dm = \int (x^2 + y^2) dm$, but $(x^2 + y^2)$ is just the distance r from the perpendicular z axis, so $I_x + I_y = \int r^2 dm = I_z$.

ASSESS   We have proven what was requested.

73.   **INTERPRET**   We find the rotational inertia of a right cylindrical cone, using the integral form for I.

DEVELOP   We divide the cone into circular slices, and by integrating over all these slices we find the total rotational inertia of the cone. The height of the cone is h and the base radius is R, so the radius of each slice is $r = \frac{R}{h}x$, where x is the distance from the apex. The volume of the cone is $V = \frac{1}{3}Ah$, where A is the area of the base, $A = \pi R^2$, so $V = \frac{\pi R^2 h}{3}$. The volume of each disk-shaped slice is $dV = \pi r^2 dx$. The cone has uniform mass density $\frac{M}{V}$, so each disk has mass $dm = \frac{M}{V}dV = \frac{3M}{\pi R^2 h}(\pi r^2 dx)$. The rotational inertia of each disk is $dI = \frac{1}{2}r^2 dm$.

EVALUATE

$$I = \int_0^h dI = \frac{1}{2}\int_0^h r^2 dm = \frac{1}{2}\int_0^h \left(\frac{Rx}{h}\right)^2 \frac{3M}{\pi R^2 h} \pi r^2 dx = \frac{3M}{2h^3}\int_0^h x^2 \left(\frac{Rx}{h}\right)^2 dx$$

$$I = \frac{3MR^2}{2h^5}\int_0^h x^4 dx = \frac{3MR^2}{2h^5}\left[\frac{1}{5}h^5\right] = \frac{3}{10}MR^2$$

ASSESS   The units are correct. The value of I is less than that of a cylinder, since more of the mass is concentrated along the axis of the cone.

75.   **INTERPRET**   We are designing a space station, and need to know the thrust of the rockets that spin the station up to speed. We know the mass and geometry of the station, so we can find the rotational inertia. We know the desired centripetal acceleration, so we can find the angular speed necessary. We know the time it takes to get up to speed, so we can use the equations for constant angular acceleration to find the force necessary.

DEVELOP   The rotational inertia is $I = MR^2$, the mass is $M = 6 \times 10^5$ kg, and the radius is $R = 50$ m. We can use $a_c = \frac{v^2}{R} = R\omega^2$ to find the desired angular speed. The rotational equivalent of Newton's second law is $\tau = I\alpha$, which we can use to find the angular acceleration α. We also use the angular acceleration in Equation 10.7, $\omega = \omega_0 + \alpha t$. The torque is provided by 4 rockets at the rim of the station, each with thrust F, so $\tau = 4RF$. Put all these pieces together to find F necessary to get to speed ω in time $t = 5$ hours $= 18 \times 10^3$ s.

EVALUATE   First, find the angular speed needed: $R\omega^2 = g \rightarrow \omega = \sqrt{\frac{g}{R}}$. Next, find the angular acceleration needed: $\omega = \omega_0 + \alpha t \rightarrow \alpha = \frac{\omega}{t} = \sqrt{\frac{g}{Rt^2}}$. Newton's second law tells us that $\tau = I\alpha \rightarrow \alpha = \frac{\tau}{I} = \frac{4RF}{MR^2} = \frac{4F}{MR}$. Equate these two expressions for α, and solve for F: $\frac{4F}{MR} = \sqrt{\frac{g}{Rt^2}} \rightarrow F = \frac{M}{4}\sqrt{\frac{Rg}{t^2}} = 0.23$ N.

ASSESS   The units in our final equation are $\text{N} = \text{kg}\sqrt{\frac{\text{m m/s}^2}{\text{s}^2}} = \text{kg}\sqrt{\frac{\text{m}^2}{\text{s}^4}} = \text{kg m/s}^2$, which is correct. This seems like a small force, but it is acting at a great distance, the time needed is long, and the final speed is not great.

77.   **INTERPRET**   Find a force from a torque and a distance. Use the definition of torque, and since no angular information is given, assume that the force is applied at 90°.

DEVELOP $\tau = RF$. We solve this for F.

EVALUATE $F = \frac{\tau}{R} = \frac{10.1 \text{ kN m}}{0.95 \text{ m}} = 10.6$ kN.

ASSESS   Since the distance is approximately 1 meter, the torque and the force have nearly the same numeric value.

11 ROTATIONAL VECTORS AND ANGULAR MOMENTUM

EXERCISES

Section 11.1 Angular Velocity and Acceleration Vectors

13. **INTERPRET** The problem asks about the angular acceleration of the wheels as the car traveling north with a speed of 70 km/h makes a 90° left turn that lasts for 25 s.

DEVELOP The speed of the car is $v_{cm} = 70$ km/h $= 19.4$ m/s. Assuming that the wheels are rolling without slipping, the magnitude of the initial angular velocity is

$$\omega = \frac{v_{cm}}{r} = \frac{19.4 \text{ m/s}}{0.31 \text{ m}} = 62.7 \text{ rad/s}$$

With the car going north, the axis of rotation of the wheels is east-west. Since the top of a wheel is going in the same direction as the car, the right-hand rule gives the direction of $\vec{\omega}_i$ as west. In unit-vector notation, we write $\vec{\omega}_i = -\omega \hat{i}$.

After making a left turn, the angular speed remains unchanged, but the direction of $\vec{\omega}_f$ is now south (see sketch). In unit-vector notation, we write $\vec{\omega}_f = -\omega \hat{j}$.

EVALUATE Using Equation 10.4, we find the angular acceleration to be

$$\vec{\alpha}_{av} = \frac{\Delta \vec{\omega}}{\Delta t} = \frac{\vec{\omega}_f - \vec{\omega}_i}{\Delta t} = \frac{-\omega \hat{j} - (-\omega \hat{i})}{\Delta t} = \frac{\omega}{\Delta t}(\hat{i} - \hat{j})$$

The magnitude of $\vec{\alpha}_{av}$ is

$$|\vec{\alpha}_{av}| = \frac{\sqrt{2}\omega}{\Delta t} = \frac{\sqrt{2}(62.7 \text{ rad/s})}{25 \text{ s}} = 3.55 \text{ rad/s}^2$$

and $\vec{\alpha}_{av}$ points in the south-east direction (in the direction of the vector $\hat{i} - \hat{j}$).

ASSESS Angular acceleration $\vec{\alpha}_{av}$ points in the same direction as $\Delta \vec{\omega}$.

15. **INTERPRET** The problem asks about the angular velocity of the wheels after an angular acceleration has been applied within a time interval.

DEVELOP Take the x-axis east and the y-axis north, with positive angles measured CCW from the x-axis. In unit-vector notation, the initial angular velocity $\vec{\omega}_i$ and the angular acceleration $\vec{\alpha}$ can be expressed as

$$\vec{\omega}_i = \omega_i \hat{i} = (140 \text{ rad/s})\hat{i}$$
$$\vec{\alpha} = \alpha(\cos\theta_\alpha \hat{i} + \sin\theta_\alpha \hat{j}) = (35 \text{ rad/s}^2)[\cos(90° + 68°)\hat{i} + \sin(90° + 68°)\hat{j}]$$
$$= (-32.45 \text{ rad/s}^2)\hat{i} + (13.11 \text{ rad/s}^2)\hat{j}$$

The final angular velocity can be found by using Equation 10.8.

EVALUATE   Using Equation 10.8, the angular velocity at $t = 5.0$ s is

$$\vec{\omega}_f = \vec{\omega}_i + \vec{\alpha}t = (140 \text{ rad/s})\hat{i} + [(-32.45 \text{ rad/s}^2)\hat{i} + (13.11 \text{ rad/s}^2)\hat{j}](5.0 \text{ s})$$
$$= (-22.3 \text{ rad/s})\hat{i} + (65.6 \text{ rad/s})\hat{j}$$

The magnitude and direction of $\vec{\omega}_f$ are

$$\omega_f = |\vec{\omega}_f| = \sqrt{(-22.3 \text{ rad/s})^2 + (65.6 \text{ rad/s})^2} = 69.2 \text{ rad/s}$$

and

$$\theta_f = \tan^{-1}\left(\frac{\omega_{f,y}}{\omega_{f,x}}\right) = \tan^{-1}\left(\frac{-22.3 \text{ rad/s}}{65.6 \text{ rad/s}}\right) = -18.8°$$

or 18.8° west of north.

ASSESS   Since the x-component of the angular acceleration is negative, $\Delta\omega_x$ is also negative. On the other hand, a positive α_y yields $\Delta\omega_y > 0$.

Section 11.2 Torque and the Vector Cross Product

17.   INTERPRET   In this problem we want to find the direction of the torque produced by an applied force. Finding the direction involves taking a cross product.

DEVELOP   The torque vector is defined as $\vec{\tau} = \vec{r} \times \vec{F}$, where $\vec{F}$ is the force vector and $\vec{r}$ is the position vector which points from the axis of rotation to the point where the force is acting. The direction of $\vec{\tau}$ is determined by the right-hand-rule.

EVALUATE   The displacement from the origin to the point of application of the force is $\vec{r} = \hat{i} + \hat{j}$ (in distance units), so the direction of the torque is the direction of the cross product $\vec{r} \times \hat{n} = (\hat{i} + \hat{j}) \times \hat{n}$ where we write $\vec{F} = F\hat{n}$ with $\hat{n}$ being a unit vector in the direction of the force.

(a) For $\hat{n} = \hat{i}$, the direction of the torque is $(\hat{i} + \hat{j}) \times \hat{i} = \hat{i} \times \hat{i} + \hat{j} \times \hat{i} = 0 - \hat{k} = -\hat{k}$.

(b) For $\hat{n} = \hat{j}$, the direction of $\vec{\tau}$ is $(\hat{i} + \hat{j}) \times \hat{j} = \hat{i} \times \hat{j} + \hat{j} \times \hat{j} = \hat{k} - 0 = \hat{k}$.

(c) For $\hat{n} = \hat{k}$, the direction of $\vec{\tau}$ is $(\hat{i} + \hat{j}) \times \hat{k} = \hat{i} \times \hat{k} + \hat{j} \times \hat{k} = -\hat{j} + \hat{i}$, which makes an angle of −45° (or 315°) CCW from the x-axis.

ASSESS   The torque vector $\vec{\tau}$ is perpendicular to both $\vec{r}$ and $\vec{F}$. It points in the direction normal to the plane formed by $\vec{r}$ and $\vec{F}$.

19.   INTERPRET   We are asked to find the torque produced by an applied force. The problem is about taking a cross product.

DEVELOP   The torque vector is defined as $\vec{\tau} = \vec{r} \times \vec{F}$, where $\vec{F}$ is the force vector and $\vec{r}$ is the position vector which points from the axis of rotation to the point where the force is acting. The direction of $\vec{\tau}$ is determined by the right-hand-rule.

EVALUATE   (a) Here we have $\vec{r} = (3 \text{ m})\hat{i}$. Therefore, with $\vec{F} = (1.3 \text{ N})\hat{i} + (2.7 \text{ N})\hat{j}$, the torque is

$$\vec{\tau} = \vec{r} \times \vec{F} = (3 \text{ m})\hat{i} \times [(1.3 \text{ N})\hat{i} + (2.7 \text{ N})\hat{j}] = (8.1 \text{ N} \cdot \text{m})\hat{k}$$

(b) Here we have $\vec{r} = (3 \text{ m})\hat{i} - [(-1.3 \text{ m})\hat{i} + (2.4 \text{ m})\hat{j}] = (4.3 \text{ m})\hat{i} - (2.4 \text{ m})\hat{j}$. Therefore, the torque is

$$\vec{\tau} = \vec{r} \times \vec{F} = [(4.3 \text{ m})\hat{i} - (2.4 \text{ m})\hat{j}] \times [(1.3 \text{ N})\hat{i} + (2.7 \text{ N})\hat{j}] = (11.6 \text{ N} \cdot \text{m})\hat{k} + (3.1 \text{ N} \cdot \text{m})\hat{k}$$
$$= (14.7 \text{ N} \cdot \text{m})\hat{k}$$

ASSESS   The torque vector $\vec{\tau}$ is perpendicular to both $\vec{r}$ and $\vec{F}$. It points in the direction normal to the plane formed by $\vec{r}$ and $\vec{F}$.

Section 11.3 Angular Momentum

21.   INTERPRET   We are given the motion of the ball and asked to find the corresponding angular momentum.

DEVELOP   The angular momentum of an object about a point is defined as (see Equation 11.3)

$$L = \vec{r} \times \vec{p}$$

where $\vec{p}$ is the linear momentum and $\vec{r}$ is the position vector of the object relative to that point. We may also express $\vec{L}$ as

$$L = \vec{r} \times \vec{p} = rp \sin\theta \hat{n}$$

where θ is the angle between $\vec{r}$ and $\vec{p}$, and $\hat{n}$ is a unit vector perpendicular to both $\vec{r}$ and $\vec{p}$.

For our problem, we assume the ball is traveling in a circle of radius r and speed v. Since the velocity of the ball, $\vec{v}$, is perpendicular to $\vec{r}$, the magnitude of the angular momentum about the center is $L = |\vec{r} \times \vec{p}| = rp = rmv$.

EVALUATE   From the problem statement, we have $r = 1.2$ m $+ 0.9$ m $= 2.1$ m and $v = 27$ m/s. Therefore,

$$L = rmv = (2.1 \text{ m})(7.3 \text{ kg})(27 \text{ m/s}) = 4.14 \text{ J·s}$$

ASSESS   The direction of $\vec{L}$ is along the axis of rotation. It is perpendicular to both $\vec{v}$ and $\vec{r}$.

23. **INTERPRET**   We are given the rotational inertia and angular velocity of the hoop and asked to find the corresponding angular momentum.

    DEVELOP   For an object rotating about a fixed axis, its angular momentum can be expressed as (see Equation 11.4)

    $$\vec{L} = I\vec{\omega}$$

    where I is the moment of inertia of the object, and $\vec{\omega}$, its angular velocity about the axis. This is the equation we shall use to calculate $\vec{L}$.

    EVALUATE   In this problem, the rotational inertia of a hoop about its central axis (perpendicular to the plane of the hoop) is $I = mr^2$. Therefore, with $\omega = 170$ rpm $= 17.8$ rad/s, the magnitude of $\vec{L}$ is

    $$L = I\omega = mr^2\omega = (0.64 \text{ kg})(0.45 \text{ m})^2(17.8 \text{ rad/s}) = 2.31 \text{ J·s}$$

    The direction of $\vec{L}$ is along the axis of rotation according to the right-hand rule.

    ASSESS   The angular momentum vector $\vec{L}$ points in the direction of $\vec{\omega}$.

Section 11.4 Conservation of Angular Momentum

25. **INTERPRET**   We want to find the angular speed of a spinning wheel after a piece of clay is dropped onto it and sticks to its surface.

    DEVELOP   If the clay is dropped vertically onto a horizontally spinning wheel, the angular momentum about the vertical spin axis is conserved. Conservation of angular momentum is expressed as

    $$L_i = L_f \quad \rightarrow \quad I_i\omega_i = I_f\omega_f$$

    EVALUATE   Given that $I_i = I_{\text{wheel}}$ and $I_f = I_{\text{wheel}} + m_{\text{clay}}r^2$, the final angular velocity is

    $$\omega_f = \frac{I_i}{I_f}\omega_i = \left(\frac{I_{\text{wheel}}}{I_{\text{wheel}} + m_{\text{clay}}r^2}\right)\omega_i = \frac{6.40 \text{ kg·m}^2}{6.40 \text{ kg·m}^2 + (2.7 \text{ kg})(0.46 \text{ m})^2}(19 \text{ rpm}) = 17.4 \text{ rpm}$$

    ASSESS   The clay increases the total rotational inertia of the system, and therefore, the angular speed is decreased, as required by angular momentum conservation.

27. **INTERPRET**   In this problem we are asked about the period of a star formed by a collapsing cloud.

    DEVELOP   If we assume there are no external torques and no mass loss during the collapse of the star-forming cloud, its angular momentum is conserved.

    $$L_i = L_f \quad \rightarrow \quad I_i\omega_i = I_f\omega_f$$

    The equation allows us to solve for ω_f, or $T_f = 2\pi/\omega_f$.

    EVALUATE   For a uniform sphere of constant mass, $I_i\omega_i = I_f\omega_f$ implies that

    $$\frac{2}{5}MR_i^2\omega_i = \frac{2}{5}MR_f^2\omega_f \quad \rightarrow \quad \frac{\omega_i}{\omega_f} = \left(\frac{R_f}{R_i}\right)^2$$

    Since the angular velocity is one rotation divided by one period, $\omega = 2\pi/T$, and

    $$T_f = T_i\left(\frac{R_f}{R_i}\right)^2 = (1.4 \times 10^6 \text{ y})\left(\frac{7 \times 10^8 \text{ m}}{10^{13} \text{ m}}\right)^2 = 6.86 \times 10^{-3} \text{ y} = 2.5 \text{ days}$$

    ASSESS   In current models of star formation, the collapsing cloud does not maintain a spherical shape, forming a flattened disk instead, and the central star retains just a fraction of the original cloud's mass.

29.   **INTERPRET**   We find the angular speed of a merry-go-round after a boy jumps onto it. Angular momentum is conserved. We will solve for ω_2, the final speed of the merry-go-round.

DEVELOP   The initial angular momentum of the merry-go-round is $\vec{L}_1 = I\vec{\omega}_1$, where we are told that $I = 240$ kg m^2 and $\omega_1 = 11$ rpm. A boy with mass $m_b = 28$ kg leaps onto the merry-go-round. His initial speed is directed towards the axis of the merry-go-round, so he has no angular momentum initially. We will use $\vec{L}_1 = \vec{L}_2$, where L_2 includes the boy rotating with the merry-go-round at a radius of $R = 1.3$ m. Since the direction of $\vec{\omega}$ does not change, we will neglect the vector nature of $\vec{L}$.

EVALUATE $L_1 = L_2 \rightarrow I\omega_1 = (I + m_b R^2)\omega_2 \rightarrow \omega_2 = \omega_1 \dfrac{I}{I + m_b R^2} = (11 \text{ rpm})(0.835) = 9.2 \text{ rpm}$

ASSESS   We did not convert the angular speed to "normal" rad/s units, but as long as we keep track of what the units are, we're okay.

PROBLEMS

31.   **INTERPRET**   The problem asks about the direction of a vector $\vec{B}$, given the directions of another vector $\vec{A}$ and their cross product $\vec{A} \times \vec{B}$.

DEVELOP   The cross product, $\vec{A} \times \vec{B}$, is perpendicular to the plane of $\vec{A}$ and $\vec{B}$. Since according to the problem statement, $\vec{A} \times \vec{B} = -A^2\hat{k}$, these vectors lie in the x-y plane. For simplicity, let's write the two vectors as

$$\vec{A} = A(\cos\theta_A \hat{i} + \sin\theta_A \hat{j})$$
$$\vec{B} = B(\cos\theta_B \hat{i} + \sin\theta_B \hat{j})$$

where $\theta_A = 30°$ and θ_B are measured counterclockwise from the x axis. Using the above expressions, the cross product $\vec{A} \times \vec{B}$ is

$$\vec{A} \times \vec{B} = AB(\cos\theta_A \hat{i} + \sin\theta_A \hat{j}) \times (\cos\theta_B \hat{i} + \sin\theta_B \hat{j}) = AB(\cos\theta_A \sin\theta_B - \sin\theta_A \cos\theta_B)\hat{k}$$
$$= -AB\sin(\theta_A - \theta_B)\hat{k}$$

Using the information given in the problem statement, the angle θ_B can be calculated.

EVALUATE   The problem states that $\vec{A} \times \vec{B} = -A^2\hat{k}$. The right-hand rule implies that the angle between $\vec{A}$ and $\vec{B}$, measured clockwise from $\vec{A}$, is less than 180°, i.e., $\theta_A - \theta_B < 180°$, or $-150° < \theta_B < 30° = \theta_A$. The magnitude of $\vec{A} \times \vec{B}$ is $AB\sin(\theta_A - \theta_B) = 2A^2\sin(\theta_A - \theta_B) = A^2$ (as given, with $B = 2A$), so

$$\sin(\theta_A - \theta_B) = \frac{1}{2}$$

or $\theta_A - \theta_B = 30°$ or $150°$. When this is combined with the given value of θ_A and the range of θ_B, one finds that $\theta_B = 0°$ or $-120°$ (i.e., along the x-axis or 120° clockwise from the x-axis).

ASSESS   The vector corresponding to $\theta_B = 0°$ can be written as $\vec{B}_1 = B\hat{i} = 2A\hat{i}$. Similarly, for $\theta_B = -120°$, we have

$$\vec{B}_2 = B[\cos(-120°)\hat{i} + \sin(-120°)\hat{j}] = 2A[(-1/2)\hat{i} - (\sqrt{3}/2)\hat{j}] = -A\hat{i} - \sqrt{3}A\hat{j}$$

With $\vec{A} = A[\cos 30°\hat{i} + \sin 30°\hat{j}] = A[(\sqrt{3}/2)\hat{i} + (1/2)\hat{j}]$, the cross products are

$$\vec{A} \times \vec{B}_1 = A[(\sqrt{3}/2)\hat{i} + (1/2)\hat{j}] \times (2A)\hat{i} = -A^2\hat{k}$$

and

$$\vec{A} \times \vec{B}_2 = A[(\sqrt{3}/2)\hat{i} + (1/2)\hat{j}] \times [-A\hat{i} - \sqrt{3}A\hat{j}] = \left(-\frac{3}{2}A^2 + \frac{1}{2}A^2\right)\hat{k} = -A^2\hat{k}$$

Both results indeed agree with the condition given in the problem statement.

33.   **INTERPRET**   In this problem we are asked to verify a vector identity: $\vec{A} \cdot (\vec{A} \times \vec{B}) = 0$.

DEVELOP   The key to the proof is to realize that the cross product $\vec{A} \times \vec{B}$ is perpendicular to $\vec{A}$ and $\vec{B}$.

EVALUATE   Let $\vec{C} = \vec{A} \times \vec{B}$. If $\vec{C}$ is perpendicular to $\vec{A}$ and $\vec{B}$, then their scalar products must vanish:

$$\vec{A} \cdot \vec{C} = \vec{A} \cdot (\vec{A} \times \vec{B}) = 0$$
$$\vec{B} \cdot \vec{C} = \vec{B} \cdot (\vec{A} \times \vec{B}) = 0$$

(Recall that $\vec{A} \cdot \vec{B} = AB\cos\theta$, where θ is the angle between $\vec{A}$ and $\vec{B}$.)

ASSESS An alternative approach is to use the component forms. Let's write the vectors as

$$\vec{A} = A_x\hat{i} + A_y\hat{j} + A_z\hat{k}, \quad \vec{B} = B_x\hat{i} + B_y\hat{j} + B_z\hat{k}$$

The cross product $\vec{A} \times \vec{B}$ is

$$\vec{A} \times \vec{B} = (A_x\hat{i} + A_y\hat{j} + A_z\hat{k})(B_x\hat{i} + B_y\hat{j} + B_z\hat{k})$$
$$= A_xB_x\hat{i}\times\hat{i} + A_xB_y\hat{i}\times\hat{j} + A_xB_z\hat{i}\times\hat{k} + A_yB_x\hat{j}\times\hat{i} + A_yB_y\hat{j}\times\hat{j}$$
$$+ A_yB_z\hat{j}\times\hat{k} + A_zB_x\hat{k}\times\hat{i} + A_zB_y\hat{k}\times\hat{j} + A_zB_z\hat{k}\times\hat{k}$$
$$= (A_yB_z - A_zB_y)\hat{i} + (A_zB_x - A_xB_z)\hat{j} + (A_xB_y - A_yB_x)\hat{k}$$

The dot product $\vec{A} \cdot (\vec{A} \times \vec{B})$ then becomes

$$\vec{A} \cdot (\vec{A} \times \vec{B}) = A_x(A_yB_z - A_zB_y) + A_y(A_zB_x - A_xB_z) + A_z(A_xB_y - A_yB_x)$$
$$= (A_yA_z - A_zA_y)B_x + (A_zA_x - A_xA_z)B_y + (A_xA_y - A_yA_x)B_z$$
$$= (\vec{A} \times \vec{A}) \cdot \vec{B} = 0$$

In general, $\vec{A} \cdot (\vec{B} \times \vec{C})$ is called the triple scalar product and $\vec{A} \cdot (\vec{B} \times \vec{C}) = (\vec{A} \times \vec{B}) \cdot \vec{C}$, i.e., the "dot" and the "cross" in the triple scalar product can be interchanged. This is equivalent to a cyclic permutation of the three vectors,

$$\vec{A} \cdot (\vec{B} \times \vec{C}) = \vec{C} \cdot (\vec{A} \times \vec{B}) = \vec{B} \cdot (\vec{C} \times \vec{A})$$

On the other hand, interchanging any two vectors introduces a minus sign,

$$\vec{A} \cdot (\vec{B} \times \vec{C}) = -\vec{C} \cdot (\vec{B} \times \vec{A}) = -\vec{B} \cdot (\vec{A} \times \vec{C}) = -\vec{A} \cdot (\vec{C} \times \vec{B})$$

35. **INTERPRET** The problem is about finding the angular momentum of the weightlifter's barbell, given its rotational motion.

DEVELOP We use Equation 11.4, $\vec{L} = I\vec{\omega}$, to compute the angular momentum. The rotational inertia of the weights and bar about the specified axis is (see Table 10.2)

$$I = 2m_{wt}\left(\frac{L}{2}\right)^2 + \frac{1}{12}m_{bar}L^2$$

EVALUATE With $\omega = 10$ rpm $= 1.05$ rad/s, the angular momentum about this axis is

$$L = I\omega = \left[2(25 \text{ kg})(0.8 \text{ m})^2 + \frac{1}{12}(15 \text{ kg})(1.6 \text{ m})^2\right](1.05 \text{ rad/s}) = 36.9 \text{ J}\cdot\text{s}$$

ASSESS The greater the angular speed, the larger the angular momentum.

37. **INTERPRET** Our system consists of two identical cars moving in the opposite direction on a highway. We want to find the angular momentum of the system about a point on the centerline of the highway.

DEVELOP The position of a particle with respect to some point can always be expressed in terms of components perpendicular and parallel to its direction of motion (i.e., momentum p). Thus, Equation 11.3 can be written as

$$\vec{L} = \vec{r} \times \vec{p} = (\vec{r}_\perp + \vec{r}_\parallel) \times \vec{p} = \vec{r}_\perp \times \vec{p}$$

since $\vec{r}_\parallel \times \vec{p} = 0$. For straight line motion, $\vec{r}_\perp$ is a constant. For the two cars which we regard as particles located at their respective centers of mass, we have $\vec{p}_1 = -\vec{p}_2$ and $\vec{r}_{1\perp} = -\vec{r}_{2\perp}$ (for any point on the center line), so the total angular momentum of their centers of mass is

$$\vec{L} = \vec{L}_1 + \vec{L}_2 = \vec{r}_{1\perp} \times \vec{p}_1 + (-\vec{r}_{1\perp})(-\vec{p}_1) = 2\vec{r}_{1\perp} \times \vec{p}_1$$

EVALUATE From the problem statement, the magnitude of $\vec{L}$ is (with $v_1 = 90$ km/h $= 25$ m/s).

$$L = 2r_{1\perp}mv_1 = 2(3 \text{ m})(1800 \text{ kg})(25 \text{ m/s}) = 2.70 \times 10^5 \text{ J}\cdot\text{s}$$

and its direction is out of the plane of Fig. 13.30.

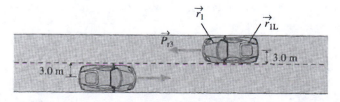

ASSESS In general, the magnitude of the particle's angular momentum about point P, for any position along its trajectory, is $L = |\vec{r} \times \vec{p}| = rp\sin\theta = mvb$, where mv is the magnitude of the particle's linear momentum and $b = r\sin\theta$ is the perpendicular distance from point P to the line of motion (called the impact parameter in a scattering experiment).

39. **INTERPRET** The problem asks about the new rotational inertia of a tire if the design reduces the angular momentum by a certain percentage, while keeping the linear speed fixed.

    DEVELOP The linear speed of the car is related to its angular speed as $v = \omega r$ (see Equation 10.3). Keeping v fixed implies

    $$\omega_1 R_1 = \omega_2 R_2$$

    Using the fact that the angular momentum is related to rotational inertia as $L = I\omega$, the new rotational inertia can be computed.

    EVALUATE The new specifications require that

    $$\frac{L_2}{L_1} = \frac{I_2\omega_2}{I_1\omega_1} = 0.7 \quad \rightarrow \quad \frac{I_2}{I_1} = 0.7\frac{\omega_1}{\omega_2}$$

    Using $\omega_1 = \omega_2 R_2/R_1$, we obtain

    $$I_2 = 0.7I_1\frac{\omega_1}{\omega_2} = 0.7I_1\frac{R_2}{R_1} = (0.7)(0.32 \text{ kg} \cdot \text{m}^2)\left(\frac{35 \text{ cm}}{38 \text{ cm}}\right) = 0.206 \text{ kg} \cdot \text{m}^2$$

    ASSESS The general condition is

    $$\frac{L_2}{L_1} = \frac{I_2\omega_2}{I_1\omega_1} = \frac{I_2 R_1}{I_1 R_2} \quad \rightarrow \quad L_2 = \left(\frac{I_2}{I_1}\right)\left(\frac{R_1}{R_2}\right)L_1$$

    A decrease in angular momentum ($L_2 < L_1$) can be achieved by either decreasing R_1/R_2 or I_2/I_1. In our problem, the ratio $R_1/R_2 = 38 \text{ cm}/35 \text{ cm} = 1.09$ actually is increased. However, this change is accompanied by a greater decrease of the rotational inertia $I_2/I_1 = 0.206/0.32 = 0.64$.

41. **INTERPRET** This problem is about the relative motion with respect to the ground, of a dog walking around a turntable. The key concept here involves conservation of angular momentum.

    DEVELOP Walking once around relative to the turntable, the dog describes an angular displacement of $\Delta\theta_D$ relative to the ground, and the turntable one of $\Delta\theta_T$ in the opposite direction, such that $\Delta\theta_D + |\Delta\theta_T| = 2\pi$. We suppose that the vertical component of the angular momentum of the dog and turntable is conserved (which was zero initially), so that

    $$L_i = L_f \quad \rightarrow \quad 0 = I_D\omega_D + I_T\omega_T = I_D\left(\frac{\Delta\theta_D}{\Delta t}\right) - I_T\left|\frac{\Delta\theta_T}{\Delta t}\right|$$

    where the angular velocities (which are in opposite directions) have been rewritten in terms of the angular displacements and the common time interval. The equation allows us to solve for $\Delta\theta_D$.

    EVALUATE The rotational inertias about the axis of rotation are

    $$I_D = mR^2 = (17 \text{ kg})(1.81 \text{ m})^2 = 55.7 \text{ kg} \cdot \text{m}^2$$

    and $I_T = 95 \text{ kg} \cdot \text{m}^2$. Eliminating $\Delta\theta_T$, we find

    $$0 = I_D\Delta\theta_D - I_T(2\pi - \Delta\theta_D) = (I_D + I_T)\Delta\theta_D - 2\pi I_T$$

    or

    $$\frac{\Delta\theta_D}{2\pi} = \frac{I_T}{I_D + I_T} = \frac{95 \text{ kg} \cdot \text{m}^2}{55.7 \text{ kg} \cdot \text{m}^2 + 95 \text{ kg} \cdot \text{m}^2} = 0.630$$

    In other words, $\Delta\theta_D$ is 63% of a full circle relative to the ground.

    ASSESS We find that $\Delta\theta_D$, the angular displacement relative to the ground, decreases with I_D. This is what we expect from angular momentum conservation.

43.   **INTERPRET**   This problem is about the rotational motion of the skaters, given their linear speed and radius of the circle they traverse. The key concept here is conservation of angular momentum.

DEVELOP   The force that abruptly stops the skater at one end exerts no torque about that skater (point P), so the total angular momentum about a vertical axis through P is conserved. Initially, the angular momentum of each of the other seven skaters about P is $L_0 = |\vec{r} \times \vec{p}| = pr\sin\theta - mv_0 b$, where b is the perpendicular distance from the original straight-line motion to the point P. For these seven skaters, $b_n = n(L/7)$, where $n = 1, 2, \ldots, 7$ and $L = 12$ m, so

$$L_i = \sum_{n=1}^{7} mv_0 n(L/7) = mv_0\,(L/7)\sum_{n=1}^{7} n = mv_0\,(L/7)\frac{7\times 8}{2} = 4mv_0 L$$

where we have used $\sum_{n=1}^{k} n = k(k+1)/2$. Once L is known, we can readily calculate the angular speed as well as the linear speeds of the skaters.

EVALUATE   **(a)** When the skaters rotate rigidly about P with angular speed ω, their angular momentum is

$$L_f = I\omega = \left(\sum_{n=1}^{7} mb_n^2\right)\omega = \frac{mL^2\omega}{49}\sum_{n=1}^{7} n^2 = \frac{mL^2\omega}{49}\frac{7\times 8\times 15}{6} = \frac{20mL^2\omega}{7}$$

where we have used $\sum_{n=1}^{k} n^2 = k(k+1)(2k+1)/6$. Since L is constant,

$$L_i = L_f \;\rightarrow\; 4mv_0 L = \frac{20mL^2\omega}{7}$$

This gives

$$\omega = \frac{7v_0}{5L} = \frac{7(4.6\text{ m/s})}{5(12\text{ m})} = 0.537\text{ rad/s}$$

(b) The linear speed of the outermost skater is

$$v_7 = \omega b_7 = \omega L = (0.537\text{ s}^{-1})(12\text{ m}) = 6.44\text{ m/s}$$

(c) The force is the centripetal force,

$$F_7 = \frac{mv_7^2}{L} = mL\omega^2 = (60\text{ kg})(12\text{ m})(0.537\text{ rad/s})^2 = 207\text{ N}$$

ASSESS   The angular speed is the same for all skaters, but the linear speed is given by

$$v_n = \omega b_n = \omega n(L/7)$$

This shows that the speed of the outermost skater is the greatest. Similarly, the centripetal force experienced by each skater is

$$F_n = \frac{mv_n^2}{b_n} = m\omega^2 b_n = m\omega^2 n(L/7)$$

Again we see that the force on the outermost skater is the greatest.

45.   **INTERPRET**   We find the angular velocity of a spinning bird feeder after a bird lands on it. The feeder is initially spinning. The bird lands tangentally to the rim of the feeder, so it must have some initial angular momentum relative to the feeder's rotational axis. We will use conservation of angular momentum. Since the bird and the feeder have opposite angular momenta, it is possible that the direction of the feeder's angular momentum will change; so we will keep track of the direction by the sign of ω.

DEVELOP   The radius and rotational inertia of the feeder are $R = 0.19$ m and $I = 0.12$ kg m^2, respectively. The initial angular velocity is $\omega_1 = 5.6$ rpm $= 0.586$ rad/s, and we will define the initial angular velocity of the feeder to be in the positive direction. The bird, with mass $m_b = 0.140$ kg and speed $v_b = 1.1$ m/s, lands on the rim with angular momentum $\omega_b = -\frac{v_b}{R}$, where the negative sign indicates that the bird is moving opposite the feeder. We use conservation of angular momentum, $L_1 = L_2$.

EVALUATE

$$L_1 = L_2 \rightarrow I\omega_1 + (m_b R^2)\left(-\frac{v_b}{R}\right) = \omega_2(I + m_b R^2)$$

$$\rightarrow I\omega_1 - m_b v_b R = \omega_2(I + m_b R^2) \rightarrow \omega_2 = \frac{I\omega_1 - m_b v_b R}{I + m_b R^2}$$

$$\rightarrow \omega_2 = 0.329 \text{ rad/s} = 3.1 \text{ rpm}$$

ASSESS   The sign of the final answer is the same as the initial sign of the angular momentum, so angular momentum does not change direction in this case.

47.   **INTERPRET**   The problem is about the rotational motion of the turntable. Tossingg a piece of clay onto its surface slows down the rotation. We are asked to find the mass of the clay.

DEVELOP   There are no external torques in the direction of the turntable's axis, so the vertical angular momentum of the turntable/clay system is conserved. The horizontal forces, which cause the clay to stick to the turntable, are internal forces. If we take the sense of rotation of the turntable to define the positive direction of vertical angular momentum, then the system's initial angular momentum is

$$L_i = I\omega_i - mvb$$

where I is the rotational inertia of the turntable, v the horizontal component of the velocity of the mass, m, of clay, and b the perpendicular distance to the axis of rotation. After the clay lands, this angular momentum equals

$$L_f = (I + mb^2)\omega_f$$

Equating L_i with L_f allows us to determine m.

EVALUATE   Solving for the mass, we find

$$m = \frac{I(\omega_i - \omega_f)}{vb + b^2\omega_f} = \frac{(0.021 \text{ kg} \cdot \text{m}^2)(0.29 \text{ rad/s} - 0.085 \text{ rad/s})}{(1.3 \text{ m/s})(0.15 \text{ m}) + (0.15 \text{ m})^2(0.085 \text{ rad/s})} = 0.0219 \text{ kg} = 21.9 \text{ g}$$

where we assumed ω_f and ω_i have the same sense.

ASSESS   The clay increases the total rotational inertia of the system, and therefore, the angular speed is decreased, as required by angular momentum conservation.

Note that if the sense of rotation of the turntable were reversed after impact of the clay, i.e., $\omega_f = -0.085$ rad/s, one can show that the mass of the clay would have been $m = 40.8$ g. A bigger piece of clay would be needed to change the sense of rotation.

49.   **INTERPRET**   In this problem we are asked to estimate how much of the solar system's angular momentum about its center is associated with the Sun.

DEVELOP   The planets orbit the Sun in planes approximately perpendicular to the Sun's rotation axis, so most of the angular momentum in the solar system is in this direction. We can estimate the orbital angular momentum of a planet by mvr, where m is its mass, v its average orbital speed, and r its mean distance from the Sun.

Compared to the orbital angular momentum of the four giant planets, everything else is negligible, except for the rotational angular momentum of the Sun itself. The latter can be estimated by assuming the Sun to be a uniform sphere rotating with an average period of $\frac{1}{2}(27 + 36)$ d. (The Sun's period of rotation at the surface varies from approximately 27 days at the equator to 36 days at the poles.)

EVALUATE   The numerical data in Appendix E results in the following estimates:

ORBITAL ANGULAR MOMENTUM (mvr) %

Jupiter	19.2×10^{42} J $\cdot$ s	59.7
Saturn	7.85×10^{42} J $\cdot$ s	24.4
Uranus	1.69×10^{42} J $\cdot$ s	5.2
Neptune	2.52×10^{42} J $\cdot$ s	7.8

ROTATIONAL ANGULAR MOMENTUM ($\frac{2}{5}MR^2\omega$)

Sun	0.89×10^{42} J $\cdot$ s	2.8
Total	32.2×10^{42} J $\cdot$ s	99.9

ASSESS   With $L_{orb} \gg L_{rot}$, we find that more than 97% of the total angular momentum of the solar system comes from the orbital angular momentum. In particular, the orbital motion of Jupiter alone accounts for roughly 60% of the total angular momentum.

51.   **INTERPRET**   This problem looks just like an inelastic collision, but instead of using conservation of linear momentum, we will use conservation of *angular* momentum.

DEVELOP   The masses of disk 1 and 2 are $m_1 = 0.440$ kg and $m_2 = 0.270$ kg, respectively. The radii are $r_1 = 0.035$ m and $r_2 = 0.023$ m. The initial angular speed of disk 1 is $\omega_1 = 180$ rpm $= 18.8$ rad/s. We will use $L_1 = L_2$ to find the final angular speed of both disks stuck together, and $f = 1 - \frac{K_{final}}{K_{initial}}$, where $K = \frac{1}{2}I\omega^2$, to find the fraction of energy lost.

EVALUATE

(a) The initial angular momentum is $L_1 = I_1 \omega_{1i} + I_2 \cancel{\omega_{2i}} = \frac{1}{2}m_1 r_1^2 \omega_{1i}$. The final angular momentum is $L_2 = (I_1 + I_2)\omega_f = (\frac{1}{2}m_1 r_1^2 + \frac{1}{2}m_2 r_2^2)\omega_f$. Conservation of angular momentum tells us that

$$L_1 = L_2 \rightarrow \frac{1}{2}m_1 r_1^2 \omega_{1i} = \left(\frac{1}{2}m_1 r_1^2 + \frac{1}{2}m_2 r_2^2\right)\omega_f \rightarrow \omega_f = \omega_{1i}\left(\frac{m_1 r_1^2}{m_1 r_1^2 + m_2 r_2^2}\right)$$

$$\rightarrow \omega_f = \omega_{1i}(0.791) = 18.8 \text{ rad/s} = 142 \text{ rpm}$$

(b) The initial kinetic energy is $K_{initial} = \frac{1}{2}I_1 \omega_{1i}^2 + \frac{1}{2}I_2 \cancel{\omega_{2i}^2} = \frac{1}{4}m_1 r_1^2 \omega_{1i}^2$. The final kinetic energy is $K_{final} = \frac{1}{2}(I_1 + I_2)\omega_f^2 = \frac{1}{4}(m_1 r_1^2 + m_2 r_2^2)\omega_f^2$.

$$f = 1 - \frac{K_{final}}{K_{initial}} = 1 - \frac{m_1 r_1^2 \omega_{1i}^2}{(m_1 r_1^2 + m_2 r_2^2)\omega_f^2} = 1 - \frac{\cancel{m_1 r_1^2}\,\cancel{\omega_{1i}^2}}{\cancel{(m_1 r_1^2 + m_2 r_2^2)}\left[\cancel{\omega_{1i}}\left(\frac{m_1 r_1^2}{m_1 r_1^2 + m_2 r_2^2}\right)\right]^{\cancel{2}}}$$

$$\rightarrow f = 1 - \frac{m_1 r_1^2 + m_2 r_2^2}{m_1 r_1^2} = \frac{m_2 r_2^2}{m_1 r_1^2} = 0.265 = 26.5\%$$

ASSESS   Note that the fractional energy loss doesn't depend on the initial energy. For this particular set of disks, 26.5% of the initial energy will be lost in the collision regardless of how fast the bottom disk is spinning!

53.   **INTERPRET**   A solid spinning ball drops onto a frictional surface. At first it slides, but due to friction it will slow down its spin and increase its linear motion until it is purely rolling without sliding. We want to find the ball's angular speed when it begins purely rolling, and how long it takes.

DEVELOP   From the problem statement, we see that the ball's mass is M, its radius is R, and its initial angular velocity around the horizontal axis is ω_0. The coefficient of kinetic friction between the ball and the surface is μ_k, so the frictional force is $F_f = \mu F_n = \mu Mg$.

There is a torque on the ball due to the frictional force, acting on the edge. This torque $\tau = -\mu MgR$ serves to slow the ball's rotation: we can use $\tau = I\alpha$ to find the angular acceleration α and then use $\omega = \omega_0 + \alpha t$ to find the resulting angular speed.

The frictional force on the ball also accelerates the ball, so we can use $F = Ma$ and $v = v_0 + at$ to find the speed of the ball.

The ball is no longer sliding when $R\omega = v$.

EVALUATE $\alpha = \frac{\tau}{I} = \frac{-\mu Mg\cancel{R}}{\frac{2}{5}MR^{\cancel{2}}} = -\frac{5\mu g}{2R}$, so $\omega = \omega_0 - \frac{5\mu g}{2R}t$.

$a = \frac{F}{M} = \frac{\mu Mg}{M} = \mu g$, so $v = v_0 + \mu gt$. We set $R\omega = v \rightarrow R\omega_0 - \frac{5}{2}\mu gt = \cancel{v_0} + \mu gt$. Solving this for t gives us $R\omega_0 = \mu gt\left(1 + \frac{5}{2}\right) \rightarrow t = \frac{7}{2}\frac{R\omega_0}{\mu g}$, and plugging this value into our equation for ω gives us $\omega = \omega_0 - \frac{5\mu g}{\cancel{2}\cancel{R}}\left(\frac{\cancel{2}}{7}\frac{\cancel{R}\omega_0}{\mu g}\right) = \omega_0 - \frac{5}{7}\omega_0 = \frac{2}{7}\omega_0$.

So for part **(a)** $\omega = \frac{2}{7}\omega_0$, and for part **(b)** $t = \frac{7}{2}\frac{R\omega_0}{\mu g}$.

ASSESS   The answer to part **(a)** is surprising—it says that no matter what the size or speed of the ball, or the coefficient of friction, the angular speed of the ball when it stops sliding is $\frac{2}{7}$ of its original value! The time, in part **(b)**, does depend on the details of the situation, but **(a)** does not.

55.   INTERPRET We find the torque due to gravity for a rod that can pivot at one end, given an angle with the horizontal. We use an integral for the torque, although we might expect that this torque was equal to the torque acting on the center of mass of the rod. We also find the angle such that the initial acceleration of the end of the rod is equal to g.

DEVELOP The rod has mass M and length L, and pivots at one end. The angle it makes with the horizontal is θ. The torque $d\vec{\tau}$ on a mass element dm is $d\vec{\tau} = \vec{r} \times d\vec{F} \rightarrow d\tau = (x)(g\,dm)\cos\theta$. We integrate this over the length of the rod, where $dm = \frac{M}{L} dx$. For the second part, we use $\tau = I\alpha$ and $a = L\alpha = g$. The rotational inertia of a rod rotating around one end is $I = \frac{1}{3} ML^2$.

EVALUATE

(a) $\tau = \int_0^L xg\cos\theta\left(\frac{M}{L}\right)dx = \frac{Mg\cos\theta}{L}\int_0^L x\,dx = \frac{Mg\cos\theta}{L}\left(\frac{1}{2}L^2\right) = \frac{1}{2}MgL\cos\theta$

(b) $\alpha = \frac{\tau}{I} = \frac{\frac{1}{2}MgL\cos\theta}{\frac{1}{3}ML^2} = \frac{3}{2}\frac{g\cos\theta}{L}$. Set $a = L\alpha = g$, so

$$g = \frac{3}{2}g\cos\theta \rightarrow \theta = \cos^{-1}\left(\frac{2}{3}\right) = 48.2°$$

ASSESS As expected, the torque was the same as if the rod were a point mass located at the center of mass of the rod.

57.   INTERPRET We are asked to show that the precessional speed of a gyroscope is $\omega_p = \frac{mgr}{L}$. We do so by using $\vec{\tau} = \vec{r} \times \vec{F}$, making sure that we keep track of directions.

DEVELOP $\vec{\tau} = \vec{r} \times \vec{F} = \frac{d\vec{L}}{dt}$. We will call the change in angle of the gyroscope as it precesses $d\phi$. The radius of the circle around which the center of mass of the gyroscope precesses is $\rho = L\sin\theta$, where θ is the angle that $\vec{L}$ makes with the vertical. The torque due to gravity is $|\vec{\tau}| = |\vec{r} \times \vec{F}_g| = mgr\sin\theta$, with direction into the page in Figure 11.9. $d\vec{L}$ (shown as $\Delta\vec{L}$ in Figure 11.9) is the change in $\vec{L}$ due to the gravitational torque, so $d\phi = \frac{|d\vec{L}|}{\rho}$. Finally, $\omega_p = \frac{d\phi}{dt}$.

EVALUATE $\tau = |\vec{r} \times m\vec{g}| = mgr\sin\theta$. $\omega_p = \frac{d\phi}{dt} = \frac{1}{\rho}\frac{dL}{dt} = \frac{1}{L\sin\theta}(\tau) = \frac{1}{L\sin\theta}(mgr\sin\theta) = \frac{mgr}{L}$.

ASSESS The faster the top spins the larger is L and the slower the precessional speed.

59.   INTERPRET We use the definition of angular acceleration to find the angular acceleration, given the initial angular speed, the final angular speed, and the time. Is the value we find between the given limits?

DEVELOP $\alpha = \frac{\Delta\omega}{\Delta t}$. The initial value of ω is $\omega_1 = 5$ rad/s, and the final value is $\omega_2 = 15$ rad/s. The time it takes to change is $t = 16.6$ s.

EVALUATE $\alpha = \frac{\Delta\omega}{\Delta t} = \frac{(10 \text{ rad/s})}{(16.6 \text{ s})} = 0.60$ rad/s^2.

ASSESS This angular acceleration is below the desired limit, which was 0.75 rad/s^2.

61.   INTERPRET We are to find the angular momentum of a device moving at a given speed. We can calculate the rotational inertia given the size and shape of the device, and then find the angular momentum.

DEVELOP There are 4 small cups, each of mass $m = 0.120$ kg. These are located $r = 0.16$ m from the center, and we can ignore the mass of the supporting rods. We will use $I = \sum m_i r_i^2$, and $L = I\omega$. The angular speed is $\omega = 12$ rev/s $= 75.4$ rad/s.

EVALUATE $I = 4mr^2 = 4[(0.12 \text{ kg}) \times (0.16 \text{ m})^2] = 0.0123$ kg m^2 $\rightarrow L = 0.926$ kg m^2/s.

ASSESS I'm not sure how this information will help us design the support, but we've calculate what was required. The units work out, and the numeric value is reasonable for a small rotor such as this.

EXERCISES

Section 12.1 Conditions for Equilibrium

15. **INTERPRET** We have been told that the choice of pivot point does not matter if the sum of forces is zero. Here we will show that this is true for two different pivot points. Three forces are acting on an object, which is in equilibrium.

    DEVELOP The three forces are $\vec{F}_1 = 2\hat{i} + 2\hat{j}$ N at point $(x, y) = (2$ m, 0 m$)$, $\vec{F}_2 = -2\hat{i} - 3\hat{j}$ N at $(-1$ m, 0 m$)$, and $\vec{F}_3 = 1\hat{j}$ N at $(-7$ m, 1 m$)$. We find the torques due to these three forces around points $(3$ m, 2 m$)$ and $(-7$ m, 1 m$)$, using $\vec{\tau} = |\vec{r} \times \vec{F}|$. To find the value of $\vec{r}$ for a point other than the origin, we take the vector difference between the point where the force is applied and the point used as the pivot: $\vec{r} = (\vec{r}_{applied} - \vec{r}_{pivot})$.

    EVALUATE For point $(3$ m, 2 m$)$, the torque due to $\vec{F}_1$ is

    $$\vec{\tau}_1 = (\vec{r}_{applied} - \vec{r}_{pivot}) \times \vec{F}_1 = ((2-3)\hat{i}\,m + (0-2)\hat{j}\,m) \times (2\hat{i} + 2\hat{j})\ N$$

    $$\vec{\tau}_1 = \begin{vmatrix} \hat{i} & \hat{j} & \hat{k} \\ -1 & -2 & 0 \\ 2 & 2 & 0 \end{vmatrix}\ Nm = (-2 - (-4))\hat{k}\ Nm = 2\hat{k}\ Nm$$

    Similarly, $\vec{\tau}_2 = 8\hat{k}$ Nm and $\vec{\tau}_3 = -10\hat{k}$ Nm. The sum of these three is $\vec{\tau}_{total} = (2 + 8 - 10)\hat{k}$ Nm $= 0$.
    Around point $(-7$ m, 1 m$)$, the torques are $\vec{\tau}_1 = 20\hat{k}$ Nm, $\vec{\tau}_2 = -20\hat{k}$ Nm, and $\vec{\tau}_3 = 0$. The sum of these three is also $\vec{\tau}_{total} = 0$.

    ASSESS Note that the torque due to force 3 around the second pivot is zero, since the force acts on the pivot.

17. **INTERPRET** The problem asks for a set of conditions that must be met for the body to be in static equilibrium. This means that all the external forces and torques must be zero.

    DEVELOP Static equilibrium demands that (see Equations 12.1 and 12.2)

    $$\Sigma \vec{F}_i = 0$$
    $$\Sigma \vec{\tau}_i = \Sigma(\vec{r}_i \times \vec{F}_i) = 0$$

    Since all of the forces lie in the same plane, which includes the points O and P, there are two independent components of the force condition (Equation 12.1) and one component of the torque condition (Equation 12.2).

    EVALUATE (a) Taking the x axis to the right, the y axis up and the z axis out of the page in Figure 12.12, we have:

    $$0 = \Sigma F_x = -F_1 + F_2 \sin\phi + F_3$$
    $$0 = \Sigma F_y = -F_2 \cos\phi + F_4$$
    $$0 = (\Sigma \tau_z)_O = -L_1 F_2 - L_2 F_3 \sin\phi + L_2 F_4 \cos\phi$$

    (b) The equation for torque about point P is

    $$0 = (\Sigma \tau_z)_P = -L_2 F_1 \sin\phi + (L_2 - L_1)F_2$$

    The lever arms of all the forces about either O or P should be evident from Fig. 12.12.

    ASSESS In static equilibrium, the torque about any given pivot point must vanish. Otherwise, the body will rotate about that point.

Section 12.2 Center of Gravity

19. **INTERPRET** In this problem we are asked to find the gravitational torque about various pivot points.

DEVELOP The torque about a point is given by Equation 10.10, $\tau = rF\sin\theta = r_\perp F$, where $r_\perp = r\sin\theta$ is the lever arm. In our problem, the center of gravity (CG) is at the center of the triangle, which is at a perpendicular distance of $\frac{L}{2\sqrt{3}}$ from any side. We regard CG as the point at which all the mass is concentrated.

EVALUATE (a) The lever arm of the weight about point A is $r_{\perp,A} = L/2$. Therefore, the gravitational torque about A is

$$\tau_A = r_{\perp,A}F_g = \frac{L}{2}mg = \frac{1}{2}mgL$$

(b) The lever arm about point B is zero (the line of action passes through point B). Therefore, we have $\tau_B = 0$.
(c) The lever arm about point C is $r_{\perp,C} = L/4$ (C is halfway up from the base) so

$$\tau_C = r_{\perp,C}F_g = \frac{L}{4}mg = \frac{1}{4}mgL$$

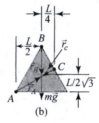

(b)

ASSESS The torques $\vec\tau_A$ and $\vec\tau_C$ are in opposite directions. If pivoted at A, the gravitational torque $\vec\tau_A$ tends to rotate the triangle CW. On the other hand, when pivoted at C, $\vec\tau_C$ gives rise to a CCW rotation.

Section 12.3 Examples of Static Equilibrium

21. **INTERPRET** In this problem we are asked to find where the child should sit on the pivot-supporting board so that the scale at the right end will read zero.

DEVELOP If we consider torques about the pivot point (so that the force exerted by the pivot does not contribute) then Equation 12.2, $\sum\vec\tau_i = \sum(\vec r_i \times \vec F_i) = 0$, is sufficient to determine the position of the child.

EVALUATE As shown on Fig. 12.15, the weight of the board (acting at its center of gravity), the weight of the child (acting a distance x from the left end), and the scale force, F_s, produce zero torque about the pivot:

$$(\Sigma\tau)_P = 0 \rightarrow 0 = F_s(1.60\text{ m}) - (60\text{ kg})(9.8\text{ m/s}^2)(0.40\text{ m}) + (40\text{ kg})(9.8\text{ m/s}^2)(0.80\text{ m} - x)$$

Therefore, the relation between the position of the child and the scale force F_s is given by

$$x = \frac{F_s(1.60\text{ m}) + 78.4\text{ kg}\cdot\text{m}^2/\text{s}^2}{(40\text{ kg})(9.8\text{ m/s}^2)} \frac{F_s(1.60\text{ m})}{(40\text{ kg})(9.8\text{ m/s}^2)} + 0.20\text{ m}$$

If we want the reading of the scale to be zero, i.e., $F_s = 0$, then the child must be at $x = 0.20$ m (distance from the left end).

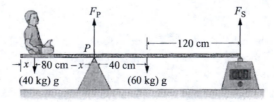

ASSESS If the child moves closer to the pivot (increasing x), then F_s will be increased, i.e., the reading of the scale will go up. One can also show that without the child, the reading would be $F_s = 147$ N.

23.   **INTERPRET**   The problem is about static equilibrium. We want to know where the concrete should be placed so that the long beam suspended by a cable and with a steelworker standing at one end will be in equilibrium.

DEVELOP   At equilibrium, the sum of the torques on the beam (taken about its center, C, so that the cable's tension and beam's weight do not enter the equation), is equal to zero:

$$(\Sigma \tau)_C = 0 \quad \rightarrow \quad m_c g x = m_s g L_s$$

where m_c and m_s are the masses of the concrete and the steelworker, and $L_s = 2.1$ m is the distance from the steelworker to C.

EVALUATE   Using the values given in the problem statement, we have

$$x = \left(\frac{m_s}{m_c}\right) L_s = \frac{65 \text{ kg}}{190 \text{ kg}}(2.1 \text{ m}) = 0.718 \text{ m}$$

on the opposite side of C from the worker.

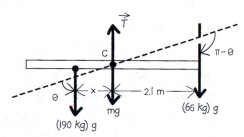

ASSESS   The lighter the concrete, the longer the lever arm (x) must be for the system to remain in static equilibrium.

Note also that since $\sin\theta = \sin(\pi - \theta)$ will cancel from the torque equation, the beam need not be horizontal to be in equilibrium; the steelworker's mental equilibrium is greatest when the beam is horizontal.

Section 12.4 Stability

25.   **INTERPRET**   The problem is about the stability of the roller coaster as it moves along the track described by a height function. We want to identify the equilibrium point and classify its stability.

DEVELOP   The potential energy of the roller coaster car, in the equivalent one-dimensional problem, is

$$U(x) = mgh(x) = mg(0.94x - 0.01x^2)$$

Equilibrium condition is given by Equation 12.3: $\frac{dU}{dx} = 0$. In addition, the equilibrium condition may be classified according to its second derivative:

$$\frac{d^2U}{dx^2} : \begin{cases} > 0, & \text{stable} \\ < 0, & \text{unstable} \\ = 0, & \text{neutral} \end{cases}$$

EVALUATE   **(a)** Equation 12.3, the condition for equilibrium, gives

$$0 = \frac{dU}{dx} = mg(0.94 - 0.02x) \quad \rightarrow \quad x = 47 \text{ m}$$

(b) Since $\frac{d^2U}{dx^2} = -0.02mg < 0$, Equation 12.5 implies that this is an unstable equilibrium.

ASSESS   The point $x = 47$ m with $U(x = 47) = 22.09mg$ corresponds to a local maximum where the potential-energy curve is concaving downward. Therefore, the point is unstable.

PROBLEMS

27.   **INTERPRET**   This problem is about static equilibrium. We want to find that tension force the left-hand bolt can withstand in order to keep the traffic signal system stable.

DEVELOP   The forces on the traffic signal structure, and their lever arms about point 0 (on the vertical member's centerline between the bolts) are shown on Fig. 12.17. The normal forces exerted by the bolts and the ground on the vertical member are designated by n_L and n_r, measured positive upward. (Of course, the ground can only make a positive contribution, and the bolts only a negative contribution, to these normal forces.)

The two conditions of static equilibrium needed to determine n_L and n_r are:

$$0 = \Sigma F_y = n_L + n_r - (9.8 \text{ m/s}^2)(320 \text{ kg} + 170 \text{ kg} + 65 \text{ kg})$$

(the vertical component of Equation 12.1, positive up), and

$$0 = (\Sigma \tau_z)_0 = (n_r - n_L)(0.38 \text{ m}) - [(170 \text{ kg})(3.5 \text{ m}) + (65 \text{ kg})(8.0 \text{ m})](9.8 \text{ m/s}^2)$$

(the out-of-the-page-component of Equation 12.2, positive CCW).

EVALUATE   The above two equations can be rewritten as

$$n_r + n_L = 5.44 \times 10^3 \text{ N}$$
$$n_r - n_L = 2.88 \times 10^4 \text{ N}$$

Thus, we find $n_L = -1.17 \times 10^4$ N, which is downward and must be exerted by the bolt.

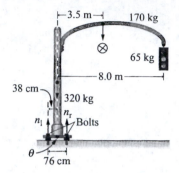

ASSESS   The reaction force on the bolt is upward and is a tensile force. Really, n_ℓ is the difference between the downward force exerted by the bolt and the upward force exerted by the ground. Tightening the bolt increases the tensile force it must withstand beyond the minimum value calculated above, under the assumption that the ground exerts no force.

29.   **INTERPRET**   The problem is about static equilibrium. The sphere is supported by a rope attached to the wall. The friction between the sphere and the wall helps keep it in equilibrium. We want to find the smallest possible value for the coefficient of friction.

DEVELOP   In equilibrium, there is no net force acting on the sphere. In addition, the sum of the torques about the center of the sphere must be zero, so the frictional force is up, as shown. Using Equations 12.1 and 12.2, the equilibrium conditions can be written as

$$\Sigma F_x = n - T \sin \theta = 0$$
$$\Sigma F_y = T \cos \theta + f - Mg = 0$$
$$(\Sigma \tau)_c = Rf - \left(\frac{R}{2}\right) T \cos \theta = 0$$

The above equations can be used to solve for the frictional force f. The coefficient of friction can then be obtained from $f \leq \mu_s n$.

EVALUATE   The x component of the force equation gives $T = n/\sin \theta$. Substituting this into the torque equation gives

$$f = \frac{1}{2} T \cos \theta = \frac{n \cos \theta}{2 \sin \theta} = \frac{n}{2} \cot \theta$$

Since $f \leq \mu_s n$, the minimum coefficient of friction is

$$\mu_s = \frac{1}{2} \cot \theta = \frac{1}{2} \cot 30° = \frac{\sqrt{3}}{2} = 0.866$$

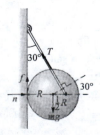

ASSESS   If the angle θ is increased, then the minimum coefficient of friction would decrease. Note that μ_s is independent of the mass and radius of the sphere; it depends only on the angle θ.

31.   **INTERPRET**   In this problem we want to find the tension in the Achilles tendon and the contact force at the ankle joint when the foot is in static equilibrium.

DEVELOP   If we approximate the bones in the foot as a massless, planar, rigid body, the equilibrium conditions for the situation depicted in Fig. 12.21 are:

$$0 = \Sigma F_x = T\sin\theta - F_{C,x}$$
$$0 = \Sigma F_y = T\cos\theta + n - F_{C,y}$$
$$0 = (\Sigma\tau)_{\text{ankle joint}} = n(12\text{ cm}) - (T\cos\theta)(7\text{ cm})$$

where $\theta = 25°$.

EVALUATE   (a) We first note that the normal force is simply equal to the weight of the person, $n = mg = (70\text{ kg})(9.8\text{ m/s}^2) = 686\text{ N}$. Substituting this into the torque equation, the tension in the Achilles tendon is

$$T = \frac{n(12\text{ cm})}{(7\text{ cm})\cos\theta} = \frac{(686\text{ N})(12\text{ cm})}{(7\text{ cm})\cos25°} = 1298\text{ N}$$

(b) Substituting the value of T into the force equations, we find

$$F_{C,x} = T\sin\theta = (1298\text{ N})\sin25° = 548\text{ N}$$
$$F_{C,y} = T\cos\theta + n = (1298\text{ N})\cos25° + 686\text{ N} = 1862\text{ N}$$

Therefore, the contact force at the ankle joint is

$$F_C = \sqrt{F_{C,x}^2 + F_{C,y}^2} = \sqrt{(548\text{ N})^2 + (1862\text{ N})^2} = 1941\text{ N}$$

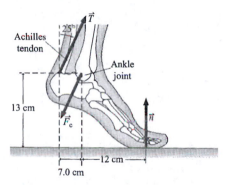

ASSESS   The tension in the Achilles tendon is almost twice the weight of the person. This is because the Achilles tendon is very close to the ankle joint, leading to a small lever arm. The contact force at ankle is roughly three times the weight of the person. The problem demonstrates that in order to maintain a static equilibrium, many parts of our body often experience forces that are greater than our own weight.

33.   **INTERPRET**   In this problem we want to find the tension in the cable supporting the boom so that the boom is in static equilibrium.

DEVELOP   For the boom to remain in static equilibrium, the forces must satisfy Equation 12.1. Since we are only asked about the tension, we can focus only on the torque about point P. The forces on the boom are shown superposed on the figure. By assumption, T is horizontal and acts at the CM of the boom. To find T, we compute the torques about P.

EVALUATE   The condition for equilibrium implies that that $(\Sigma\tau)_P = 0$, or

$$T(L/2)\sin\theta - m_b g(L/2)\cos\theta - mgL\cos\theta = 0$$

which can be solved to give

$$T = (2m + m_b)g\cot\theta = (4400\text{ kg} + 1700\text{ kg})(9.8\text{ m/s}^2)\cot50° = 5.02\times10^4\text{ N}$$

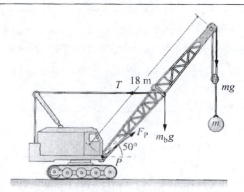

ASSESS To see that our result makes sense, let's consider the following cases: (i) $\theta = 90°$. In this limit, the boom is vertical and the tension in the cable vanishes $(\cot 90° = 0)$, as expected. (ii) $m = 0$, when no mass is hung from the end of the boom, the tension in the cable is reduced to $T = m_b g \cot \theta$. This comes from the force equations at P (with $m = 0$):

$$0 = \Sigma F_x = F_P \cos \theta - T \sin \theta$$
$$0 = \Sigma F_y = F_P \sin \theta - m_b g$$

35. **INTERPRET** In this problem we want to find the force the people apply to pull the car so that it is in static equilibrium. The applied force is equal to the tension in the rope.

DEVELOP For the car to remain in static equilibrium, the forces must satisfy Equation 12.1. Three forces act on the car, as shown added to Fig. 12.24. The unknown force, F_P, exerted by the edge of the embankment, does not contribute to Equation 12.2 (positive torques CCW) if evaluated about point P, so the tension necessary to keep the car in equilibrium can be found directly. Thus, our plan is to compute the torques about P.

EVALUATE The condition for equilibrium implies that that $(\Sigma \tau)_P = 0$, or

$$0 = Mg(L - l)\cos \theta - Tl \sin \theta$$

where $L = 2.4$ m and $l = 1.8$ m. This gives

$$T = Mg\left(\frac{L - l}{l}\right)\cot \theta$$
$$= (1250 \text{ kg})(9.8 \text{ N})\left(\frac{2.4 \text{ m} - 1.8 \text{ m}}{1.8 \text{ m}}\right)\cot 34°$$
$$= 6.05 \times 10^3 \text{ N}$$

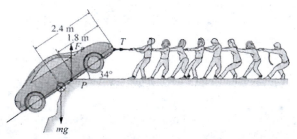

ASSESS Note that if $l \geq L$, (when the CM lies above the edge of the embankment), then the tension becomes zero. In this situation, the car would remain in equilibrium.

37. **INTERPRET** This problem is about static equilibrium. The hanging sign is being supported by a rod and a cable. Given the maximum tension in the cable, we want to find the minimum height above the pivot for anchoring the cable to the wall so that the sign remains in static equilibrium.

DEVELOP Suppose that the sign is centered on the rod, so that its CM lies under the center of the rod. Then the total weight may be considered to act through the center of the rod, as shown. For the sign to remain in static equilibrium, the forces and the torques must satisfy Equations 12.1 and 12.2. In this situation, since we are only interested in the minimum height h and not F_P, it is sufficient to focus only on the torques about point P.

EVALUATE   In equilibrium, we have $(\sum \tau)_P = 0$. Note that the pivot force F_P makes no contribution to the torque about P. The equation implies

$$0 = (\sum \tau)_P = TL \sin \theta - Mg(L/2)$$

Therefore, we find the tension to be

$$T = \frac{Mg}{2 \sin \theta} = \frac{1}{2} Mg \sqrt{1 + \frac{L^2}{h^2}}$$

where we have used $\tan \theta = \frac{h}{L}$ and the identity $1 + \cot^2 \theta = \csc^2 \theta$. Solving for h, we obtain

$$h = L \left[\left(\frac{2T}{Mg} \right)^2 - 1 \right]^{-1/2}$$

Given the maximum tension that could be withstood in the cable, T_{max}, the condition that must be met by h for the sign to remain in static equilibrium is

$$h \geq L \left[\left(\frac{2T_{max}}{Mg} \right)^2 - 1 \right]^{-1/2} = (2.3 \text{ m}) \left[\left(\frac{2(800 \text{ N})}{(66 \text{ kg} + 8.2 \text{ kg})(9.8 \text{ m/s}^2)} \right)^2 - 1 \right]^{-1/2} = 1.17 \text{ m}$$

So the minimum height is $h_{min} = 1.17$ m.

ASSESS   When the maximum tension which the cable can withstand is such that $T_{max} < Mg/2$, static equilibrium would not be possible!

39.   **INTERPRET**   This problem is an equilibrium problem, as long as the log does not slip! We use the conditions for equilibrium to find the maximum mass of the climber at the end of the log.

DEVELOP   We start by drawing a free-body diagram showing all of the forces on the log, as shown in the figure. The maximum frictional force on the end of the log is $F_f = \mu_s F_N$. This frictional force is balanced by the normal force from the wall, F_N', so if the force from the wall is *greater* than the maximum frictional force, then the log will slip. We can find the force from the wall by using $\sum \tau = 0$ with the left end of the log as the pivot point. We also see from the figure that $F_N = (M + m)g$, where $M = 340$ kg is the mass of the log, and m is the unknown mass of the climber. The length of the log is $L = 6.3$ m, and the log forms an angle $\theta = 27°$ with the horizontal. The center of mass of the log is located $\frac{L}{3}$ from the left end, and the coefficient of friction between the left end of the log and the ground is $\mu = 0.92$.

EVALUATE $\sum \vec{\tau} = mg L \cos \theta + Mg \frac{L}{3} \cos \theta - F_N' L \sin \theta = 0 \rightarrow F_N' = \frac{\left(m + \frac{M}{3}\right) g \cos \theta}{\sin \theta}$

This normal force from the wall must be balanced by the static frictional force on the left end of the log:

$$F_f = F_N' \rightarrow \mu F_N = \mu(m + M) g = \frac{\left(m + \frac{M}{3}\right) g \cos \theta}{\sin \theta} = \frac{\left(m + \frac{M}{3}\right)}{\tan \theta}$$

$$\rightarrow \mu m + \mu M = \frac{m}{\tan \theta} + \frac{M}{3 \tan \theta}$$

$$\rightarrow m \left(\mu - \frac{1}{\tan \theta} \right) = M \left(\frac{1}{3 \tan \theta} - \mu \right)$$

$$\rightarrow m = M \left(\frac{\mu - \frac{1}{\tan \theta}}{\frac{1}{3 \tan \theta} - \mu} \right) = M \left(\frac{3\mu \tan \theta - 3}{1 - 3 \tan \theta} \right) = 3.0M \approx 1000 \text{ kg}$$

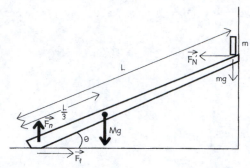

ASSESS   This mass limit is large enough that any climber can safely cross.

41.   **INTERPRET**   In this problem we are given a block that makes an angle θ with the horizontal, and we want to find out the values of θ which make it stable.

DEVELOP   Let the block be tilted in a plane perpendicular to its thickness, as shown in the figure. The condition for equilibrium implies that the torque about the pivot point is zero, i.e., $(\sum \tau)_p = 0$. Since there's no external force acting on the block, the only force that contributes to the torque is the weight of the block.

EVALUATE   To satisfy the condition that $(\sum \tau)_p = 0$, we require that the weight force passes through the pivot point so that the lever arm vanishes. One can readily show that $\theta = 0°$ is a possibility and it represents a stable equilibrium position. On the other hand, $\theta = 90°$ is a metastable one. Finally, $\theta + \alpha = 90°$ is an unstable one (α is the angle a diagonal makes with the longer side, as shown). Since $\tan \alpha = \frac{L}{2L} = \frac{1}{2}$, the unstable equilibrium is at

$$\theta = 90° - \tan^{-1}\left(\frac{1}{2}\right) = 63.4°$$

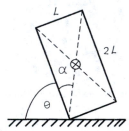

ASSESS   For a rectangular block of length L and width W, one may show that the general expression for the torque about the pivot point is given by

$$\tau = r_\perp F = \frac{1}{2}(L\cos\theta - W\sin\theta)mg$$

Thus, the condition that $\tau = 0$ implies that

$$\tan\theta = \frac{L}{W}$$

In our case, we have $L/W = 2$, which gives $\theta = 63.4°$.

43.   **INTERPRET**   In this problem a cubical block is placed on an incline. Given the coefficient of friction between the block and the incline, we'd like to find out whether the block first slides or tips when the angle of the incline is increased.

DEVELOP   We suppose that the block is oriented with two sides parallel to the direction of the incline, and that its CM is at the center. The condition for sliding is that

$$mg\sin\theta > f_s^{\max} = \mu_s n = \mu_s mg\cos\theta$$

or $\tan\theta > \mu_s$. The condition for tipping over is that the CM lie to the left of the lower corner of the block (see sketch). Thus $\theta > \alpha$, where

$$\alpha = \tan^{-1}\left(\frac{w}{h}\right)$$

is the diagonal angle of the block.

EVALUATE   For a cubical block, $w = h$, and the cube will tip over when $\theta > \alpha = \tan^{-1}(w/h) = \tan^{-1}(1) = 45°$ but will slide when

$$\theta > \tan^{-1}\mu_s = \tan^{-1} 0.95 = 43.5°$$

It thus slides before tipping.

ASSESS   We find that sliding happens first if $\mu_s < w/h$. This makes sense because when the coefficient of friction is small, the block has a greater tendency to slide. On the other hand, when the coefficient of friction is large ($\mu_s > w/h$), we'd expect tipping to take place first.

45.   **INTERPRET**   In this problem a ladder is leaning against the wall and we want to find the mass of the heaviest person who can climb to the top of the ladder while keeping it in static equilibrium.

DEVELOP   The forces on the uniform ladder are shown in the sketch, with the force exerted by the (frictionless) wall horizontal. The person is up the ladder a fraction α of its length. Equilibrium conditions require:

$$0 = \Sigma F_x = f - F_{wall}$$
$$0 = \Sigma F_y = n - (m_L + m)g$$
$$0 = (\Sigma\tau)_A = F_{wall}L\cos\theta - m_L g(L/2)\sin\theta - mg\alpha L\sin\theta$$

The ladder will not slip if $f \le \mu_s n$. Using the equations above, this condition can be rewritten as

$$f = F_{wall} = \left(\frac{1}{2}m_L + \alpha m\right)g\tan\theta \le \mu_s n = \mu_s(m_L + m)g$$

or

$$\alpha \le \frac{\mu_s(m_L + m)\cot\theta - m_L/2}{m} = \mu_s\cot\theta + \frac{m_L}{m}\left(\mu_s\cot\theta - \frac{1}{2}\right)$$

Here, we used the horizontal force equation to find f, the torque equation to find F_{wall}, and the vertical force equation to find n.

EVALUATE   For a person at the top of the ladder, $\alpha = 1$ and the condition for no slipping becomes

$$m \le m_L\left(\frac{\mu_s\cot\theta - 1/2}{1 - \mu_s\cot\theta}\right)$$

With the data given for the ladder (note that $\cot\theta = \cot(\pi/2 - 66°) = \tan 66°$), we obtain

$$m \le (9.5\,\text{kg})\frac{(0.42)\tan 66° - 1/2}{1 - (0.42)\tan 66°} = 74.3\,\text{kg}$$

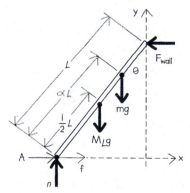

ASSESS   The above equation shows that when the coefficient of friction becomes too small, $\mu_s\cot\theta < 1/2$, or $\mu_s < \tan\theta/2$ (see Example 12.2), slipping will occur and it's no longer possible for the ladder to remain in static equilibrium. In this situation, nobody can climb up to the top of the ladder without making the ladder slip, regardless of his or her mass.

47.   **INTERPRET**   We want to find the energy required to bring a cube to an unstable equilibrium. The problem is equivalent to finding the increase in potential energy of the system.

DEVELOP   When resting in a stable equilibrium position, the CM of a uniform cube of side s is at a distance $y_0 = s/2$ above the tabletop. When balancing on a corner, the CM is now a distance

$$y = \sqrt{(s/2)^2 + (s/2)^2 + (s/2)^2} = \frac{\sqrt{3}}{2}s$$

above the corner resting on the tabletop.

EVALUATE   From the above, the potential energy difference is

$$\Delta U = mg\ \Delta y_{\text{cm}} = mgs\left(\frac{\sqrt{3}}{2} - \frac{1}{2}\right) = 0.366\ mgs$$

This is the energy required to bring the cube to an unstable equilibrium.

ASSESS   Raising the vertical distance of the CM increases the potential energy of the cube. In general, the stability of a system decreases as its potential energy is increased.

49.   **INTERPRET**   This problem is about a ladder leaning against the wall and we want to verify the condition under which any person (with any mass) can climb to the top, and also the one in which nobody can.

DEVELOP   The forces on the uniform ladder are shown in the sketch, with the force exerted by the (frictionless) wall horizontal. The person is up the ladder a fraction α of its length. Equilibrium conditions require:

$$0 = \Sigma F_x = f - F_{\text{wall}}$$
$$0 = \Sigma F_y = n - (m_L + m)g$$
$$0 = (\Sigma \tau)_A = F_{\text{wall}}L\cos\theta - m_\ell g(L/2)\sin\theta - mg\alpha L\sin\theta$$

The ladder will not slip if $f \leq \mu_s n$. Using the equations above, this condition can be rewritten as

$$f = F_{\text{wall}} = \left(\frac{1}{2}m_L + \alpha m\right)g\tan\theta \leq \mu_s n = \mu_s(m_L + m)g$$

or

$$\alpha \leq \frac{\mu_s(m_L + m)\cot\theta - m_L/2}{m} = \mu_s\cot\theta + \frac{m_L}{m}\left(\mu_s\cot\theta - \frac{1}{2}\right)$$

Here, we used the horizontal force equation to find f, the torque equation to find F_{wall}, and the vertical force equation to find n.

EVALUATE   For a person at the top of the ladder, $\alpha = 1$ and the condition for no slipping becomes

$$m \leq m_L\left(\frac{\mu_s\cot\theta - 1/2}{1 - \mu_s\cot\theta}\right) = m_L\left(\frac{\mu_s - \tan\theta/2}{\tan\theta - \mu_s}\right)$$

Since m is positive, this condition cannot be fulfilled if $\mu_s \leq \frac{1}{2}\tan\theta$, i.e., no one can climb to the top without causing the ladder to slip, whereas if $\mu_s = \tan\theta$, the limit is ∞, so anyone can climb to the top.

ASSESS   When the coefficient of friction becomes too small, $\mu_s\cot\theta < 1/2$, or $\mu_s < \tan\theta/2$ (see Example 12.2), slipping will occur and it's no longer possible for the ladder to remain in static equilibrium. In this situation, nobody can climb up to the top of the ladder without making the ladder slip, regardless of his or her mass.

51.   **INTERPRET**   In this problem a wheel has been placed on a slope. We want to apply a horizontal force at its highest point to keep it from rolling down.

DEVELOP   Consider the conditions for static equilibrium of the wheel, under the action of the forces shown. Here F_{app} is the applied horizontal force, F_c is the contact force of the incline, normal plus friction, and we assumed that the CM is at the center.

For the wheel to remain in static equilibrium, the forces must satisfy Equation 12.1. Our plan is to compute the torques about P using Equation 12.2. Note that the contact force, F_c does not contribute.

EVALUATE   The torques about the point of contact sum to zero, or

$$0 = (\Sigma \tau)_P = F_{\text{app}}R(1 + \cos\theta) - MgR\sin\theta$$

Therefore, the applied force is $F_{\text{app}} = Mg\frac{\sin\theta}{1+\cos\theta} = Mg\tan\left(\frac{\theta}{2}\right)$.

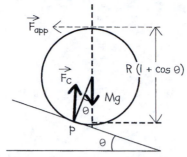

ASSESS The applied force vanishes when $\theta = 0$ (flat surface), and become maximum when $\theta = 90°$. In this limit, $F_{app} = Mg$, and points vertically upward.

53. **INTERPRET** In this problem a rectangular block is placed on an incline. Given the coefficient of friction between the block and the incline, we'd like to find out under what condition the block would slide first.

DEVELOP We suppose that the block is oriented with two sides parallel to the direction of the incline, and that its CM is at the center. The condition for sliding is that

$$mg\sin\theta > f_s^{max} = \mu_s n = \mu_s mg\cos\theta$$

or $\tan\theta > \mu_s$. The condition for tipping over is that the CM lie to the left of the lower corner of the block (see sketch). Thus $\theta > \alpha$, where

$$\alpha = \tan^{-1}\left(\frac{w}{h}\right)$$

is the diagonal angle of the block.

EVALUATE For the rectangular block with $w = h/2$, it tips over when $\theta > \alpha = \tan^{-1}(w/h) = \tan^{-1}(1/2)$ but will slide when $\theta > \tan^{-1}\mu_s$. Thus, if $\mu_s < \tan\alpha = 1/2$, the block in the Problem 52 will slide before tipping.

ASSESS We find that sliding happens first if $\mu_s < w/h$. This makes sense because when the coefficient of friction is small, the block has a greater tendency to slide. On the other hand, when the coefficient of friction is large ($\mu_s > w/h$), we'd expect tipping to take place first.

55. **INTERPRET** In this problem we want to verify the statement that the choice of pivot point does not matter when applying the conditions for static equilibrium.

DEVELOP With reference to Fig. 12.29, we follow the hint given in the problem statement and write

$$\vec{\tau}_P = \Sigma\vec{r}_{Pi} \times \vec{F}_i = \Sigma(\vec{r}_{Oi} + \vec{R}) \times \vec{F}_i = \Sigma\vec{r}_{Oi} \times \vec{F}_i + \vec{R} \times \Sigma\vec{F}_i = \vec{\tau}_O + \vec{R} \times \vec{F}_{net}$$

EVALUATE When the system is in static equilibrium, the total force and torque acting on the system vanish: $\vec{F}_{net} = 0$ and $\vec{\tau}_P = \vec{0}$. Therefore, we have $\vec{\tau}_P = \vec{\tau}_O = \vec{0}$, i.e., the total torque about any two points is the same.

ASSESS If the angle θ is increased, then the corresponding coefficient of friction must also be increased in order to keep the pole from slipping.

57. **INTERPRET** This problem is about static equilibrium. The forces acting on the pole are the tension in the rope, gravity acting at the CM at its center, and the contact force of the incline (perpendicular component n and parallel component f). We want to find the minimum coefficient of friction that will keep the pole from slipping.

DEVELOP Consideration of Equation 12.2 about the CM shows that a frictional force f must be acting up the plane if the rod is to remain in static equilibrium. Since the weight of the rod, mg, and the normal force, n, contribute no torques about the CM, there must be a force to oppose the torque of the tension, T. The equations for static equilibrium (parallel and perpendicular components of Equation 12.1, and CCW-positive component of Equation 12.2) are:

$$0 = \Sigma F_\parallel = f + T\cos\theta - mg\sin\theta$$
$$0 = \Sigma F_\perp = n - T\sin\theta - mg\cos\theta$$
$$0 = (\Sigma\tau)_{cm} = T(L/2)\cos\theta - f(L/2)$$

The solutions for the forces are $f = \frac{1}{2}mg\sin\theta$, $T = \frac{1}{2}mg\tan\theta$, and

$$n = \frac{1}{2}mg\left(2\cos\theta + \frac{\sin^2\theta}{\cos\theta}\right)$$

subject to the condition that $f \le \mu n$. Therefore,

$$\sin\theta \le \mu\left(2\cos\theta + \frac{\sin^2\theta}{\cos\theta}\right) \rightarrow \mu \ge \frac{\tan\theta}{2 + \tan^2\theta}$$

EVALUATE   By use of the identities $\sin 2\theta = 2\sin\theta\cos\theta$, $\cos 2\theta = \cos^2\theta - \sin^2\theta$, and $\sin^2\theta = 1 - \cos^2\theta$, this may be rewritten as

$$\mu \ge \frac{\sin 2\theta}{3 + \cos 2\theta}$$

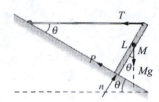

ASSESS   If the angle θ is increased, then the corresponding coefficient of friction must also be increased in order to keep the pole from slipping.

59.   **INTERPRET**   We use equilibrium methods to find the horizontal component of force on a bookshelf bracket tab. The bookshelf is in equilibrium, so the sum of forces and the sum of torques are both zero. Since the sum of forces is zero, we may use any point as the pivot for calculating the torques. We would expect that the horizontal force is much larger than the weight of the books, since the books have more leverage than the bracket.

DEVELOP   We start by drawing a diagram showing the forces and their approximate locations, as shown in the figure. The mass of books is $m = 32$ kg, the distance $y = 4.5$ cm, and the distance $x = 12$ cm.

Since we know nothing about the force $\vec{F}_b$ acting on the bottom corner of the bracket, we will use that point as our pivot point. The sum of torques around this point must be zero; use this to find F_h.

EVALUATE $\sum\tau = 0 = F_g x - F_h y \rightarrow F_h = F_g \frac{x}{y} = mg\frac{x}{y} = 836$ N.

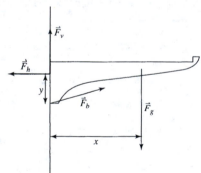

ASSESS   The force is much larger than the weight of the books, as we expected.

61.   **INTERPRET**   We have in this problem a disk with an additional off-center mass, and we need to find the angle at which it will balance on a slope. We do this by finding the torques on the disk, and setting the angle so that the sum of torques is zero. We will use the point of contact between the disk and the ramp as the pivot point for calculating torques.

DEVELOP   As we can see from the figure, the magnitude of the clockwise torque around the contact point is $\tau_1 = F_{gM} R\sin\theta$. The magnitude of the counterclockwise torque is $\tau_2 = F_{gm}(R\cos\phi - R\sin\theta)$. In order to balance, the two torques must have the same magnitude.

EVALUATE

$$\tau_1 = \tau_2 \rightarrow F_{gM} R\sin\theta = F_{gm}R(\cos\phi - \sin\theta) \rightarrow M\,g\!\!\!/R\sin\theta = m\,g\!\!\!/R(\cos\phi - \sin\theta)$$
$$\rightarrow (M + m)\sin\theta = m\cos\phi$$
$$\rightarrow \phi = \cos^{-1}\left(\frac{M + m}{m}\sin\theta\right)$$

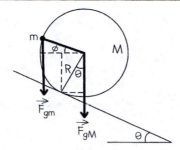

ASSESS   We can check this equation for some special cases. If m is zero, then the disk will balance anywhere as long as θ is zero also. If θ is zero, then the disk will balance at $\phi = \pm 90°$.

63.   **INTERPRET**   This problem involves two masses, separated by a massless rod, hovering far above the Earth. We need to find the net gravitational force on these masses, the net torque around the center of mass, and the center of gravity. We will use the equation for gravitational force, as well as the equation for torque. The center of gravity is not the same as the center of mass, in this case, since there is a change in gravitational field between the two masses.

DEVELOP   We start with a diagram, as shown in the figure. We use $\vec{F}_g = G\frac{m_E m}{r^2}$ to find the force on each end of the spacecraft, then find the vector sum of the forces to obtain the net gravitational force. Use the forces on each end, and the angle found from the figure, to find the torque on each end and thus the net torque around the center of mass. The center of gravity is the point on the spacecraft at which the torques will be zero. We have labeled the distance from the left end of the spacecraft to the center of gravity x.

EVALUATE

(a) The gravitational force on the left mass is $\vec{F}_1 = -G\frac{m_E m}{(2R_E)^2}\hat{j}$. The gravitational force on the right mass

is $F_2 = -G\frac{m_E m}{(2R_E)^2 + (R_E)^2}$, directed at an angle $\theta = \tan^{-1}\left(\frac{R_E}{2R_E}\right) = \tan^{-1}\left(\frac{1}{2}\right) = 26.7°$. We note in passing that $\sin\theta = \frac{1}{\sqrt{5}}$

and $\cos\theta = \frac{2}{\sqrt{5}}$. We break F_2 into x and y components: $\vec{F}_2 = -G\frac{m_E m}{(2R_E)^2 + (R_E)^2}(\sin\theta\hat{i} + \cos\theta\hat{j}) = -G\frac{m_E m}{5R_E^2}\left(\frac{1}{\sqrt{5}}\hat{i} + \frac{2}{\sqrt{5}}\hat{j}\right)$.

Now we add the vector components to find the total force: $\vec{F} = -G\frac{m_E m}{R_E^2}\left[\left(\frac{1}{4} + \frac{2}{\sqrt{5}}\right)\hat{j} + \left(\frac{1}{\sqrt{5}}\right)\hat{i}\right]$. The magnitude of this

force is $|\vec{F}| = G\frac{m_E m}{R_E^2}[1.229]$, and the direction of the force is $\tan^{-1}\left(\frac{\frac{1}{\sqrt{5}}}{\frac{1}{4} + \frac{2}{\sqrt{5}}}\right) = 21.3°$ left of the negative y axis.

(b) The net torque around the center of mass is

$$\tau = -F_1\frac{R_E}{2} + F_{2j}\frac{R_E}{2} = -G\frac{m_E m}{4R_E^2}\left(\frac{R_E}{2}\right) + G\frac{m_E m}{5R_E^2}\frac{2}{\sqrt{5}}\left(\frac{R_E}{2}\right)$$

$$\tau = G\frac{m_E m}{R_E}\left(-\frac{1}{8} + \frac{1}{5\sqrt{5}}\right) = G\frac{m_E m}{R_E}(-0.0356)$$

(c) To find the center of gravity, repeat the calculation for (b) but use x for the left-hand distance and $(R_E - x)$ for the right-hand distance, setting $\tau = 0$. Solving this for x will give the distance of the center of gravity from the left end, which we can compare to $\frac{R_E}{2}$.

$$0 = -F_1 x + F_{2j}(R_E - x) = -G\frac{m_E m}{4R_E^2}(x) + G\frac{m_E m}{5R_E^2}\frac{2}{\sqrt{5}}(R_E - x)$$

$$0 = -\frac{1}{4}x + \frac{2}{5\sqrt{5}}(R_E - x) \rightarrow x\left(\frac{1}{4} + \frac{2}{5\sqrt{5}}\right) = R_E\frac{2}{5\sqrt{5}}$$

$$\rightarrow x = 0.417R_E$$

$$\rightarrow \frac{R_E}{2} - x = 0.083$$

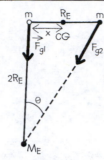

ASSESS The torque is negative, which in this case means counterclockwise as we would expect. The center of gravity in this case is not at the center of mass, due to the decrease in gravitational force with altitude.

65. **INTERPRET** We analyze the forces on a roof rafter, as shown in Figure 12.37. We use equilibrium techniques, with the sum of torques equaling zero, to determine whether the tie beam will withstand the force.

    DEVELOP We start with a diagram showing the relevant forces, as shown in the figure. We will analyze just half of the roof, since it is symmetric. We use the point where the tie beam attaches as the pivot point.
    $F_{wall} = F_{snow} = \frac{m}{2} g$, where m is the mass of snow and building materials, $m = 170$ kg. We use half this mass because the other half is on the other half of the roof. The angle of the roof is $\theta = \tan^{-1}\left(\frac{y_1 + y_2}{\frac{1}{2}w}\right)$, where the width of the roof is $w = 9.6$ m, so $\theta = 39.8°$. This allows us to calculate values of the torque arms for the snow and wall forces:

    $$\tan\theta = \frac{y_1}{x_1} \rightarrow x_1 = \frac{y_1}{\tan\theta} = 0.96 \text{ m, and similarly } x_2 = 3.84 \text{ m.}$$

    We see from the figure also that the force on the tie beam is equal to the horizontal force at the peak due to the other half of the roof. So we can use $\sum\tau = 0$ to find F_{roof}, and if this has a magnitude greater than $F = 7500$ N then the roof collapses.

    EVALUATE

    $$\sum\tau = 0 \rightarrow F_{wall}x_2 + F_{snow}x_1 = F_{roof}y_1 \rightarrow F_{roof} = \frac{F_{wall}x_2 + F_{snow}x_1}{y_1}$$

    $$\rightarrow F_{roof} = 4170 \text{ N}$$

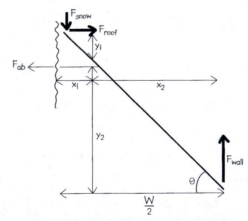

    ASSESS The roof holds. (You were rather hoping your professor's roof would collapse, weren't you?) Note that the force increases with decreasing values of y_1, as you would expect.

13

OSCILLATORY MOTION

EXERCISES

Section 13.1 Describing Oscillatory Motion

17. **INTERPRET** The question here is about the oscillatory behavior of the violin string. Given the frequency of oscillation, we are asked to find the period.

    DEVELOP The relationship between period and frequency is given by Equation 13.1, $T = \frac{1}{f}$.

    EVALUATE Using Equation 13.1, we obtain

    $$T = \frac{1}{f} = \frac{1}{440\ \text{Hz}} = 2.27 \times 10^{-3}\ \text{s}$$

    ASSESS The period is the oscillation is the inverse of the frequency. Note that the unit of frequency is the hertz: $1\ \text{Hz} = 1\ \text{s}^{-1}$.

19. **INTERPRET** The problem involves simple harmonic motion. We want to write down an expression that characterizes the oscillation, given its amplitude, frequency, and the function at $t = 0$.

    DEVELOP The general expression of the position of an object undergoing simple harmonic motion is given by Equation 13.8:

    $$x(t) = A\cos(\omega t + \phi)$$

    where A is the amplitude, ω is the angular frequency, and ϕ is the phase constant. By taking the time derivative of $x(t)$, we obtain the corresponding velocity as a function of time (see Equation 13.9):

    $$v(t) = \frac{dx(t)}{dt} = -A\omega\sin(\omega t + \phi)$$

    EVALUATE (a) Since the displacement is a maximum at $t = 0$, the phase constant is zero: $\phi = 0$. Use Equation 13.8 with $A = 10$ cm, $\omega = 2\pi(5\ \text{Hz}) = 10\pi\,\text{s}^{-1}$, the displacement is found to be

    $$x(t) = A\cos(\omega t + \phi) = (10\ \text{cm})\cos[(10\pi\,\text{s}^{-1})t]$$

    (b) Equation 13.9 shows that the maximum (positive) velocity occurs at $t = 0$ if $\sin\phi = -1$ or $\phi = -\pi/2$. Therefore, with $A = 2.5$ cm, $\omega = 5\ \text{s}^{-1}$, we have

    $$x(t) = A\cos(\omega t + \phi) = (2.5\ \text{cm})\cos[(5\ \text{s}^{-1})t - \pi/2] = (2.5\ \text{cm})\sin[(5\ \text{s}^{-1})t]$$

    where we have used $\cos(\omega t - \pi/2) = \sin\omega t$.

    ASSESS For a system undergoing simple harmonic motion, once the position as a function of time is given in the form of Equation 13.8, physical quantities such as velocity, acceleration, angular frequency and period can all be readily determined.

Section 13.2 Simple Harmonic Motion

21. **INTERPRET** In this problem a mass attached to a spring undergoes simple harmonic motion. Given the mass, the spring constant, and the amplitude, we are asked to compute the frequency and the period of oscillation, the maximum velocity, and the maximum force in the spring.

    DEVELOP Given the spring constant k and the mass, the frequency and period of the oscillation may be obtained from Equations 13.8b and c:

    $$f = \frac{1}{2\pi}\sqrt{\frac{k}{m}} \qquad T = \frac{1}{f} = 2\pi\sqrt{\frac{m}{k}}$$

Using Equations 13.9 and 13.10, the maximum speed and the maximum force are:

$$v_{max} = \omega A$$
$$F_{max} = ma_{max} = m\omega^2 A$$

EVALUATE   **(a)** For a mass on a spring, Equation 13.7a gives

$$\omega = \sqrt{\frac{k}{m}} = \sqrt{\frac{5.6 \text{ N/m}}{0.2 \text{ kg}}} = 5.29 \text{ s}^{-1}$$

Using Equation 13.7b, the frequency of oscillation is

$$f = \frac{\omega}{2\pi} = \frac{5.29 \text{ s}^{-1}}{2\pi} = 0.842 \text{ Hz}$$

(b) The period of oscillation is the inverse of the frequency (see Equation 13.7c):

$$T = \frac{1}{f} = 1.19 \text{ s}$$

(c) From Equation 13.9, we find the maximum speed to be

$$v_{max} = \omega A = (5.29 \text{ s}^{-1})(0.25 \text{ m}) = 1.32 \text{ m/s}$$

(d) Similarly, using Equation 13.10, the maximum force in the spring is

$$F_{max} = ma_{max} = m\omega^2 A = (0.2 \text{ kg})(5.29 \text{ s}^{-1})^2(0.25 \text{ m}) = 1.40 \text{ N}$$

ASSESS   The force in the spring can also be calculated by using Equation 13.2: $F = kx$. In fact,

$$F_{max} = ma_{max} = m\omega^2 A = kx_{max} = kA = (5.6 \text{ N/m})(0.25 \text{ m}) = 1.40 \text{ N}$$

23.   **INTERPRET**   In this problem a mass attached to a spring executes simple harmonic motion. Given the mass and the spring constant, we are asked to compute the frequency and the period of the oscillation.
       DEVELOP   Once the spring constant k and the mass m are known, the frequency and period of the oscillation may be obtained from Equations 13.8b and c:

$$f = \frac{1}{2\pi}\sqrt{\frac{k}{m}} \qquad T = \frac{1}{f} = 2\pi\sqrt{\frac{m}{k}}$$

       EVALUATE   The above equations give

$$f = \frac{1}{2\pi}\sqrt{\frac{k}{m}} = \frac{1}{2\pi}\sqrt{\frac{26,000 \text{ N/m}}{1900 \text{ kg}}} = 0.589 \text{ Hz} \qquad T = \frac{1}{f} = 1.70 \text{ s}$$

       ASSESS   Given the large spring constant and mass involved, a period of 1.70 s for the automobile suspension is reasonable.

25.   **INTERPRET**   In this problem a mass which is being attached to a spring undergoes simple harmonic motion. Given the maximum speed and the maximum acceleration of the mass, we want to find the angular frequency, the spring constant, and the amplitude of the oscillation.
       DEVELOP   The maximum speed and the maximum acceleration of the mass are given by Equations 13.9 and 13.10:

$$v_{max} = \omega A \qquad a_{max} = \omega^2 A$$

From a_{max} and v_{max}, the angular frequency may be obtained as $\omega = \frac{a_{max}}{v_{max}}$. Once ω is known, the spring constant and the amplitude of the oscillation can be calculated using:

$$k = m\omega^2 \qquad A = \frac{a_{max}}{\omega^2} = a_{max}\left(\frac{v_{max}}{a_{max}}\right)^2 = \frac{v_{max}^2}{a_{max}}$$

       EVALUATE   **(a)** The angular frequency is

$$\omega = \frac{a_{max}}{v_{max}} = \frac{15 \text{ m/s}^2}{3.5 \text{ m/s}} = 4.286 \text{ s}^{-1} \approx 4.29 \text{ s}^{-1}$$

       (b) The spring constant is $k = m\omega^2 = (0.05 \text{ kg})(4.29 \text{ s}^{-1})^2 = 0.918 \text{ N/m}$.
       (c) Similarly, the amplitude of the motion is

$$A = \frac{v_{max}^2}{a_{max}} = \frac{(3.5 \text{ m/s})^2}{15 \text{ m/s}^2} = 0.817 \text{ m}$$

ASSESS   To check that our results are correct, let's compute the amplitude A using the value of ω found in **(a)**. In either way, we have

$$A = \frac{v_{max}}{\omega} = \frac{3.5 \text{ m/s}}{4.286 \text{ s}^{-1}} = 0.817 \text{ m}$$

$$A = \frac{a_{max}}{\omega^2} = \frac{15 \text{ m/s}^2}{(4.286 \text{ s}^{-1})^2} = 0.817 \text{ m}$$

The results indeed agree.

27.   **INTERPRET**   The problem here is about simple harmonic motion of a particle. Given its maximum speed and maximum acceleration, we want to find the angular frequency, the period, and the amplitude of the oscillation.

DEVELOP   The maximum speed and the maximum acceleration of the mass are given by Equations 13.9 and 13.10:

$$v_{max} = \omega A \qquad a_{max} = \omega^2 A$$

From a_{max} and v_{max}, the angular frequency may be obtained as $\omega = \frac{a_{max}}{v_{max}}$. Once ω is known, the period and the amplitude of the oscillation can be computed using

$$T = \frac{1}{f} = \frac{2\pi}{\omega} \qquad A = \frac{a_{max}}{\omega^2} = a_{max}\left(\frac{v_{max}}{a_{max}}\right)^2 = \frac{v_{max}^2}{a_{max}}$$

EVALUATE   **(a)** The angular frequency is

$$\omega = \frac{a_{max}}{v_{max}} = \frac{3.1 \text{ m/s}^2}{1.4 \text{ m/s}} = 2.21 \text{ s}^{-1}$$

(b) The period of the motion is $T = \frac{2\pi}{\omega} = \frac{2\pi}{2.21 \text{ s}^{-1}} = 2.84 \text{ s}$.

(c) Similarly, the amplitude of the motion is

$$A = \frac{v_{max}^2}{a_{max}} = \frac{(1.4 \text{ m/s})^2}{3.1 \text{ m/s}^2} = 0.632 \text{ m}$$

ASSESS   To check that our results are correct, let's compute the amplitude A using the value of ω found in **(a)**:

$$A = \frac{v_{max}}{\omega} = \frac{1.4 \text{ m/s}}{2.21 \text{ s}^{-1}} = 0.63 \text{ m}$$

$$A = \frac{a_{max}}{\omega^2} = \frac{3.1 \text{ m/s}^2}{(2.21 \text{ s}^{-1})^2} = 0.63 \text{ m}$$

The results indeed agree.

Section 13.3 Applications of Simple Harmonic Motion

29.   **INTERPRET**   This problem is about the simple harmonic motion of a grandfather clock. We want to find the time interval between successive ticks.

DEVELOP   The period of a simple pendulum is given by Equation 13.15:

$$T = 2\pi\sqrt{\frac{L}{g}}$$

We note that the clock ticks twice each period of oscillation.

EVALUATE   Using Equation 13.15, the time between ticks is

$$\Delta t = \frac{T}{2} = \pi\sqrt{\frac{L}{g}} = \pi\sqrt{\frac{1.45 \text{ m}}{9.8 \text{ m/s}^2}} = 1.21 \text{ s}$$

ASSSESS   One tick every 1.21 s seems reasonable. Note that if we increase the length of the pendulum, then the period and the time between ticks will increase as well.

31.   **INTERPRET**   The problem is about a physical pendulum—a meter stick suspended from one end. We want to know its period of oscillation.

DEVELOP   The period of a physical pendulum can be obtained from Equation 13.13.

$$T = \frac{2\pi}{\omega} = 2\pi\sqrt{\frac{I}{mgL}}$$

The meter stick is a physical pendulum whose CM is $L = L_0/2 = 0.50$ m below the point of suspension through one end. The rotational inertia of the stick about one end is $I = \frac{1}{3}mL_0^2 = \frac{1}{3}m(1 \text{ m})^2$.

EVALUATE   Using Equation 13.13, the period of the meter stick is

$$T = 2\pi\sqrt{\frac{I}{mgL}} = 2\pi\sqrt{\frac{mL_0^2/3}{mgL_0/2}} = 2\pi\sqrt{\frac{2L_0}{3g}} = 2\pi\sqrt{\frac{2(1 \text{ m})}{3(9.8 \text{ m/s}^2)}} = 1.64 \text{ s}$$

ASSESS   A simple experiment can be carried out to verify that a period of 1.64 s (or roughly 5 complete oscillations in 8 seconds) for the meter stick is reasonable.

Section 13.4 Circular and Harmonic Motion

33.   **INTERPRET**   The problem here is about a body undergoing harmonic motion in two dimensions, with a different angular frequency in each dimension.

DEVELOP   Let the angular frequencies be ω_x and ω_y, such that

$$\frac{\omega_x}{\omega_y} = \frac{1.75}{1} = \frac{7}{4}$$

Since $T = 2\pi/\omega$, the ratio of the periods for the x and y components of the motion is then equal to

$$\frac{T_x}{T_y} = \frac{2\pi/\omega_x}{2\pi/\omega_y} = \frac{\omega_y}{\omega_x} = \frac{4}{7}$$

EVALUATE   The above equation gives $7T_x = 4T_y$, which means that seven oscillations in the x direction are completed when four are completed in the y direction.

ASSESS   Since $\omega_x = 1.75\,\omega_y > \omega_y$, we have $T_x = T_y/1.75 < T_y$. The smaller the period, the greater the number of oscillations completed in a given time interval.

Section 13.5 Energy in Simple Harmonic Motion

35.   **INTERPRET**   In this problem we have a mass undergoing a simple harmonic motion. We'd like to know its amplitude of oscillation, if the total energy of the mass-spring system is given.

DEVELOP   The total energy stored in the system is

$$E = U + K = \frac{1}{2}kx^2 + \frac{1}{2}mv^2 = \frac{1}{2}k(A\cos\omega t)^2 + \frac{1}{2}m(-A\omega\sin\omega t)^2$$

$$= \frac{1}{2}kA^2\cos^2\omega t + \frac{1}{2}m\omega^2 A^2\sin^2\omega t = \frac{1}{2}kA^2(\cos^2\omega t + \sin^2\omega t)$$

$$= \frac{1}{2}kA^2$$

where we have used Equations 13.8 and 13.9 for $x(t)$ and $v(t)$, with $\phi = 0$. The equation relates the total energy to the amplitude of the oscillation.

EVALUATE   From the above expression for E, we find the amplitude to be

$$A = \sqrt{\frac{2E}{m\omega^2}} = \sqrt{\frac{2E}{m(2\pi f)^2}} = \sqrt{\frac{2(0.51 \text{ J})}{(0.45 \text{ kg})(2\pi \times 1.2 \text{ Hz})^2}} = 0.20 \text{ m}$$

ASSESS   The total energy stored in the system, E, is found to be proportional to A^2, and is independent of time, as required by energy conservation, even though both K and U vary with time.

Sections 13.6 and 13.7 Damped Harmonic Motion and Resonance

37.   **INTERPRET**   The problem is about damped harmonic motion. We are interested in finding out how long it takes for vibration amplitude of the piano string to drop to half its initial value.

DEVELOP   The solution to the damped harmonic motion is given by Equation 13.17:

$$x(t) = Ae^{-bt/2m}\cos(\omega t + \phi)$$

Thus, the amplitude is half the initial value when $e^{-bt/2m} = \frac{1}{2}$.

EVALUATE   Taking the natural logarithms of both sides of the equation above gives $bt/2m = \ln 2$. Thus, the time for the amplitude to be halved is

$$t = \frac{2m}{b} \ln 2 = \frac{\ln 2}{2.8 \text{ s}^{-1}} = 0.248 \text{ s}$$

ASSESS   Since the amplitude drops by half in just 0.248 s, we conclude that the damping must be rather strong.

39.   **INTERPRET**   In this problem we want to find the speed of the car that leads to a maximum vibration amplitude.

DEVELOP   The peak amplitude occurs at resonance, when $\omega_d = \omega_0$, so the car receives an impulse from the bumps once each period. It therefore travels the distance between bumps in one period.

EVALUATE   The condition for resonance is $L_0 = 40 \text{ m}/vT_0$, or

$$v = \frac{L_0}{T_0} = L_0 f_0 = (40 \text{ m})(0.45 \text{ Hz}) = 18 \text{ m/s} = 64.8 \text{ km/h}$$

ASSESS   If the spacing between bumps increases, then the car speed must also go up to meet the rather unpleasant resonance condition!

PROBLEMS

41.   **INTERPRET**   The problem involves two identical mass-spring systems undergoing simple harmonic motion. We want to find the time elapsed between releasing the masses so that the two oscillations differ in phase by $\pi/2$.

DEVELOP   Suppose that both masses are released from their maximum positive displacements, the first at $t = 0$ and the second at $t = t_0$. Then, since the phase (the argument of the cosine in Equation 13.8) of each motion is zero at release, $\phi_{10} = 0$ and $\omega_2 t_0 + \phi_{20} = 0$. The difference in phase is

$$\Delta\phi = \phi_1 - \phi_2 = (\omega_1 t + \phi_{10}) - (\omega_2 t + \phi_{20}) = (\omega_1 - \omega_2)t + (\phi_{10} - \phi_{20}) = -\phi_{20}$$

where $\omega_1 = \omega_2 = \sqrt{k/m}$ for identical mass-spring systems. This allows us to solve for t_0.

EVALUATE   With $\Delta\phi = \pi/2$, we have

$$t_0 = \frac{-\phi_{20}}{\omega_2} = \frac{\pi/2}{\sqrt{k/m}} = \frac{\pi/2}{\sqrt{(2.2 \text{ N/m})/(0.43 \text{ kg})}} = 0.694 \text{ s}$$

ASSESS   In terms of the period, the time is $t_0 = \frac{-\phi_{20}}{2\pi} T = \frac{\pi/2}{2\pi} T = \frac{1}{4} T$. This makes sense since the phase change in one period is 2π.

43.   **INTERPRET**   We want to show that $x(t) = A \sin \omega t$ is a solution to the differential equation given in Equation 13.3.

DEVELOP   Equation 13.3, $m\frac{d^2 x}{dt^2} = -kx$, describes a system undergoing simple harmonic motion in one dimension. To show that $x(t) = A \sin \omega t$ is a solution to the differential equation, we simply compute the second time derivative of x, and show that the left-hand side is the same as the right-hand side.

EVALUATE   By differentiating $x = A \sin \omega t$ with respect to t twice, we obtain

$$\frac{dx}{dt} = \frac{d}{dt}(A \sin \omega t) = \omega A \cos \omega t$$

$$\frac{d^2 x}{dt^2} = \frac{d}{dt}(\omega A \cos \omega t) = -\omega^2 A \sin \omega t = -\omega^2 x$$

Substituting into Equation 13.3, we find $m(-\omega^2 x) = -kx$, which is satisfied if $\omega^2 = k/m$.

ASSESS   Equation 13.3 describes a system with periodic motion. Since the cosine function is also periodic, it can be used as a solution to the differential equation.

45.   **INTERPRET**   In this problem we want to find the period of a pendulum inside a rocket in various accelerating conditions. It may be helpful to think of an elevator instead of a rocket in this problem.

DEVELOP   Let a be the acceleration of the pendulum relative to the rocket, and let a_0 be the acceleration of the rocket relative to the ground (assumed to be an inertial system). Then Newton's second law is

$$\sum \vec{F} = m(\vec{a} + \vec{a}_0) \quad \rightarrow \quad \sum \vec{F} - m\vec{a}_0 = m\vec{a}$$

The rotational analog of this equation is the appropriate generalization of Equation 13.11 for a simple pendulum in an accelerating frame. The "fictitious" torque (about the point of suspension), $\vec{r} \times (-m\vec{a}_0)$, can be combined with the torque of gravity, $\vec{r} \times m\vec{g}$, if we replace g by an effective gravitational acceleration $|\vec{g} - \vec{a}_0|$, while the right-hand side, $|\vec{r} \times m\vec{a}| = I\alpha$, is the same as Equation 13.11. For small oscillations about the equilibrium position (which depends on $\vec{a}_0$), the period is

$$T = 2\pi \sqrt{\frac{L}{|\vec{g} - \vec{a}_0|}}$$

EVALUATE   (a) If $\vec{a}_0 = 0$, then $T = 2\pi\sqrt{L/g}$ as before.

(b) Take the y-axis positive up. Then $\vec{a}_0 = \frac{1}{2}g\hat{j}$ and $\vec{g} = -g\hat{j}$. This gives

$$T = 2\pi \sqrt{\frac{L}{g + \frac{1}{2}g}} = 2\pi \sqrt{\frac{2L}{3g}}$$

(c) If $\vec{a}_0 = -\frac{1}{2}g\hat{j}$, then

$$T = 2\pi \sqrt{\frac{L}{g - \frac{1}{2}g}} = 2\pi \sqrt{\frac{2L}{g}}$$

(d) If $\vec{a}_0 = g\hat{j}$, then $T = \infty$, since there is no restoring torque and the pendulum does not oscillate.

ASSESS   The period of a swinging pendulum increases as the effective gravitational acceleration decreases.

47.   **INTERPRET**   The problem is about simple harmonic motion of the mass-spring system in the vertical direction.

DEVELOP   We note that at the highest point, there is no spring force, since the spring is unstretched. Therefore, the acceleration is just g (downward). This is also the maximum acceleration during the simple harmonic motion, since a_{max} occurs where the displacement is a maximum. Thus, using Equation 13.10, we have $a_{max} = g = \omega^2 A$. Once ω is known, the period of oscillation can be obtained by using Equation 13.7b.

EVALUATE   The peak-to-peak displacement is $2A = 5.8$ cm. Thus, using Equation 13.7b, the period is

$$T = \frac{1}{f} = \frac{2\pi}{\omega} = 2\pi \sqrt{\frac{A}{g}} = 2\pi \sqrt{\frac{0.029 \text{ m}}{9.8 \text{ m/s}^2}} = 0.342 \text{ s}$$

ASSESS   Our result confirms that the greater the amplitude, the longer the period of oscillation.

49.   **INTERPRET**   In this problem we want to find the radius of the solid disk such that its vertical oscillation will have the same period as torsional oscillation.

DEVELOP   If the periods are the same, then the angular frequencies must be the same as well. The angular frequencies for the vertical and torsional oscillations are given by Equations 13.7a and 13.12:

$$\omega = \sqrt{\frac{k}{m}} \quad \text{(vertical)}$$

$$\omega = \sqrt{\frac{\kappa}{I}} \quad \text{(torsional)}$$

EVALUATE   The rotational inertia of the solid disk is $I = mR^2/2$. Equating the angular frequencies for vertical and torsional oscillations, we find

$$\frac{k}{m} = \frac{\kappa}{I} = \frac{\kappa}{mR^2/2} \quad \rightarrow \quad R = \sqrt{\frac{2\kappa}{k}}$$

ASSESS   The torsional constant κ has units N · m while the spring constant has units N/m. Thus, the quantity $\sqrt{\kappa/k}$ has units of meters. The result makes sense because if the radius is increased, then the rotational inertia also increases. To keep the period the same would require a greater torsional constant.

51.   **INTERPRET**   In this problem we want to find the mass of the tire valve stem, given the period of oscillation of the bicycle wheel.

DEVELOP   The bicycle wheel may be regarded as a physical pendulum, with rotational inertia $I = MR^2 + mR^2 = (M + m)R^2$ about its central axle, where $M = 600$ g is the mass of the wheel (thin ring) and m is the mass of the valve stem (a circumferential point mass).

The distance of the CM from the axle is given by $(M + m)L = mR$ (this is just Equation 9.2, with origin at the center of the wheel so $x_1 = 0$ for M, $x_2 = R$ for m, and $x_{cm} = L$).

For small oscillations, the angular frequency is given by Equation 13.13, where $M + m$ is the total mass. Knowing the angular frequency, or the period allows us to determine m, the mass of the valve stem.

EVALUATE   The period of oscillation is

$$T = \frac{2\pi}{\omega} = 2\pi\sqrt{\frac{I}{(M + m)gL}} = 2\pi\sqrt{\frac{(M + m)R^2}{mgR}} = 2\pi\sqrt{\left(\frac{M}{m} + 1\right)\frac{R}{g}}$$

Solving for m, we find

$$m = M\left[\frac{g}{R}\left(\frac{T}{2\pi}\right)^2 - 1\right]^{-1} = (600 \text{ g})\left[\frac{9.8 \text{ m/s}^2}{0.3 \text{ m}}\left(\frac{12 \text{ s}}{2\pi}\right)^2 - 1\right]^{-1} = 5.04 \text{ g}$$

ASSESS   A mass of 5.04 g is reasonable for the valve stem. Note that this value is much smaller than the mass of the wheel, as we expect.

53.   **INTERPRET**   This problem is about simple harmonic motion of a pendulum. We want to know what its period is when it is treated as a simple pendulum or as a physical pendulum.

DEVELOP   The periods of a simple pendulum and a physical pendulum are given by

$$T_{simple} = 2\pi\sqrt{\frac{L}{g}} \qquad T_{phys} = 2\pi\sqrt{\frac{I}{mgL}}$$

EVALUATE   When treated as a simple pendulum, its period is

$$T_{simple} = 2\pi\sqrt{\frac{L}{g}} = 2\pi\sqrt{\frac{0.875 \text{ m}}{9.8 \text{ m/s}^2}} = 1.877 \text{ s}$$

On the other hand, if we regard the pendulum as a physical pendulum with rotational inertia $I = \frac{2}{5}mR^2 + mL^2$, then its period becomes

$$T_{phys} = 2\pi\sqrt{\frac{I}{mgL}} = 2\pi\sqrt{\frac{\frac{2}{5}mR^2 + mL^2}{mgL}} = 2\pi\sqrt{\frac{L}{g}}\sqrt{1 + \frac{2R^2}{5L^2}} = T_{simple}\sqrt{1 + \frac{2R^2}{5L^2}}$$

$$= T_{simple}\sqrt{1 + \frac{2}{5}\left(\frac{0.075 \text{ m}}{0.875 \text{ m}}\right)^2} = (1.00147)T_{simple} = 1.880 \text{ s}$$

The fractional error is $\frac{T_{phys} - T_{simple}}{T_{phys}} = 0.147\%$.

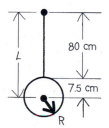

ASSESS   The period found by assuming the pendulum to be simple is slightly less than if we treat it as a physical pendulum.

55.   **INTERPRET**   This problem is about simple harmonic motion of a hoop. We treat it as a physical pendulum, and solve for its period.

DEVELOP   Using Equation 13.13, the period of a physical pendulum is

$$T_{phys} = 2\pi\sqrt{\frac{I}{mgL}}$$

where I is the rotational inertia of the hoop. We shall calculate I by using the parallel-axis theorem.

EVALUATE   Using parallel-axis theorem, the rotational inertia of the hoop about the pivot is

$$I = I_0 + mh^2 = mR^2 + mR^2 = 2mR^2$$

Thus, using Equation 13.13, we find the period to be

$$T_{\text{phys}} = 2\pi\sqrt{\frac{I}{mgL}} = 2\pi\sqrt{\frac{2mR^2}{mgR}} = 2\pi\sqrt{\frac{2R}{g}}$$

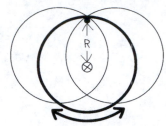

ASSESS   The period of our oscillating hoop is the same as a simple pendulum of mass m and length $L = 2R$.

57.   **INTERPRET**   This problem is about simple harmonic motion in two dimensions. We want to verify that the motion is elliptical if the two components have different amplitudes and are $\pi/2$ out of phase.

DEVELOP   Simple harmonic motions in the x and y directions, with different amplitudes and $\pi/2$ out of phase, are

$$x = a\cos(\omega t + \phi)$$
$$y = b\cos(\omega t + \phi \pm \pi/2) = \mp b\sin(\omega t + \phi)$$

EVALUATE   The above equation can be rewritten as

$$\left(\frac{x}{a}\right)^2 + \left(\frac{y}{b}\right)^2 = \cos^2(\omega t + \phi) + \sin^2(\omega t + \phi) = 1$$

Thus, we see that the equations describe an elliptical path with semi-major or minor axis equal to the amplitudes, a or b.

ASSESS   In cases where $b = a$, the path would be circular with radius a.

59.   **INTERPRET**   In this problem we want to recover the Newton's second law for simple harmonic motion by differentiating the energy equation.

DEVELOP   Newton's second law for the simple harmonic motion of the mass-spring system gives $m\frac{d^2x}{dt^2} = -kx$ (Equation 13.3). To recover this expression from the energy equation, we shall make use of $v = \frac{dx}{dt}$ and $\frac{dv}{dt} = \frac{d^2x}{dt^2}$.

EVALUATE   Since E is a constant, we have

$$0 = \frac{dE}{dt} = \frac{d}{dt}\left(\frac{1}{2}kx^2 + \frac{1}{2}mv^2\right) = kx\frac{dx}{dt} + mv\frac{dv}{dt} = kxv + mv\frac{dv}{dt}$$
$$= v\left(kx + m\frac{dv}{dt}\right) = v\left(kx + m\frac{d^2x}{dt^2}\right)$$

The above expression implies $m\frac{d^2x}{dt^2} = -kx$, which is Equation 13.3.

ASSESS   Since there is no damping term in the equation of motion, the total energy in the system must be conserved.

61.   **INTERPRET**   The motion of the mass sliding on the track is periodic. Given the vertical height y as a function of the position, we can determine its period of oscillation.

DEVELOP   The potential energy, relative to the bottom of the track, is

$$U(x) = mgy = mgax^2$$

which is analogous to $\frac{1}{2}kx^2$. Thus, we suspect that the x component of the motion is simple harmonic motion with "spring constant" $k = 2mga$.

EVALUATE   Using Equation 13.7c, we find the period of the motion to be

$$T = 2\pi\sqrt{\frac{m}{k}} = 2\pi\sqrt{\frac{m}{2mga}} = \frac{2\pi}{\sqrt{2ga}}$$

ASSESS   Indeed, we find that

$$F_x = -\frac{dU}{dx} = -2mgax = m\frac{d^2x}{dt^2}$$

represents simple harmonic motion with $\omega = \sqrt{2ga}$. (The y component of the motion, however, is not simple harmonic motion.)

63.   **INTERPRET**   This problem is about an underdamped oscillator. We would like to show that the amplitude of oscillation has a maximum at some driving frequency less than the natural frequency of undamped motion.

DEVELOP   The amplitude of the driven oscillation is given by Equation 13.19:

$$A(\omega) = \frac{F_0/m}{\sqrt{\left(\omega_d^2 - \omega_0^2\right)^2 + b^2\omega_d^2/m^2}}$$

From the equation, we see that A is a maximum when the denominator of the right-hand side is a minimum. The condition for this is

$$\frac{d}{d\omega_d}\left[\left(\omega_d^2 - \omega_0\right)^2 + \frac{b^2\omega_d^2}{m^2}\right] = 2\left(\omega_d^2 - \omega_0^2\right)(2\omega_d) + \frac{2\omega_d b^2}{m^2} = 2\omega_d\left[2\left(\omega_d^2 - \omega_0^2\right) + \frac{b^2}{m^2}\right] = 0$$

EVALUATE   Since $\omega_d \neq 0$, the driving frequency that satisfies the above condition is

$$\omega_d^* = \sqrt{\omega_0^2 - b^2/2m^2}$$

Evidently, we have $\omega_d^* < \omega_0$.

ASSESS   Although the motion is underdamped for $b < 2m\omega_0$, the amplitude A has a maximum in the physical region ($\omega_d > 0$, $b > 0$) only for $b < \sqrt{2}m\omega_0$, and A has sharp resonance-type behavior for $b \ll 2m\omega_0$.

65.   **INTERPRET**   The problem is about a physical pendulum—a meter stick suspended from one end. We want to know its period of oscillation.

DEVELOP   The period of a physical pendulum can be obtained from Equation 13.13:

$$T = \frac{2\pi}{\omega} = 2\pi\sqrt{\frac{I}{mgL}}$$

The meter stick is a physical pendulum whose CM is at the 50-cm mark, or $L = 0.25$ m below the point of suspension through one end. The rotational inertia of the stick about the pivot point is

$$I = \frac{1}{12}m(1\text{ m})^2 + m(0.25\text{ m})^2 = \frac{7}{3}m(0.25\text{ m})^2.$$

EVALUATE   Using Equation 13.13, the period of the meter stick is

$$T = 2\pi\sqrt{\frac{I}{mgL}} = 2\pi\sqrt{\frac{7m(0.25\text{ m})^2/3}{mg(0.25\text{ m})}} = 2\pi\sqrt{\frac{7(1\text{ m})}{12g}} = 2\pi\sqrt{\frac{7(1\text{ m})}{12(9.8\text{ m/s}^2)}} = 1.53\text{ s}$$

ASSESS   The result can be compared with that obtained in Exercise 31. The rotational inertia I about the pivot point decreases as the pivot moves closer to the center of mass. As I decreases, the period decrease as well.

67.   **INTERPRET**   In this problem we are given two mass-spring systems with same mass but different frequencies, and we want to compare their energies and acceleration.

DEVELOP   As shown in Section 13.5, the energy of a mass-spring system is $E = \frac{1}{2}m\omega^2 A^2$. Also, from Equation 13.10, the maximum acceleration is $a_{max} = \omega^2 A$.

EVALUATE   **(a)** If m and A are the same but $\omega_1 = 2\omega_2$, we have

$$\frac{E_1}{E_2} = \frac{m\omega_1^2 A/2}{m\omega_2^2 A/2} = \left(\frac{\omega_1}{\omega_2}\right)^2 = (2)^2 = 4 \quad \Rightarrow \quad E_1 = 4E_2$$

(b) Comparing the maximum accelerations of the two systems, we find

$$\frac{a_{max,1}}{a_{max,2}} = \frac{\omega_1^2 A}{\omega_2^2 A} = \left(\frac{\omega_1}{\omega_2}\right)^2 = (2)^2 = 4 \quad \Rightarrow \quad a_{max,1} = 4a_{max,2}$$

ASSESS   Both the energy and the maximum acceleration of the mass-spring system increase with ω^2. Doubling the frequency quadruples E and a_{max}.

69.   **INTERPRET**   Given the maximum tension that could be sustained by a thread, we want to know what is the maximum allowable amplitude for the pendulum motion.

DEVELOP   It is shown in Example 13.3 that the greatest tension in a simple pendulum occurs at the bottom of its swing, where $T_{max} = mg(1 + A^2)$, and A is the angular amplitude.

EVALUATE   For the thread in this problem, the maximum tension is $T_{max} = 6.0$ N. Therefore, the maximum allowable amplitude is

$$A_{max} = \sqrt{\frac{T_{max}}{mg} - 1} = \sqrt{\frac{6.0 \text{ N}}{(0.50 \text{ kg})(9.8 \text{ m/s}^2)} - 1} = 0.474 \text{ rad} = 27.16°$$

ASSESS   If the maximum tension the thread can sustain is only equal to the weight of the object, $T_{max} = mg$, then $A = 0$, which implies that any pendulum motion will break the thread.

71.   **INTERPRET**   The motion of the ball rolling on the track is simple harmonic. Given the vertical height y as a function of the position, we can determine its period of oscillation.

DEVELOP   The potential energy of the ball, relative to the bottom of the track, is

$$U(x) = mgy = mgax^2$$

and its kinetic energy from rolling (without slipping) is

$$K = \frac{1}{2}mv^2 + \frac{1}{2}I_{cm}\omega^2 = \frac{1}{2}mv^2 + \frac{1}{2}\left(\frac{2}{5}mR^2\right)\left(\frac{v}{R}\right)^2 = \frac{7}{10}mv^2$$

Since the total mechanical energy, $E = U + K$ is constant, we have

$$0 = \frac{dE}{dt} = \frac{d}{dt}\left(mgax^2 + \frac{7}{10}mv^2\right) = 2mgax\frac{dx}{dt} + \frac{7}{5}mv\frac{dv}{dt}$$

If we assume that $v = \sqrt{(dx/dt)^2 + (dy/dt)^2} \approx dx/dt$, i.e., the vertical displacement is small, then the above expression can be simplified to

$$0 = 2mgax\frac{dx}{dt} + \frac{7}{5}mv\frac{dv}{dt} = 2mgaxv + \frac{7}{5}mv\frac{d^2x}{dt^2} = mv\left(2gax + \frac{7}{5}\frac{d^2x}{dt^2}\right)$$

or

$$\frac{d^2x}{dt^2} = -\frac{10}{7}gax$$

The expression allows us to solve for the angular frequency, and hence the period.

EVALUATE   Comparing the above expression with Equation 13.3 and with the help of Equation 13.7a, we readily see that the angular frequency of the oscillation is

$$\omega = \sqrt{\frac{10ga}{7}}$$

Therefore, the period is

$$T = \frac{2\pi}{\omega} = 2\pi\sqrt{\frac{7}{10ga}}$$

ASSESS   In Problem 61 the period of the sliding point mass is found to be $\frac{2\pi}{\sqrt{2ga}}$. Therefore, we see that the period for the rolling ball is longer. We expect this to be so, since increasing rotational inertia increases the period.

73.   **INTERPRET**   This problem involves a completely inelastic collision between a moving block and a mass-spring system. We'd like to know the frequency, amplitude, and phase constant after the two masses stick and oscillate together.

DEVELOP   The simple harmonic motion with just the first block on the spring can be described by Equation 13.8 and the given amplitude and phase constant;

$$x(t) = (10 \text{ cm})\cos(\omega_1 t - \pi/2) = (10 \text{ cm})\sin\omega_1 t$$

where

$$\omega_1 = \sqrt{\frac{k}{m_1}} = \sqrt{\frac{23 \text{ N/m}}{1.2 \text{ kg}}} = 4.38 \text{ s}^{-1}$$

This equation holds up to the time of the collision, i.e., for $t < t_c$, where $t_c = \frac{\pi}{2\omega_1}$, since for the rightmost point of oscillation, $\sin \omega_1 t_c = 1$ or $\omega_1 t_c = \pi/2$. (This specifies the original zero of time appropriate to the given phase constant of $-\pi/2$.)

Equation 13.8 also describes the simple harmonic motion after the collision;

$$x(t) = A\cos(\omega_2 t + \phi)$$

for $t > t_c$, where $\omega_2 = \sqrt{k/(m_1 + m_2)} = 3.39 \text{ s}^{-1}$ is the angular frequency when both blocks oscillate on the spring. It follows from this that

$$v(t) = -\omega_2 A\sin(\omega_2 t + \phi)$$

The amplitude A and phase constant ϕ can be determined from these two equations evaluated just after the collision, essentially at t_c, if we assume that the collision takes place almost instantaneously; then conservation of momentum during the collision can be applied (see Equation 9.11). Just after the collision, $x(t_c) = 10$ cm (given) and

$$v(t_c) = \frac{m_1 v_1 + m_2 v_2}{m_1 + m_2}$$

where just before the collision, $v_1 = 0$ (given m_1 at rightmost point of its original motion) and $v_2 = -1.7$ m/s (also given).

EVALUATE The frequency of oscillation after the inelastic collision is

$$f_2 = \frac{\omega_2}{2\pi} = \frac{1}{2\pi}\sqrt{\frac{k}{m_1 + m_2}} = \frac{3.39 \text{ s}^{-1}}{2\pi} = 0.540 \text{ Hz}$$

To solve for the amplitude, we note that numerically,

$$v(t_c) = v(t_c) = \frac{m_1 v_1 + m_2 v_2}{m_1 + m_2} = \frac{0 + (0.8 \text{ kg})(-1.7 \text{ m/s})}{1.2 \text{ kg} + 0.8 \text{ kg}} = -0.68 \text{ m/s}$$

Thus, with

$$x(t_c) = 10 \text{ cm} = A\cos(\omega_2 t_c + \phi),$$
$$v(t_c) = -68 \text{ cm/s} = -\omega_2 A\sin(\omega_2 t_c + \phi)$$

we solve for A (using $\sin^2 + \cos^2 = 1$) and find

$$A = \sqrt{x(t_c)^2 + [-v(t_c)/\omega_2]^2} = \sqrt{(10 \text{ cm})^2 + (68 \text{ cm}/3.39)^2} = 22.4 \text{ cm}$$

To find the phase, we first note that

$$\omega_2 t_c = \omega_2\left(\frac{\pi}{2\omega_1}\right) = \frac{\pi}{2}\frac{\sqrt{k/(m_1 + m_2)}}{\sqrt{k/m_1}} = \frac{\pi}{2}\sqrt{\frac{m_1}{m_1 + m_2}} = \frac{\pi}{2}\sqrt{\frac{1.2 \text{ kg}}{1.2 \text{ kg} + 0.8 \text{ kg}}} = 1.22 \text{ radians}$$
$$= 69.7°$$

Solving for ϕ (using $\sin\alpha/\cos\alpha = \tan\alpha$), we find $\tan(\omega_2 t_c + \phi) = \frac{-v(t_c)/\omega_2}{x(t_c)}$, or

$$\phi = \tan^{-1}\left(\frac{-v(t_c)}{\omega_2 x(t_c)}\right) - \omega_2 t_c = \tan^{-1}\left(\frac{68 \text{ cm/s}}{(3.39 \text{ s}^{-1})(10 \text{ cm})}\right) - 69.7° = 63.5° - 69.7° = -6.22°$$
$$= -0.109 \text{ rad}$$

ASSESS The solution for A is equivalent to calculating the various energies in the second simple harmonic motion, since just after the collision,

$$K(t_c) = \frac{1}{2}(m_1 + m_2)v(t_c)^2 = \frac{1}{2}(2 \text{ kg})(-0.68 \text{ m/s})^2 = 0.462 \text{ J}$$

$$U(t_c) = \frac{1}{2}kx(t_c)^2 = \frac{1}{2}(23 \text{ N/m})(0.1 \text{ m})^2 = 0.115 \text{ J}$$

$$E = K(t_c) + U(t_c) = 0.577 \text{ J} = \frac{1}{2}kA^2$$

or $A = \sqrt{2(0.577 \text{ J})/(23 \text{ N/m})} = 22.4$ cm. Once A is known, ϕ can also be found from either expression for $x(t_c)$ or $v(t_c)$, e.g.,

$$\omega_2 t_c + \phi = \cos^{-1}\left(\frac{10 \text{ cm}}{22.4 \text{ cm}}\right) = \sin^{-1}\left(\frac{68 \text{ cm/s}}{(3.39 \text{ s}^{-1})(22.4 \text{ cm})}\right)$$

75.   **INTERPRET**   In this problem, we want to show, by substitution, that $x = A\cos(\omega_d t + \phi)$ is a solution to the differential equation given in Equation 13.18, with A given by Equation 13.19.

DEVELOP   The algebra involved in this problem is straightforward but somewhat tedious.

EVALUATE   When $x = A\cos(\omega_d t + \phi)$ is substituted into Equation 13.18, one obtains

$$m\left[-\omega_d^2 A\cos(\omega_d t + \phi)\right] = -kA\cos(\omega_d t + \phi) - b(-\omega_d A\sin(\omega_d t + \phi))$$
$$+F_0[\cos(\omega_d t + \phi)\cos\phi + \sin(\omega_d t + \phi)\sin\phi]$$

where we let $\omega_d t = \omega_d t + \phi - \phi$ in the F_0-term, and used a trigonometric identity. This equation is true if the coefficient of the $\sin(\omega_d t + \phi)$ and $\cos(\omega_d t + \phi)$ terms on each side are equal, respectively, that is,

$$-m\omega_d^2 A = -kA + F_0\cos\phi$$
$$0 = b\omega_d A + F_0\sin\phi$$

Let $\omega_0^2 = \frac{k}{m}$ and these equations become

$$F_0\cos\phi = -m\left(\omega_d^2 - \omega_0^2\right)A$$
$$F_0\sin\phi = -b\omega_d A$$

Squaring and adding, we get Equation 13.19:

$$A(\omega) = \frac{F_0/m}{\sqrt{\left(\omega_d^2 - \omega_0^2\right)^2 + b^2\omega_d^2/m^2}}$$

ASSESS   By substitution, we have verified that the function $x = A\cos(\omega_d t + \phi)$ indeed is a solution to the differential equation

$$m\frac{d^2x}{dt^2} = -kx - b\frac{dx}{dt} + F_0\cos\omega_d t$$

77.   **INTERPRET**   We show that $x(t) = a\cos(\omega t) - b\sin(\omega t)$ is equivalent to $x(t) = A\cos(\omega t + \phi)$, where $A = \sqrt{a^2 + b^2}$ and $\phi = \tan^{-1}(\frac{b}{a})$. We use trigonometric identities.

DEVELOP   We use the trig identity $\cos(\alpha + \beta) = \cos\alpha\cos\beta - \sin\alpha\sin\beta$, using $x(t) = A\cos(\omega t + \phi)$ as a starting point.

EVALUATE

$$x(t) = A\cos(\omega t + \phi) = A[\cos\omega t\cos\phi - \sin\omega t\sin\phi]$$
$$x(t) = (A\cos\phi)\cos\omega t - (A\sin\phi)\sin\omega t$$

Now we draw an arbitrary right triangle, as shown in the figure. We can see from the figure that $a = A\cos\phi$ and $b = A\sin\phi$, and also that $A = \sqrt{a^2 + b^2}$ and $\tan\phi = \frac{b}{a}$. So $x(t) = a\cos\omega t - b\sin\omega t$.

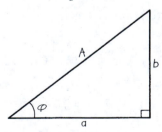

ASSESS   This is as much a trigonometry problem as a physics problem—it allows us to see that simple harmonic motion can be expressed in many ways.

79.   **INTERPRET**   A pendulum at rest on Earth has a given frequency. The frequency becomes higher when accelerating. We use a vector sum to find the effective value of g in the simple pendulum equation and calculate the acceleration. We will use the simple pendulum approximation.

DEVELOP   We start with a sketch, as shown in the figure. From the ratio of the initial period $T = 2\pi\sqrt{\frac{l}{g}} = \frac{60\text{ s}}{90\text{ cycles}} = 0.667$ s and the final period $T' = 2\pi\sqrt{\frac{l}{g_{eff}}} = \frac{60\text{ s}}{91\text{ cycles}} = 0.659$ s, we can calculate g_{eff}. Once we know g_{eff}, we can use $g_{eff}^2 = g^2 + a^2$ to find a.

EVALUATE $\frac{T}{T'} = \sqrt{\frac{g_{eff}}{g}} \rightarrow g_{eff} = g\frac{T^2}{T'^2}$. $a = \sqrt{g_{eff}^2 - g^2} = \sqrt{g^2\frac{T^4}{T'^4} - g^2} = g\sqrt{\frac{T^4}{T'^4} - 1} = 0.21g = 2.1 \text{ m/s}^2$.

ASSESS   It turns out that our use of ratios eliminates the need to know much of anything about the physical setup. In fact, we get the same result even if we use the equation for the physical pendulum, rather than for the simple pendulum.

81.   **INTERPRET**   In this problem, we want to show, by substitution, that $x = A\cos(\omega_d t + \phi)$ is a solution to the differential equation given in Equation 13.18, with A given by Equation 13.19.
  DEVELOP   The algebra involved in this problem is straightforward but somewhat tedious.
  EVALUATE   When $x = A\cos(\omega_d t + \phi)$ is substituted into Equation 13.18, one obtains

$$m\left[-\omega_d^2 A\cos(\omega_d t + \phi)\right] = -kA\cos(\omega_d t + \phi) - b(-\omega_d A\sin(\omega_d t + \phi))$$
$$+ F_0[\cos(\omega_d t + \phi)\cos\phi + \sin(\omega_d t + \phi)\sin\phi]$$

where we let $\omega_d t = \omega_d t + \phi - \phi$ in the F_0-term, and used a trigonometric identity. This equation is true if the coefficients of the $\sin(\omega_d t + \phi)$ and $\cos(\omega_d t + \phi)$ terms on each side are equal, respectively, that is,

$$-m\omega_d^2 A = -kA + F_0\cos\phi$$
$$0 = b\omega_d A + F_0\sin\phi$$

Let $\omega_0^2 = \frac{k}{m}$ and these equations become

$$F_0\cos\phi = -m(\omega_d^2 - \omega_0^2)A$$
$$F_0\sin\phi = -b\omega_d A$$

Squaring and adding, we get Equation 13.19:

$$A(\omega) = \frac{F_0/m}{\sqrt{(\omega_d^2 - \omega_0^2)^2 + b^2\omega_d^2/m^2}}$$

ASSESS   By substitution, we have verified that the function $x = A\cos(\omega_d t + \phi)$ indeed is a solution to the differential equation

$$m\frac{d^2x}{dt^2} = -kx - b\frac{dx}{dt} + F_0\cos\omega_d t$$

WAVE MOTION

14

EXERCISES

Section 14.1 Waves and Their Properties

17. **INTERPRET** This problem is about wave propagation. Given the speed and frequency of the ripples, we are asked to compute the period and the wavelength.

    DEVELOP Equation 14.1 relates the speed of the wave to its period, frequency, and wavelength:

    $$v = \frac{\lambda}{T} = \lambda f$$

    This is the equation we shall use to solve the problem.

    EVALUATE Equation 14.1 gives (a) $T = \frac{1}{f} = \frac{1}{5.2\ \text{Hz}} = 0.192$ s, and (b) $\lambda = \frac{v}{f} = \frac{34\ \text{cm/s}}{5.2\ \text{Hz}} = 6.54$ cm.

    ASSESS The unit of frequency is Hz, with $1\ \text{Hz} = 1\ \text{s}^{-1}$. If the frequency is kept fixed, then increasing the wavelength will increase the speed of propagation.

19. **INTERPRET** This problem is about wave propagation. Given the speed and frequency of various electromagnetic waves, we are asked to compute their wavelength.

    DEVELOP Equation 14.1 relates the speed of the wave to its period, frequency, and wavelength:

    $$v = \frac{\lambda}{T} = \lambda f \quad \rightarrow \quad \lambda = \frac{v}{f}$$

    This is the equation we shall use to solve the problem.

    EVALUATE Since the speed of propagation of electromagnetic waves in vacuum is simply equal to the speed of light, $v = c = 3.0 \times 10^8$ m/s, Equation 14.1 gives

    (a) $\lambda = \frac{c}{f} = \frac{3 \times 10^8\ \text{m/s}}{10^6\ \text{Hz}} = 300$ m;

    (b) $\lambda = \frac{c}{f} = \frac{3 \times 10^8\ \text{m/s}}{190 \times 10^6\ \text{Hz}} = 1.58$ m;

    (c) $\lambda = \frac{c}{f} = \frac{3 \times 10^8\ \text{m/s}}{10^{10}\ \text{Hz}} = 0.03$ m = 3 cm;

    (d) $\lambda = \frac{c}{f} = \frac{3 \times 10^8\ \text{m/s}}{4 \times 10^{13}\ \text{Hz}} = 7.5 \times 10^{-6}$ m = 7.5 μm;

    (e) $\lambda = \frac{c}{f} = \frac{3 \times 10^8\ \text{m/s}}{6 \times 10^{14}\ \text{Hz}} = 5.0 \times 10^{-7}$ m = 500 nm;

    (f) $\lambda = \frac{c}{f} = \frac{3 \times 10^8\ \text{m/s}}{1.0 \times 10^{18}\ \text{Hz}} = 3.0 \times 10^{-10}$ m = 3Å (See Appendix C on units.)

    ASSESS If the speed of propagation is kept fixed, then a higher frequency means a shorter wavelength.

Section 14.2 Wave Math

21. **INTERPRET** This problem is about the ultrasound wave. Given its frequency, and wavelength, we want to find its angular frequency, wave number, and wave speed.

    DEVELOP The relationships between the speed of the wave, its wave number, frequency, and wavelength are given by Equations 13.6, 14.1, and 14.2:

    $$f = \frac{\omega}{2\pi} \quad v = \frac{\lambda}{T} = \lambda f, \quad k = \frac{2\pi}{\lambda}$$

EVALUATE   (a) Equation 13.6 gives $\omega = 2\pi f = 2\pi(4.8 \text{ MHz}) = 3.02 \times 10^7 \text{ s}^{-1}$.

(b) Equation 14.2 gives $k = \frac{2\pi}{\lambda} = \frac{2\pi}{0.31 \text{ mm}} = 2.03 \times 10^4 \text{ m}^{-1}$.

(c) Using Equation 14.1, the speed of the ultrasound wave is

$$v = f\lambda = (4.8 \times 10^6 \text{ Hz})(0.31 \times 10^{-3} \text{ m}) = 1.49 \times 10^3 \text{ m/s}$$

ASSESS   The speed of the wave can also be computed as

$$v = \frac{\omega}{k} = \frac{3.02 \times 10^7 \text{ s}^{-1}}{2.03 \times 10^4 \text{ m}^{-1}} = 1.49 \times 10^3 \text{ m/s}$$

Thus, we see that the pairs f, λ and ω, k are equivalent ways to describe the same wave.

23.   **INTERPRET**   We are given a function that describes a traveling sinusoidal wave, and asked to compute various physical quantities associated with the wave.

DEVELOP   Consider a traveling wave of the form given in Equation 14.3:

$$y(x,t) = A\cos(kx \pm \omega t)$$

The amplitude of the wave is A; its wavelength is given by Equation 14.2: $\lambda = 2\pi/k$; its period is given by Equation 13.5: $T = 2\pi/\omega$. The speed of propagation is $v = \lambda f = \omega/k$ and the direction of propagation is $+x$ if the argument is $kx - \omega t$, and $-x$ if the argument is $kx + \omega t$.

EVALUATE   (a) Comparing $y = 1.3\cos(0.69x + 31t)$ with Equation 14.3, we find the amplitude to be **(a)**
(b) Since $A = y_{\max} = 1.3$ cm. **(b)** Since $k = 0.69$ cm^{-1}, the wavelength is

$$\lambda = \frac{2\pi}{k} = \frac{2\pi}{0.69 \text{ cm}^{-1}} = 9.11 \text{ cm}$$

(c) Equation 13.5 gives $T = \frac{2\pi}{\omega} = \frac{2\pi}{31 \text{ s}^{-1}} = 0.203$ s^{-1}.

(d) The speed of the wave is $v = \frac{\omega}{k} = \frac{31 \text{ s}^{-1}}{0.69 \text{ cm}} = 44.9$ cm/s.

(e) A phase of the form $kx + \omega t$ describes a wave propagating in the negative x direction.

ASSESS   This problem demonstrates that the wave function of the form given in Equation 14.3 contains all the information about the amplitude, the wavelength, the period, the phase and the speed of propagation of a wave. So once the wave function is given, all these quantities can be readily calculated.

Section 14.3 Waves on a String

25.   **INTERPRET**   Given the tension in the cable and the linear mass density (mass per unit length) of the cable, we want to find the speed of propagation of a transverse wave

DEVELOP   The relations among the speed of propagation, the tension and the mass per unit length of the medium is given by Equation 14.5, $v = \sqrt{F/\mu}$. Given F and μ, we can calculate v.

EVALUATE   The speed of the transverse wave in the cables is

$$v = \sqrt{\frac{F}{\mu}} = \sqrt{\frac{2.5 \times 10^8 \text{ N}}{4100 \text{ kg/m}}} = 247 \text{ m/s}$$

ASSESS   Increasing the tension results in a greater acceleration of the disturbed cables, and hence the wave propagates more rapidly.

27.   **INTERPRET**   In this problem we are given the mass per unit length of the spring and the speed of propagation of the wave, and we would like to know the tension in the spring.

DEVELOP   The relations among the speed of propagation, the tension, and the mass per unit length of the medium is given by Equation 14.5, $v = \sqrt{F/\mu}$. Given v and μ, we can calculate F.

EVALUATE   Using Equation 14.5, we find the tension in the spring to be

$$F = \mu v^2 = (0.17 \text{ kg/m})(6.7 \text{ m/s})^2 = 7.63 \text{ N}$$

ASSESS   Our result indicates that, keeping μ fixed, increasing the tension will result in a greater propagation speed.

29.   **INTERPRET**   This problem is about the average power carried by a wave propagating along a rope.

DEVELOP   The average power transmitted by transverse traveling waves in a string is given by Equation 14.6:
$\bar{P} = \frac{1}{2}\mu\omega^2 A^2 v$. The speed of propagation can be obtained by using Equation 14.5: $v = \sqrt{F/\mu}$.

EVALUATE   Using the values given in the problem statement, we find the average power to be

$$\bar{P} = \frac{1}{2}\mu\omega^2 A^2 v = \frac{1}{2}\mu\omega^2 A^2\sqrt{\frac{F}{\mu}} = \frac{1}{2}(0.28 \text{ kg/m})(2\pi \times 3.3 \text{ Hz})^2(0.061 \text{ m})^2\sqrt{\frac{550 \text{ N}}{0.28 \text{ kg/m}}}$$
$$= 9.93 \text{ W}$$

ASSESS   The wave power is proportional to the speed of propagation. It is also proportional to the square of the amplitude and the square of the angular frequency.

Section 14.4 Sound Waves

31.   **INTERPRET**   We apply Equation 14.8 to find the speed of sound with a given air pressure and density.

DEVELOP   We use Equation 14.8, $v = \sqrt{\frac{\gamma P}{\rho}}$ relates sound speed v to pressure P and density ρ. For air, $\gamma = \frac{7}{5}$.

EVALUATE $v = \sqrt{\frac{\gamma P}{\rho}} = 343$ m/s.

ASSESS   This is the accepted value for the speed of sound at standard temperature and pressure.

33.   **INTERPRET**   This problem is about finding the speed of sound in a given medium.

DEVELOP   The speed of sound in a medium is given by Equation 14.8:

$$v = \sqrt{\frac{\gamma P}{\rho}}$$

where γ is a constant characteristic of the gas, P is the pressure, and ρ is the density of the medium.

EVALUATE   Using the given in the problem, we find the speed of sound in the NO_2 medium to be

$$v = \sqrt{\frac{\gamma P}{\rho}} = \sqrt{\frac{1.29(4.8 \times 10^4 \text{ N/m}^2)}{0.35 \text{ kg/m}^3}} = 421 \text{ m/s}$$

ASSESS   The speed of sound at room temperature (20°C) is about 343 m/s. The speed depends on the thermodynamic properties of the medium.

35.   **INTERPRET**   In this problem we are asked about the frequency of a sound wave in a medium—the underwater habitat.

DEVELOP   To compute the frequency, we first calculate the speed of sound in the underwater habitat using Equation 14.8, $v = \sqrt{\gamma P/\rho}$. Once v is known, we use Equation 14.1, $v = \lambda f$, to find the frequency.

EVALUATE   The speed of sound is

$$v = \sqrt{\frac{\gamma P}{\rho}} = \sqrt{\frac{1.61(6.2 \times 10^5 \text{ N/m}^2)}{4.5 \text{ kg/m}^3}} = 471 \text{ m/s}$$

Therefore, the frequency of 0.5 m wavelength sound waves is

$$f = \frac{v}{\lambda} = \frac{471 \text{ m/s}}{0.50 \text{ m}} = 942 \text{ Hz}$$

ASSESS   In "normal air," the frequency would be about $f = \frac{v}{\lambda} = \frac{343 \text{ m/s}}{0.50 \text{ m}} = 686$ Hz.

Section 14.5 Interference

37.   **INTERPRET**   This problem is about wave interference. Given the condition for the second calm region where waves interfere destructively, we want to compute the wavelength of the ocean wave.

DEVELOP   The condition for destructive interference is a phase difference of $k_2\Delta r = (2\pi/\lambda_2)\Delta r = 3\pi$, or an odd multiple of $\pi = 180°$. The second node occurs when the path difference is three half-wavelengths, or $\Delta r \equiv AP - BP = \frac{3}{2}\lambda_2$.

EVALUATE From Example 14.5, we have $\Delta r = 8.1$ m, so the wavelength is

$$\lambda_2 = \frac{2}{3}\Delta r = \frac{2}{3}(8.1 \text{ m}) = 5.4 \text{ m}$$

ASSESS Comparing with Example 14.5, we expect the wavelength in this case to be shorter since at the same distance away from the source the calm region encountered here is the second one.

Section 14.7 Standing Waves

39. **INTERPRET** This problem is about standing-wave modes in a string that is clamped at both ends.

DEVELOP Since the string is clamped at both ends, the amplitudes there must be zero. If L is the length of the string, then the standing waves must satisfy the condition given in Equation 14.12:

$$L = \frac{m\lambda}{2} \quad m = 1, 2, 3, \ldots$$

It follows that the frequencies of the standing-wave modes of a string fixed at both ends are all the (positive) integer multiples of the fundamental frequency,

$$f_m = \frac{v}{\lambda_m} = m\left(\frac{v}{2L}\right) = mf_1 \quad m = 1, 2, 3, \ldots$$

where $f_1 = \frac{v}{2L}$ is the fundamental frequency. On the other hand, if only one end of the string is fixed, then from Figure 14.29, we may show that the relation between λ and L is given by

$$L = \frac{m\lambda}{4} \quad m = 1, 3, 5, \ldots$$

and the corresponding frequencies are

$$f_m = \frac{v}{\lambda_m} = m\left(\frac{v}{4L}\right) = mf_1, \quad m = 1, 3, 5, \ldots$$

where $f_1 = \frac{v}{4L}$ is the fundamental frequency.

EVALUATE (a) With both ends fixed, the next higher frequency is

$$f_2 = 2f_1 = 2(140 \text{ Hz}) = 280 \text{ Hz}$$

(b) The fundamental frequency for the string fixed at one end is

$$f_1 = \frac{v}{4L} = \frac{1}{2}(140 \text{ Hz}) = 70 \text{ Hz}$$

i.e., one half the fundamental frequency of the string fixed at both ends.

(c) In this case, the standing-wave frequencies are only the odd multiples of the fundamental frequency, therefore the second standing-wave mode has frequency

$$f_2 = 3f_1 = 3(70 \text{ Hz}) = 210 \text{ Hz}$$

ASSESS When the string is clamped at both ends, it can accommodate an integer number of half-wavelength. However, if it's clamped only at one end, then the string can accommodate only an odd number of quarter-wavelengths.

41. **INTERPRET** This problem is about standing-wave modes in the human vocal tract. We assume it to be a cylinder with one end open.

DEVELOP If only one end of the cylinder of length L is closed, then from Figure 14.29, we may show that the relation between λ and L is given by

$$L = \frac{m\lambda}{4}, \quad m = 1, 3, 5, \ldots$$

That is, the tract can accommodate only an odd number of quarter-wavelengths. The corresponding frequencies are

$$f_m = \frac{v}{\lambda_m} = m\left(\frac{v}{4L}\right) = mf_1, \quad m = 1, 3, 5, \ldots$$

where $f_1 = \frac{v}{4L}$ is the fundamental frequency.

EVALUATE The fundamental frequency is

$$f_1 = \frac{v}{4L} = \frac{340 \text{ m/s}}{4(0.15 \text{ m})} = 567 \text{ Hz}$$

ASSESS   The typical value of fundamental frequency ranges from 80–200 Hz for males and 150–350 Hz for females. So the estimate based on the cylinder model is a bit high.

Section 14.8 The Doppler Effect and Shock Waves

43.   INTERPRET   This problem is about Doppler effect. We want to find the frequency perceived by a moving observer.
DEVELOP   The Doppler-shifted frequency perceived by the firefighter moving toward the siren is given by Equation 14.15:

$$f' = f\left(1 + \frac{u}{v}\right)$$

EVALUATE   With $u = 120$ km/h $= 33.3$ m/s, the frequency the firefighter perceived is

$$f' = f\left(1 + \frac{u}{v}\right) = (85 \text{ Hz})\left(1 + \frac{33.3 \text{ m/s}}{343 \text{ m/s}}\right) = 93.3 \text{ Hz}$$

ASSESS   As expected, since the firefighter is moving toward the sound source, the frequency he perceives is higher than when he is at rest, i.e., $f' > f$. On the other hand, $f' < f$ if he were to move away from the source.

45.   INTERPRET   This problem is about using Doppler effect for light to deduce the galaxy's motion relative to Earth.
DEVELOP   The formula for the Doppler shift for light is different than for sound, but when the relative velocity of the source and observer, u, is very small compared to the wave speed, $v = c$ for light, the result is the same as Equations 14.13a and b.
EVALUATE   For the galaxy described in this problem, the observed wavelength is greater (red-shifted) than the laboratory wavelength, so the galaxy is receding with speed

$$\frac{u}{c} = \frac{\lambda'}{\lambda} - 1 = \frac{708 \text{ nm}}{656 \text{ nm}} - 1 = 7.93 \times 10^{-2} \quad \rightarrow \quad u = 0.0793c \approx 2.38 \times 10^7 \text{ m/s}$$

ASSESS   The red shift observed in light from distant galaxies is an indication that the universe is expanding, as suggested by the Big Bang theory.

PROBLEMS

47.   INTERPRET   In this problem we want to find out how a change in the tension in the spring affects the speed of wave propagation.

DEVELOP   The relations among the speed of propagation, the tensions and the mass per unit length of the medium is given by Equation 14.5, $v = \sqrt{F/\mu}$. With μ kept fixed, we see that increasing F will increase v.

EVALUATE   Using Equation 14.5, obtain $\mu = \frac{F_1}{v_1^2} = \frac{F_2}{v_2^2}$, which gives

$$v_2 = \sqrt{\frac{F_2}{F_1}} v_1 = \sqrt{\frac{40 \text{ N}}{14 \text{ N}}} (18 \text{ m/s}) = 30.4 \text{ m/s}$$

ASSESS   Our result indicates that, keeping μ fixed, increasing the tension will result in a greater propagation speed.

49.   INTERPRET   The problem is about the total energy carried by the traveling wave, given the string tension F, the wave amplitude A, and the wavelength λ.

DEVELOP   The average wave energy, $d\bar{E}$, in a small element of string of length dx, is transmitted in time, dt, at the same speed as the waves, $v = dx/dt$. From Equation 14.6,

$$d\bar{E} = \bar{P}dt = \frac{1}{2}\mu\omega^2 A^2 v\, dt = \frac{1}{2}\mu\omega^2 A^2 dx$$

Thus, the average linear energy density is $\frac{d\bar{E}}{dx} = \frac{1}{2}\mu\omega^2 A^2$. The total average energy in a wave train of length $L = 2\lambda$ is $\bar{E} = \left(\frac{d\bar{E}}{dx}\right)L = \frac{1}{2}\mu\omega^2 A^2 (2\lambda)$.

EVALUATE   In terms of the quantities specified in this problem (see Equations 14.1 and 14.5), the total energy in the wave train is

$$\bar{E} = \frac{1}{2}\mu\omega^2 A^2 (2\lambda) = \frac{1}{2}\left(\frac{F}{v^2}\right)\left(\frac{2\pi v}{\lambda}\right)^2 A^2 (2\lambda) = \frac{4\pi^2 F A^2}{\lambda}$$

ASSESS   The relation derived can be written as $\bar{P} = (d\bar{E}/dx)v$. For a one-dimensional wave, $\bar{P}$ is the intensity, so the average intensity equals the average energy density times the speed of wave energy propagation. This is a general wave property.

51.   **INTERPRET**   This problem is about the power output from a light bulb, given its intensity at a certain distance.

DEVELOP   The intensity at a distance r from a light bulb that has an average power output P is given by Equation 14.7, $I = P/(4\pi r^2)$. This equation is what we shall utilize to solve for P.

EVALUATE   From Equation 14.7, we have

$$P = 4\pi r^2 I = 4\pi (3.3 \text{ m})^2 (0.73 \text{ W/m}^2) = 99.9 \text{ W} \approx 100 \text{ W}$$

ASSESS   The light source is a 100-watt light bulb. Note that the intensity decreases with the square of the distance.

53.   **INTERPRET**   This problem is wave superposition. We want to show that the superposition of two harmonic waves results in a wave which is also harmonic.

DEVELOP   Using the identity

$$\cos\alpha + \cos\beta = 2\cos\left(\frac{\alpha - \beta}{2}\right)\cos\left(\frac{\alpha + \beta}{2}\right)$$

we find

$$y = y_1 + y_2 = A\cos(kx - \omega t) + A\cos(kx - \omega t + \phi) = 2A\cos\left(\frac{\phi}{2}\right)\cos\left(kx - \omega t + \tfrac{1}{2}\phi\right)$$

EVALUATE   Writing $y = A_s \cos(kx - \omega t + \phi_s)$ and comparing with the expression above, we find the amplitude and the phase to be

$$A_s = 2A\cos\left(\frac{\phi}{2}\right) \qquad \phi_s = \frac{\phi}{2}$$

ASSESS   Let's check our results by considering the following limits: **(i)** $\phi = 0$. In this case, $y_1 = y_2$, and the resultant amplitude is simply $A_s = A + A = 2A$. **(ii)** $\phi = \pi$. In this case, we have (using the identity $\cos(\theta + \pi) = -\cos\theta$),

$$y_2 = A\cos(kx - \omega t + \pi) = -A\cos(kx - \omega t) = -y_1$$

and therefore, $y = y_1 + y_2 = 0$.

55.   **INTERPRET**   This problem is about how the propagation speed of a transverse wave on the spring is affected as the spring is stretched.

DEVELOP   The speed of propagation can be obtained by using Equation 14.5: $v = \sqrt{F/\mu}$. We regard the spring as a stretched string with tension, $F = k\Delta x = k(L - L_0)$. In addition, its linear mass density is $\mu = m/L$.

EVALUATE   Equation 14.5 gives the speed of transverse waves as

$$v = \sqrt{\frac{F}{\mu}} = \sqrt{\frac{k\Delta x}{m/L}} = \sqrt{\frac{k(L - L_0)}{m/L}} = \sqrt{\frac{kL(L - L_0)}{m}}$$

ASSESS   There are two effects here that affect the speed of propagation. The first one is the amount of stretching, Δx. This makes the speed of propagation go up as $(\Delta x)^{1/2}$. The second effect is the change in mass per unit length, characteristic of the inertia of the spring. The mass density decreases as the spring is stretched. This makes it easier for the wave to propagate through.

57.   **INTERPRET**   This problem is about how the intensity of sound waves varies with distance.

DEVELOP   The intensity of spherical waves from a point source is given by Equation 14.7:

$$I = \frac{P}{A} = \frac{P}{4\pi r^2}$$

Since the power output by the source remains constant, the intensity decreases as we move further away from the source.

EVALUATE   From Equation 14.7, we have

$$P = 4\pi r_1^2 I_1 = 4\pi r_2^2 I_2 = \text{constant}$$

which gives $r_2 = \sqrt{\frac{I_1}{I_2}} r_1 = \sqrt{\frac{750 \text{ mW/m}^2}{270 \text{ mW/m}^2}} (15 \text{ m}) = 25 \text{ m}$. Thus, the person needs to walk a distance

$d = r_2 - r_1 = 25 \text{ m} - 15 \text{ m} = 10 \text{ m}$ from where he is currently standing.

ASSESS   The intensity falls off as the inverse square of distance. The further you walk away from the source, the weaker the intensity.

59.   **INTERPRET**   This problem is about the relation between the propagation speed of a transverse wave on the spring and the amount the spring is stretched.

DEVELOP   The speed of propagation can be obtained by using Equation 14.5: $v = \sqrt{F/\mu}$. We regard the spring as a stretched string with tension, $F = k\Delta x = k(L_1 - L_0)$. In addition, its linear mass density is $\mu = m/L_1$. Therefore, the speed of transverse waves is

$$v_1 = \sqrt{\frac{F}{\mu}} = \sqrt{\frac{k\Delta x}{m/L_1}} = \sqrt{\frac{k(L_1 - L_0)}{m/L_1}} = \sqrt{\frac{kL_1(L_1 - L_0)}{m}}$$

This is the equation we shall use to solve for the unstretched length, L_0.

EVALUATE   The fact that the speed triples as the length is doubled allows us to write

$$v_2 = 3v_1 = \sqrt{\frac{k(2L_1)(2L_1 - L_0)}{m}}$$

Combining the two equations gives $9L_1(L_1 - L_0) = 2L_1(2L_1 - L_0)$, or $L_0 = 5L_1/7$.

ASSESS   There are two effects here that affect the speed of propagation. The first one is the amount of stretching, Δx. This makes the speed of propagation go up as $(\Delta x)^{1/2}$. The second effect is the change in mass per unit length, which is characteristic of the inertia of the spring. The mass density decreases as the spring is stretched. This makes it easier for the wave to propagate through.

61.   **INTERPRET**   This problem is about finding the speed of the airplane, given its apparent location as the source of sound.

DEVELOP   The travel time of the sound from the airplane, reaching you along a line making an angle of $\theta = 35°$ with the vertical (from the apparent sound source), is $\Delta t = d/v$. During this time, the airplane moved a horizontal distance $\Delta x = d\sin\theta$. Once Δt and Δx are known, we can calculate the speed of the airplane as $u = \Delta x/\Delta t$.

EVALUATE   Using the values given in the problem statement, the speed of the airplane is

$$u = \frac{\Delta x}{\Delta t} = \frac{d\sin\theta}{d/v} = v\sin\theta = (330 \text{ m/s})\sin 35° = 189 \text{ m/s}$$

or, approximately 420 mi/h.

ASSESS   The value is reasonable for an airplane's speed. Note that the airplane's altitude, $5.2 \text{ km} = d\cos 35°$, was not needed in this calculation.

63.   **INTERPRET**   This problem is about sound intensity measured in decibel level.

DEVELOP   The sound intensity level in decibels is given by Equation 14.9:

$$\beta = (10 \text{ db})\log\left(\frac{I}{I_0}\right)$$

where I is the intensity (measured in W/m²), and $I_0 = 10^{-12}$ W/m² is the reference level.

EVALUATE   If the sound intensity is doubled, $I' = 2I$. Equation 14.9 shows that

$$\beta' = (10 \text{ dB})\log\left(\frac{I'}{I_0}\right) = (10 \text{ dB})\log\left(\frac{2I}{I_0}\right) = (10 \text{ dB})\log\left(\frac{I}{I_0}\right) + (10 \text{ dB})\log 2$$

$$= \beta + 3.01 \text{ dB}$$

Thus, the decibel level increases by about 3 dB.

ASSESS The problem demonstrates that doubling the intensity corresponds to a 3 dB increase. Human ears, however, do not respond linearly to the intensity change. For each 10-dB increase, you perceive an increase in loudness by roughly a factor of 2.

65. **INTERPRET** This problem is about how the sound intensity measured in decibel level varies with distance. We want to find out the distance one must move away from the sound source for the loudness to drop by half.

    DEVELOP The sound intensity level in decibels is given by Equation 14.9:

$$\beta = (10 \text{ db}) \log \left(\frac{I}{I_0} \right)$$

where I is the intensity (measured in W/m²), and $I_0 = 10^{-12}$ W/m² is the reference level. A change of 10 dB (i.e., $\beta' - \beta = -10$ dB) corresponds to the intensity decreasing by a factor of one tenth (i.e., $I' = I/10$). To see this, note that Equation 14.9 may be written as

$$\beta' - \beta = (10 \text{ db}) \log \left(\frac{I'}{I_0} \right) - (10 \text{ db}) \log \left(\frac{I}{I_0} \right) = (10 \text{ dB}) \log \left(\frac{I'}{I} \right)$$

or $\frac{I'}{I} = 10^{(\beta' - \beta)/10 \text{ dB}}$. Thus, if $\beta' - \beta = -10$ dB, then $I' = I/10$.

    EVALUATE For an isotropic point source of sound, the intensity falls inversely with the square of the distance (Equation 14.7). Therefore, $r'^2 = 10r^2$ or

$$r' = \sqrt{10}\,r = \sqrt{10}\,(2 \text{ m}) = 6.32 \text{ m}$$

    ASSESS To summarize, we find that for the loudness perceived to go down by half, the intensity I must decrease by 10 dB, or a factor of 10. Since I decreases as $1/r^2$, the distance the person has to be standing from the sound source is $r' = \sqrt{10}\,r$.

67. **INTERPRET** This problem is about the standing-wave condition and the requirement it imposes on the time it takes for the wave to complete one round trip in the medium.

    DEVELOP Since the string is clamped at both ends, the amplitudes there must be zero. If L is the length of the string, then the standing waves must satisfy the condition given in Equation 14.12:

$$L = \frac{m\lambda}{2} \quad m = 1, 2, 3, \ldots$$

    On the other hand, the round-trip time for waves on a string of length L, clamped at both ends, is

$$t = \frac{2L}{v}$$

    EVALUATE Substituting the first equation to the second gives

$$t = \frac{2L}{v} = \frac{m\lambda}{v} = \frac{m(vT)}{v} = mT \quad m = 1, 2, 3, \ldots$$

    Therefore, we see that t is a multiple of the wave period.

    ASSESS The conclusion can also be drawn by examining Figure 14.28. For example, the wavelength of the fundamental harmonic is $\lambda_1 = 2L$. This gives $t_1 = \lambda_1/v = T$.

69. **INTERPRET** This problem is about the Doppler effect in sound from a moving source.

    DEVELOP Suppose a wave source approaches you at constant speed, and you measure a wave frequency f_1, and as the source passes and then recedes, you measure frequency f_2. From Equation 14.14, we find that

$$f_1 = \frac{f}{1 - u/v} \qquad f_2 = \frac{f}{1 + u/v}$$

    The two equations can be combined to solve for u, the speed of the source.

    EVALUATE The above expressions for f_1 and f_2 imply

$$f = (1 - u/v)f_1 = (1 + u/v)f_2$$

which can be solved to give

$$\frac{u}{v} = \frac{f_1 - f_2}{f_1 + f_2} = \frac{1100 \text{ Hz} - 950 \text{ Hz}}{1100 \text{ Hz} + 950 \text{ Hz}} = 0.0732$$

For sound waves in "normal" air, this implies a truck speed of

$$u = 0.0732v = 0.0732(343 \text{ m/s}) = 25.1 \text{ m/s} = 90.4 \text{ km/h}$$

ASSESS A speed of 90.4 km/h is reasonable for a truck on a freeway. Note that an increase in the speed of the truck would result in a greater difference between f_1 and f_2.

71. **INTERPRET** In this problem we want to find the altitude of a supersonic plane, given its speed and the time you hear the sonic boom.

DEVELOP Since the plane is moving at a supersonic speed, shock waves are formed and the Mach angle is given by $\sin\theta = \frac{u}{v}$. The altitude and distance traveled in level flight are related by $h = (u\Delta t)\tan\theta$ (the shock front moves with the same speed as the aircraft).

EVALUATE The Mach angle is $\theta = \sin^{-1}(1/2.2) = 27.0°$ for the plane. Its altitude is

$$h = (u\Delta t)\tan\theta = (2.2 \times 340 \text{ m/s})(19 \text{ s})\tan 27.0° = 7.25 \text{ km}$$

ASSESS The altitude is greater than the typical flying altitude of commercial aircraft (about 10,000 ft, or 3000 m). However, a high-altitude surveillance plane (such as Lockheed U-2) can fly at an altitude greater than 21 km!

73. **INTERPRET** This problem is about the Doppler effect, with the source receding from the observer.

DEVELOP The wavelength measured by an observer as the wave source recedes is given by Equation 14.13b:

$$\lambda' = \lambda\left(1 + \frac{u}{v}\right)$$

where v is the speed of the wave, and u is the speed of the source.

EVALUATE The above equation gives

$$v = \frac{u}{(\lambda'/\lambda) - 1} = \frac{8.2 \text{ m/s}}{1.2 - 1} = 41 \text{ m/s}$$

ASSESS The Doppler-shifted wavelength increases as the source recedes from the observer. However, the Doppler-shifted frequency decreases, and is given by Equation 14.14, $f' = f/(1 + u/v)$.

75. **INTERPRET** The problem is about the Doppler effect in a radar signal. As the signal is reflected off a moving car, it's frequency is shifted.

DEVELOP The car receives waves at the frequency of an observer moving toward a stationary source, so $f' = f(1 + u/v)$. The reflected waves are re-emitted by the moving car at this frequency, f', and so are received by the radar (the original stationary source) at frequency

$$f'' = \frac{f'}{1 - u/v} = \left(\frac{1 + u/v}{1 - u/v}\right)f$$

The overall frequency shift is

$$\Delta f = f'' - f = \left(\frac{1 + u/v}{1 - u/v} - 1\right)f = \left(\frac{2u/v}{1 - u/v}\right)f$$

EVALUATE In our case, $u \ll v = c$, and the above expression simplifies to $\Delta f \approx \frac{2u}{c}f$. Substituting the values given in the problem statement, with $u = 120 \text{ km/h} = 33.3 \text{ m/s}$, we find

$$\Delta f \approx \frac{2u}{c}f = \frac{2(33.3 \text{ m/s})}{3 \times 10^8 \text{ m/s}}(7 \times 10^{10} \text{ Hz}) = 15.6 \text{ kHz}$$

ASSESS Two Doppler shifts are involved in this problem. The first instance involves the stationary source (police radar) and a moving observer (the car). The second instance involves a moving source (the car reflecting the wave) and a stationary observer (the police radar). Both effects must be taken into consideration in computing the total frequency shift.

77. **INTERPRET** We use beat frequency to determine the rotational speed of one motor, knowing the speed of another.

DEVELOP The beat frequency is equal to the difference in frequencies of the two motors. The beat period is 30 seconds, so the beat frequency is $f_{beat} = \frac{1}{T} = 0.033$ Hz. The speed of the slower motor is $f = 3600$ rpm $= 60$ Hz.

EVALUATE The difference in speeds is the beat frequency, $f_{beat} = 0.033$ Hz. The faster motor is going at $f + f_{beat} = 60.033$ Hz.

ASSESS Beat frequencies are a highly precise method of tuning things ranging from twin-engine aircraft to pianos.

79. **INTERPRET** How does sound level change with distance? We are given a sound level, in dB, at a close distance and are asked to find the sound level at some point much further away.

DEVELOP $\beta = 10\log\left(\frac{I}{I_0}\right)$. The intensity I is in watts per meter squared, and $I_0 = 10^{-12}$ W/m^2. We can use the sound level reading at $d = 10$ m to find the intensity there, then calculate the intensity at distance $d' = 2600$ m and convert that intensity back to dB. We will assume that the sound is a spherical wave, so $I \propto \frac{1}{r^2}$. The sound level at d is $\beta = 80$ dB.

EVALUATE The intensity at d is $\beta = 10\log(\frac{I}{I_0}) \rightarrow I = 10^{\left(\frac{\beta}{10}\right)}I_0$. The distance d' is 260 times greater, so the intensity at d' is $I' = \frac{I}{(260)^2} = \frac{I_0}{67600}10^{\left(\frac{\beta}{10}\right)}$. The sound level at d' is then

$$\beta' = 10\log\left(\frac{I}{I_0}\right) = 10\log\left(\frac{\frac{I_0}{67600}10^{\left(\frac{\beta}{10}\right)}}{I_0}\right) = 10\log\left(\frac{10^8}{67600}\right) = 31.7 \text{ dB}.$$

ASSESS This, in human hearing perception, is about twice the volume of a loud whisper. Note that in the process of solving the problem we never did bother to find what I was, or use the value of I_0. This is the great power of using ratios in solving problems!

15

FLUID MOTION

EXERCISES

Section 15.1 Density and Pressure

17. **INTERPRET** This problem is about the volume fraction of water that's made up of the atomic nuclei.

DEVELOP The average density of a mixture of two substances, with definite volume fractions, is

$$\rho_{av} = \frac{m_{tot}}{V_{tot}} = \frac{m_1 + m_2}{V_1 + V_2} = \frac{\rho_1 V_1 + \rho_2 V_2}{V_1 + V_2} = \rho_1\left(\frac{V_1}{V_1 + V_2}\right) + \rho_2\left(\frac{V_2}{V_1 + V_2}\right)$$

where $V_{tot} = V_1 + V_2$ is the total volume. The density of water is approximately the average density $(\rho_{av} = 10^3 \text{ kg/m}^3)$ of the nuclei $(\rho_1 = 10^{17} \text{ kg/m}^3)$ and empty space $(\rho_2 = 0)$, provided we neglect the mass of the atomic electrons.

EVALUATE The volume fraction of nuclei in water is

$$\frac{V_1}{V_{tot}} = \frac{\rho_{av}}{\rho_1} = \frac{10^3}{10^{17}} = 10^{-14}$$

ASSESS The result agrees with the fact that almost all the mass of an atom is concentrated in its nucleus which is made up of protons and neutrons that are much more massive than the electrons.

19. **INTERPRET** This problem is about expressing pressure in SI units, using suitable conversion factors.

DEVELOP The pressure at a depth h in an incompressible fluid of uniform density ρ is given by Equation 15.3:

$$p = p_0 + \rho g h$$

where p_0 is the pressure at the surface of the liquid. By definition, one torr is the pressure that will support a column of mercury 1 mm high: 1 torr $\equiv \rho_{Hg} g(1 \text{ mm})$.

EVALUATE From the definition above, we obtain

$$1 \text{ torr} \equiv \rho_{Hg} g(1 \text{ mm}) = (1.36 \times 10^4 \text{ kg/m}^3)(9.81 \text{ m/s}^2)(10^{-3} \text{ m}) = 133 \text{ Pa}$$

Similarly, 1 in (or 25.4 mm) of Hg is

$$25.4 \text{ torr} \equiv \rho_{Hg} g(25.4 \text{ mm}) = (1.36 \times 10^4 \text{ kg/m}^3)(9.81 \text{ m/s}^2)(25.4 \times 10^{-3} \text{ m}) = 3.39 \text{ kPa}$$

ASSESS One atmospheric pressure (1 atm) of 101.3 kPa supports a column of Hg 760 mm (or 29.92 in) high. So 1 torr is 1/760 of 1 atm.

21. **INTERPRET** This problem is about computing the weight of a column of air in the atmosphere.

DEVELOP As shown in Equation 15.1, pressure measures the normal force per unit area exerted by a fluid, $p = F/A$. In this problem the fluid is the air. Since the atmospheric pressure supports the entire weight of the air column, we have $F_g = m_{air} g = p_{atm} A$.

EVALUATE The weight of the air is

$$F_g = m_{air} g = p_{atm} A = (101.3 \text{ kPa})(1 \text{ m}^2) = 101.3 \text{ kN}$$

ASSESS This is the force that's pushing down on the 1-m^2 cross-sectional area. It is enormous!

23. **INTERPRET** This problem is about finding the force must be exerted on the paper clip to produce a certain pressure.

DEVELOP As shown in Equation 15.1, pressure measures the normal force per unit area exerted by a fluid, $p = F/A$. The cross-sectional area of the paper clip is $A = \pi(d/2)^2 = \pi d^2/4$, where d is the diameter.

EVALUATE   An average pressure of 120 atm over the area results in a force of

$$F = pA = (120 \times 101.3 \text{ kPa})\pi(1.5 \times 10^{-3} \text{ m})^2/4 = 21.5 \text{ N}$$

ASSESS   Pressure is inversely proportional to the area the force is exerted. The smallness of the cross section of the paper clip is what makes the pressure so large.

Section 15.2 Hydrostatic Equilibrium

25. **INTERPRET**   This problem is about hydrostatic equilibrium. Given the maximum pressure the submarine can withstand, we want to find its maximum allowable depth below water.

    DEVELOP   The external pressure, p, at a depth, h, below the surface of water (where the pressure is atmospheric pressure, $P_0 = 1$ atm) is given by Equation 15.1:

    $$p = p_0 + \rho g h$$

    A typical density for open ocean seawater (which varies with salinity) is $\rho = 1027 \text{ kg/m}^3$.

    EVALUATE   The depth corresponding to $p = 50$ MPa is

    $$h = \frac{p - p_0}{\rho g} = \frac{50 \text{ MPa} - 101.3 \text{ kPa}}{(1027 \text{ kg/m}^3)(9.8 \text{ m/s}^2)} = 4.96 \text{ km}$$

    ASSESS   The pressure the submarine can withstand is quite high (more than $490\, p_{\text{atm}}$). Note that our result is not exact because water at this depth is slightly compressible.

27. **INTERPRET**   We have an open tube filled with water on the bottom and oil on the top of water. The two fluids do not mix. We want to find the gauge pressure at the bottom of the tube.

    DEVELOP   The pressure at the top of the tube is atmospheric pressure, p_{atm}. The absolute pressure at the interface of the oil and water is $p_i = p_{\text{atm}} + \rho_{\text{oil}} g h_{\text{oil}}$ and at the bottom is, using Equation 15.3,

    $$p = p_i + \rho_{\text{water}} g h_{\text{water}} = p_{\text{atm}} + \rho_{\text{oil}} g h_{\text{oil}} + \rho_{\text{water}} g h_{\text{water}}$$

    EVALUATE   Therefore, the gauge pressure at the bottom is

    $$\begin{aligned}\Delta p = p - p_{\text{atm}} &= (\rho_{\text{oil}} h_{\text{oil}} + \rho_{\text{water}} h_{\text{water}})g \\ &= [(0.82 \times 10^3 \text{ kg/m}^3)(0.05 \text{ m}) + (1.00 \times 10^3 \text{ kg/m}^3)(0.05 \text{ m})](9.8 \text{ m/s}^2) \\ &= 892 \text{ Pa}\end{aligned}$$

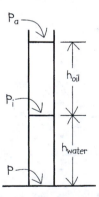

    ASSESS   Oil is less dense than water. Therefore it flows on top of the water. The gauge pressure at the bottom of the tube is due to the weight of both the oil and the water.

29. **INTERPRET**   In this problem we are given two barometric pressures, one at the eye of the hurricane, another at a fair-weather region. We want to compare their level of ocean surface.

    DEVELOP   The pressure difference between two points differed by an altitude h can be calculated by using Equation 15.3: $p - p_0 = \rho g h$.

    EVALUATE   With $p = 0.91$ atm and $p_0 = p_{\text{atm}}$, we obtain

    $$h = \frac{p - p_0}{\rho g} = \frac{(0.91 \text{ atm} - 1 \text{ atm})}{(1000 \text{ kg/m}^3)(9.8 \text{ m/s}^2)} \frac{1.013 \times 10^5 \text{ Pa}}{1 \text{ atm}} = -0.93 \text{ m}$$

ASSESS   The negative sign means that the fair weather region is 0.93 m *under* the hurricane eye. Note that pressure increases with increasing depth of the water.

Section 15.3 Archimedes' Principle and Buoyancy

31.   **INTERPRET**   Given the apparent weight of a jewel when submerged in water, we want to know whether the jewel is diamond or not. To answer this question, we need to calculate the density of the jewel.

DEVELOP   According to Archimedes' principle, the buoyant force on an object is equal to the weight of the fluid displaced by the object:

$$F_b = m_{water}g = \rho_{water}V_{jewel}g = \rho_{water}\frac{m_{jewel}}{\rho_{jewel}}g = \left(\frac{\rho_{water}}{\rho_{jewel}}\right)W_{jewel}$$

where $W_{jewel} = m_{jewel}g$ is the actual weight of the jewel. The apparent weight of the jewel is

$$W_{apparent} = W_{jewel} - F_b = W_{jewel} - \left(\frac{\rho_{water}}{\rho_{jewel}}\right)W_{jewel} = \left(1 - \frac{\rho_{water}}{\rho_{jewel}}\right)W_{jewel}$$

EVALUATE   Using the above equation, we obtain

$$\rho_{jewel} = \frac{\rho_{water}}{1 - W_{apparent}/W_{jewel}} = \frac{1000 \text{ kg/m}^3}{1 - (32\times10^{-3} \text{ N})/(0.0054 \text{ kg} \cdot 9.8 \text{ m/s}^2)} = 2.53\times10^3 \text{ kg/m}^3$$

The value is too small for the specific gravity of diamond ($\rho_{diamond} = 3.52\times10^3 \text{ kg/m}^3$).

ASSESS   Knowing the apparent weight of a submerged object allows us to determine the density of the object.

33.   **INTERPRET**   In this problem we are given the volume and mass of a steel drum and asked to determine whether or not it will float in water when filled with water or with gasoline.

DEVELOP   An object will float in water if its average density is less than the density of water. This follows from Archimedes' principle, since the volume of water displaced by an object floating on the surface is less than its total volume, i.e., $V_{dis} < V$. Since the buoyant force equals the weight of a floating object,

$$F_b = \rho_{H_2O}gV_{dis} = W = \rho_{av}gV$$

this implies $\rho_{av} < \rho_{H_2O}$.

EVALUATE   **(a)** When the drum is filled with water, since $\rho_{steel} > \rho_{H_2O}$, we have $\rho_{av} > \rho_{H_2O}$ and the drum will sink. This is because the average density of a composite object is always greater than the smallest density of its components. **(b)** When the drum is filled with gasoline, its average density is

$$\rho_{av} = \frac{M_{steel} + M_{gas}}{V_{steel} + V_{gas}}$$

If we neglect the volume occupied by the steel compared to the volume of the drum ($V_{steel} \ll V_{gas}$), then $M_{gas} = \rho_{gas}V$ and

$$\rho_{av} \approx \frac{M_{steel} + M_{gas}}{V_{gas}} = \frac{M_{steel}}{V_{gas}} + \rho_{gas} = \frac{16 \text{ kg}}{0.23 \text{ m}^3} + 860 \text{ kg/m}^3 = 930 \text{ kg/m}^3$$

which is less than ρ_{H_2O}. Therefore, the drum will float in this case.

ASSESS   Even though steel has greater density than water, the effective volume of the steel drum includes all of the liquid inside it. Thus, when filled with gasoline, its average density is then less than that of water, and hence it will float.

Sections 15.4 and 15.5 Fluid Dynamics and Applications

35.   **INTERPRET**   This problem deals with fluid dynamics. We are asked to sketch a streamline pattern of the flow of a fluid.

DEVELOP   The flow velocity of a fluid can be characterized by drawing continuous lines called streamlines. They are lines that are everywhere tangent to the local flow direction.

EVALUATE   In order to maintain a constant volume rate of flow, in an incompressible fluid, streamlines must be closer together (smaller cross-section of tube of flow) where the velocity is greater. The figure is a sketch of the streamlines.

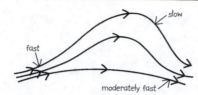

ASSESS   The spacing between streamlines gives a measure of the flow speed; region with higher speed has streamlines that are closely spaced, while the spacing is larger in regions with lower speed.

37.   INTERPRET   This problem deals with fluid flow rate. The key concepts involved are mass conservation and continuity equation.

DEVELOP   The mass flow rate is given by Equation 15.4, $R_m = \rho vA$, while the volume flow rate is given by Equation 15.5: $R_V = vA$. The speed of the flow can be determined once the flow rate and the cross-sectional area are known.

EVALUATE   (a) The volume flow rate for the Mississippi River is

$$R_V = \rho v = \frac{R_m}{\rho} = \frac{1.8 \times 10^7 \ \text{kg/s}}{10^3 \ \text{kg/m}^3} = 1.8 \times 10^4 \ \text{m}^3/\text{s}$$

(b) At a point in the river where the cross-sectional area is given, the average speed of flow is

$$v = \frac{R_V}{A} = \frac{1.8 \times 10^4 \ \text{m}^3/\text{s}}{2 \times 10^3 \times 6.1 \ \text{m}^2} = 1.48 \ \text{m/s} = 5.31 \ \text{km/h} = 3.30 \ \text{mph}$$

ASSESS   The flow speed we find is reasonable. Note that the actual flow rate of any river varies with the season, local weather and vegetation conditions, and human water consumption.

39.   INTERPRET   This problem deals with flow speed of a fluid, which in this case is the blood in the artery. The key concepts involved are mass conservation and the continuity equation.

DEVELOP   The continuity equation, $vA =$ constant, as given in Equation 15.5, is a reasonable approximation for blood circulation in an artery.

EVALUATE   The continuity equation gives $v' = vA/A'$. If the cross-sectional area is reduced by 80%, then $A' = 0.20A$, and

$$v' = v\left(\frac{A}{A'}\right) = (35 \ \text{cm/s})\frac{A}{0.20A} = 1.75 \ \text{m/s}$$

ASSESS   The flow speed of blood increases in the region where the cross-sectional area of the artery has been reduced due to clotting.

PROBLEMS

41.   INTERPRET   Given the mass and the gauge pressure of the tires, we want to find the total tire area that's in contact with the road.

DEVELOP   As shown in Equation 15.1, pressure measures the normal force per unit area exerted by a fluid. Our fluid here is the air. The force the tires exert on the road is the weight of the car, $F_g = mg$.

EVALUATE   With a gauge pressure of $\Delta p = 230$ kPa, the area of contact is

$$A = \frac{F_g}{\Delta p} = \frac{mg}{\Delta p} = \frac{(1950 \ \text{kg})(9.8 \ \text{m/s}^2)}{230 \times 10^3 \ \text{Pa}} = 0.0831 \ \text{m}^2 = 831 \ \text{cm}^2$$

ASSESS   Our result implies that the area of contact for each wheel would be about 200 cm², or the area of a 25 cm × 8 cm rectangular surface. This seems reasonable.

43.   INTERPRET   We have an open tube filled with water on the bottom and oil on the top of water. The two fluids do not mix. We want to find the gauge pressures at the interface as well as at the bottom of the tube.

DEVELOP   The pressure at the top of the tube is atmospheric pressure, p_{atm}. The gauge pressure at the interface of the oil and water is $\Delta p = p_i - p_{atm} = \rho_{oil} g h_{oil}$, and the gauge pressure at the bottom is the total weight of fluid divided by the cross-sectional area of the tube.

EVALUATE   (a) To find h_{oil}, note that

$$m_{oil} = \rho_{oil} V_{oil} = \rho_{oil} A_{tube} h_{oil} = 5 \text{ g}$$

where V_{oil} is the volume of oil and A_{tube} the cross-sectional area of the tube. Thus, the gauge pressure at the interface is

$$\Delta p = p_i - p_{atm} = \rho_{oil} g h_{oil} = \frac{m_{oil} g}{A_{tube}} = \frac{(0.005 \text{ kg})(9.8 \text{ m/s}^2)}{\pi (0.01 \text{ m})^2 / 4} = 624 \text{ Pa (gauge)}$$

(b) The gauge pressure at the bottom is the total weight of fluid divided by the cross-sectional area of the tube, which equals twice the answer to part **(a)**, or 1.25 kPa.

ASSESS   Oil flows on top of water since its density is lower than that of water. The gauge pressure at the bottom of the tube is due to the weight of both the oil and the water. The absolute pressure there would be equal to the sum of the gauge pressure and the atmospheric pressure.

45.   **INTERPRET**   The U-tube contains two liquids, oil and water, in hydrostatic equilibrium. We want to find their height difference.

DEVELOP   Let's consider the pressure at the level of the interface. By considering the right side, we find the pressure to be

$$p_{right} = p_{atm} + \rho_{oil} g l$$

where $l = 2.0$ cm. On the other hand, by considering the left side, we find the pressure to be

$$p_{left} = p_{atm} + \rho_{H_2O} g (l - h)$$

From Equation 15.3, $p = p_0 + \rho g h$, we see that the pressure at points at the same level in the water is the same, $p_{right} = p_{left}$.

EVALUATE   Equating the two pressures leads to $\rho_{oil} g l = \rho_{H_2O} g(l - h)$, or

$$h = \left(1 - \frac{\rho_{oil}}{\rho_{H_2O}}\right) l = (1 - 0.82)(2 \text{ cm}) = 3.6 \text{ mm}$$

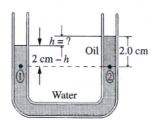

ASSESS   Note that the final answer does not depend on the atmospheric pressure, p_{atm}. The U-tube can be used to measure the density of a fluid, if we know the height difference h and the density of the other fluid.

47.   **INTERPRET**   This problem involves Pascal's law. We want to know the maximum mass the hydraulic lift can support.

DEVELOP   If we neglect the variation of pressure with height in the hydraulic system (which is usually small compared to the applied pressure), the fluid pressure is the same throughout, for either the small or large cylinders:

$$p = \frac{F_1}{A_1} = \frac{F_2}{A_2}$$

where F_1 is applied force and F_2 is the output force.

EVALUATE   The above equation gives

$$F_2 = pA_2 = (500 \text{ kPa}) \frac{\pi (0.45 \text{ m})^2}{4} = 79.5 \text{ kN}$$

which corresponds to a mass-load of $m_2 = F_2/g = 8112$ kg, or 8.112 tonnes (metric tons).

ASSESS   Through the constant fluid pressure, the input force multiplies by the area ratios ($A_2/A_1 > 1$) to give the output force. Energy, however, is conserved in this process.

49.   INTERPRET   This problem is about applying Archimedes' principle to find the change of the weight of the beer in a bottle.

DEVELOP   Archimedes' principle implies that the weight of the beer swallowed equals the difference in the weight of water displaced by the bottle, before and after. Therefore, the change of the weight of the beer is

$$\Delta w_{beer} = (\Delta m_{beer})g = \rho_{H_2O} g \, \Delta V$$

The difference in the volume of water displaced is $\Delta V = A\Delta h$, where A is the cross-sectional area of the bottle and $\Delta h = 28$ mm is the height change.

EVALUATE   If we ignore the thickness of the walls of the bottle, then

$$\Delta m_{beer} = \rho_{H_2O}\Delta V = \rho_{H_2O} A\Delta h = (1 \text{ g/cm}^3)\frac{\pi(5.2 \text{ cm})^2}{4}(2.8 \text{ cm}) = 59.5 \text{ g}$$

ASSESS   As expected, the amount of beer consumed is proportional to the height change, Δh.

51.   INTERPRET   This problem involves applying Archimedes' principle to find the minimum water depth for the load-carrying ship to navigate.

DEVELOP   If the sides of the hull are vertical, and its bottom flat, by Archimedes' principle, the buoyancy force is equal to the weight of the water displaced:

$$F_b = \Delta m_{H_2O} g = \rho_{H_2O} g \Delta V$$

and the volume it displaces is proportional to its draft (depth in the water), i.e., $\Delta V = A\Delta y$, where A is the cross-sectional area.

EVALUATE   Since the total mass of the full supertanker is three times that when empty, the draft when full is simply $\Delta y_{full} = 3\Delta y_{empty} = 3(9 \text{ m}) = 27 \text{ m}$.

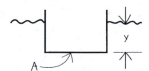

ASSESS   Our result is independent of the mass of the supertanker. The heavier the supertanker plus the load, the deeper it will submerge.

53.   INTERPRET   This problem is about the buoyancy force provided by the helium balloon and the hot-air balloon.

DEVELOP   For the balloon to lift off, the buoyancy force must exceed the weight of the load (mass M, including the balloon) plus the gas (mass m),

$$F_b \geq (M + m)g$$

where the buoyancy force is simply equal to $F_b = \rho_{air} g V$. If we neglect the volume of the balloon's skin etc. compared to that of the gas it contains, then $V = m/\rho_{gas}$. Therefore,

$$m = \rho_{gas}V = \rho_{gas}\frac{F_b}{\rho_{air}g} \geq \frac{\rho_{gas}}{\rho_{air}}(M + m) \quad \text{or} \quad m \geq \left(\frac{\rho_{gas}}{\rho_{air} - \rho_{gas}}\right)M$$

EVALUATE   **(a)** When the gas is helium, the density ratio is $\rho_{air}/\rho_{He} = (1.2 \text{ kg/m}^3)/(0.18 \text{ kg/m}^3) = 6.67$, and

$$m \geq \left(\frac{\rho_{gas}}{\rho_{air} - \rho_{gas}}\right)M = \left(\frac{1}{6.67 - 1}\right)(280 \text{ kg}) = 49.4 \text{ kg}$$

(b) For hot air, $\rho_{gas} = 0.9\rho_{air}$, and

$$m \geq \left(\frac{\rho_{gas}}{\rho_{air} - \rho_{gas}}\right)M = \left(\frac{0.9}{1 - 0.9}\right)(280 \text{ kg}) = 2520 \text{ kg}$$

ASSESS These masses correspond to gas volumes of 275 m^3 for helium and 2330 m^3 for hot air. These volumes are reasonable for a helium-filled balloon and a hot-air balloon.

55. **INTERPRET** This problem deals with flow speed of a fluid, which in this case is the blood in the artery. The key point involved here is Bernoulli's equation.

    DEVELOP The continuity equation, $vA = \text{constant}$, as given in Equation 15.5, is a reasonable approximation for blood circulation in an artery. Neglecting any pressure differences due to height, we find, from Bernoulli's equation, that

    $$p + \frac{1}{2}\rho v^2 = p' + \frac{1}{2}\rho v'^2$$

    EVALUATE The continuity equation gives $v' = vA/A'$. If the cross-sectional area is reduced by 80%, then $A' = 0.20A$, and

    $$v' = v\left(\frac{A}{A'}\right) = (0.35 \text{ m/s})\frac{A}{0.20A} = 1.75 \text{ m/s}$$

    From Bernoulli's equation, the gauge pressure is

    $$p' = p + \frac{1}{2}\rho(v^2 - v'^2) = p + \frac{1}{2}\rho v^2(1 - 25) = 16 \text{ kPa (gauge)} - 12(1060 \text{ kg/m}^2)(0.35 \text{ m/s})^2$$
    $$= 14.4 \text{ kPa (gauge)}$$

    ASSESS The flow speed of blood increases in the region where the cross-sectional area of the artery has been reduced due to clotting. Since $p + \frac{1}{2}\rho v^2 = \text{constant}$, the gauge pressure must decrease.

57. **INTERPRET** This problem involves flow of water, an incompressible fluid. We apply Bernoulli's equation to find the maximum height reached by the water coming out from the hose.

    DEVELOP The flow of water in the hose can be described by Bernoulli's equation (Equation 15.6):

    $$p + \frac{1}{2}\rho v^2 + \rho gy = \text{constant}$$

    The pressure, velocity, and height of the water in the hose (point 1) are $p_1 = p_{atm} + \Delta p_1 = p_{atm} + 140 \text{ kPa}$, $v_1 \approx 0$ and $y_1 = 0$, while at the highest point of a jet of water from a hole (point 2), $p_2 = p_{atm}$, $v_2 \approx 0$, and $y_2 = h$.

    EVALUATE Using Bernoulli's equation we have $p_1 = p_2$, which gives

    $$h = \frac{\Delta p_1}{\rho g} = \frac{140 \text{ kPa}}{(1000 \text{ kg/m}^3)(9.8 \text{ m/s}^2)} = 14.3 \text{ m}$$

    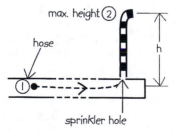

    ASSESS At the maximum height, all the work done by pressure has been converted to potential energy of the fluid. Energy is conserved in the process.

59. **INTERPRET** A narrower section is placed in a pipe carrying an incompressible fluid. We are to find the flow rate in the pipe and the volume flow rate, given the pressure difference between the fluid in the pipe and the fluid in the narrow section. We will assume that the flow is non-turbulent, and use Bernoulli's equation. The velocity is related to the cross-sectional area by the continuity equation.

    DEVELOP The pressure difference between the venturi and the unrestricted pipe is $\Delta P = 17 \text{ kPa}$. The radius of the unconstricted pipe is $r_1 = 0.010 \text{ m}$, and the radius of the constricted region is $r_2 = 0.005 \text{ m}$. We will assume that any height changes are negligible, and use the density of water as $\rho = 1000 \text{ kg/m}^3$. Bernoulli's equation is $P + \rho gh + \frac{1}{2}\rho v^2 = \text{constant}$, and the continuity equation for incompressible fluids such as water is $v_1 A_1 = v_2 A_2$.

EVALUATE

(a) We set the unrestricted pipe equal to the restricted pipe, using Bernoulli's equation with $h_1 = h_2$:

$P_1 + \rho g h_1 + \frac{1}{2}\rho v_1^2 = P_2 + \rho g h_2 + \frac{1}{2}\rho v_2^2$. We substitute $v_2 = v_1 \frac{A_1}{A_2}$ from the continuity equation and solve for v_1:

$$P_1 + \frac{1}{2}\rho v_1^2 = P_2 + \frac{1}{2}\rho\left(v_1\frac{A_1}{A_2}\right)^2 \rightarrow P_1 - P_2 = \frac{1}{2}\rho v_1^2\left(\frac{A_1^2}{A_2^2}-1\right)$$

$$\rightarrow v_1^2 = \frac{2\Delta P}{\rho\left(\frac{A_1^2}{A_2^2}-1\right)} = \frac{2\Delta P}{\rho\left(\left(\frac{r_1}{r_2}\right)^4-1\right)} = \frac{2\Delta P}{\rho(2^4-1)}$$

$$\rightarrow v_1 = \sqrt{\frac{2\Delta P}{\rho(2^4-1)}} = 1.5 \text{ m/s}$$

(b) To find the volume flow rate, in m³/s, we multiply the speed by the area of the pipe. $VFR = v_1 A_1 = v_1 \pi r_1^2 = 4.7 \times 10^{-4}$ m³/s $= 0.47$ liters/second.

ASSESS Both the flow speed and volume flow rate seem reasonable for a small pipe such as this.

61. **INTERPRET** This problem involves flow of water, an incompressible fluid, in a pipe. Pressure is reduced due to obstruction. We apply the continuity equation and Bernoulli's equation to solve for the fraction of area obstructed.
    DEVELOP We assume horizontal flow of water in a narrow pipe so there is no dependence on height. For steady incompressible fluid flow, the continuity equation (Equation 15.5) is $v_1 A_1 = v_2 A_2$, and Bernoulli's equation reads

    $$p_1 + \frac{1}{2}\rho v_1^2 = p_2 + \frac{1}{2}\rho v_2^2$$

    where subscripts 1 and 2 refer to without and with the obstruction. The two equations can be combined to solve for the fraction of area obstructed.
    EVALUATE Since the pressure at the obstruction is 5% less,

    $$\Delta p = p_1 - p_2 = 0.05 p_1 = \frac{1}{2}\rho\left(v_2^2 - v_1^2\right) = \frac{1}{2}\rho v_1^2\left[\left(\frac{A_1}{A_2}\right)^2 - 1\right]$$

    where we have eliminated v_2. Thus,

    $$\frac{A_1}{A_2} = \sqrt{1 + \frac{2\Delta p}{\rho v_1^2}} = \sqrt{1 + \frac{2(0.05)(230 \text{ kPa})}{(1000 \text{ kg/m}^3)(1.5 \text{ m/s})^2}} = 3.35$$

    The fraction of area obstructed is

    $$\text{frac} = \frac{A_1 - A_2}{A_1} = 1 - \frac{A_2}{A_1} = 1 - \frac{1}{3.35} = 0.701 = 70.1\%$$

    ASSESS Increasing Δp will lead to a greater A_1/A_2, and hence a larger fraction of area obstructed.

63. **INTERPRET** This problem involves flow of juice which we take to be an incompressible fluid. We apply both the continuity equation and Bernoulli's equation to solve the problem.
    DEVELOP For steady incompressible fluid flow, the continuity equation (Equation 15.5) is

    $$v_1 A_1 = v_2 A_2 \quad \rightarrow \quad v_1\left(\frac{\pi D_1^2}{4}\right) = v_2\left(\frac{\pi D_2^2}{4}\right)$$

    and Bernoulli's equation reads

    $$p_1 + \frac{1}{2}\rho v_1^2 + \rho g y_1 = p_2 + \frac{1}{2}\rho v_2^2 + \rho g y_2$$

    Eliminating v_2 using the first equation, we obtain

    $$\Delta p = p_1 - p_2 = \frac{1}{2}\rho v_1^2\left[\left(\frac{A_1}{A_2}\right)^2 - 1\right] + \rho g(y_2 - y_1) = \frac{1}{2}\rho v_1^2\left[\left(\frac{D_1}{D_2}\right)^4 - 1\right] + \rho g(y_2 - y_1)$$

EVALUATE   (a) We assume the juice density to be that of water. When $y_2 - y_1 = 6.5$ cm, the pressure difference is

$$\Delta p = \frac{1}{2}\rho v_1^2 \left[\left(\frac{D_1}{D_2} \right)^4 - 1 \right] + \rho g(y_2 - y_1)$$

$$= \frac{1}{2}(1000 \text{ kg/m}^3)(0.002 \text{ m/s})^2 \left[\left(\frac{0.08 \text{ m}}{0.003 \text{ m}} \right)^4 - 1 \right] + (1000 \text{ kg/m}^3)(9.8 \text{ m/s}^2)(0.065 \text{ m})$$

$$= 1.65 \times 10^3 \text{ Pa} = (1.63\%) p_{atm}$$

(b) For a constant pressure difference, $y_2 - y_1$ attains its maximum value when $v_1 = 0$. Thus,

$$(y_2 - y_1)_{max} = \frac{\Delta p}{\rho g} = \frac{1.65 \times 10^3 \text{ Pa}}{(1000 \text{ kg/m}^3)(9.8 \text{ m/s}^2)} = 16.8 \text{ cm}$$

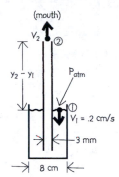

ASSESS   As the juice level drops, the pressure difference and/or flow speed may change.

65.   **INTERPRET**   This problem is about finding the force on a suction cup due to atmospheric pressure.
DEVELOP   The normal force on the suction cup is the result of atmospheric pressure, where we assume a perfect vacuum inside the cup. The frictional force that supports the object must be such that
$f = mg \leq f_{max} = \mu n = \mu p_{atm} A = \mu p_{atm} (\pi d^2/4)$.
EVALUATE   The maximum mass that can be supported by the suction cup is

$$m_{max} = \frac{\mu p_{atm} \pi d^2}{4g} = \frac{(0.72)(101.3 \text{ kPa})\pi(0.05 \text{ m})^2}{4(9.8 \text{ m/s}^2)} = 14.6 \text{ kg}$$

ASSESS   The maximum value of 14.6 kg is about the mass of a small toddler. The force on the cup due to the atmospheric pressure is quite large.

67.   **INTERPRET**   This problem is about how the speed of the wind turbine affects its maximum power output.
DEVELOP   The maximum rate of energy extraction per unit area of turbine is

$$R_{max} = \frac{P_{max}}{A} = \frac{8}{27}\rho v^3$$

as discussed in the Application to wind energy in Section 15.5. A 65 m diameter blade sweeps through an area of $A = \frac{1}{4}\pi d^2 = \frac{1}{4}\pi(65 \text{ m})^2 = 3.32 \times 10^3 \text{ m}^2$.
EVALUATE   (a) For $v = 10$ m/s, we have

$$P_{max} = R_{max} A = \frac{8}{27}\rho v^3 A = \frac{8}{27}(1.2 \text{ kg/m}^3)(10 \text{ m/s})^3(3.32 \times 10^3 \text{ m}^2) = 1.18 \text{ MW}$$

(b) For $v = 15$ m/s, we have

$$P_{max} = R_{max} A = \frac{8}{27}\rho v^3 A = \frac{8}{27}(1.2 \text{ kg/m}^3)(15 \text{ m/s})^3(3.32 \times 10^3 \text{ m}^2) = 3.98 \text{ MW}$$

(c) For $v = 12$ m/s, each turbine actually produces only $P = (0.50)R_{max} A = 0.5\frac{8}{27}\rho v^3 A =$
$(0.50)\frac{8}{27}(1.2 \text{ kg/m}^3)(12 \text{ m/s})^3(3.32 \times 10^3 \text{ m}^2) = 1.02 \text{ MW}$ and thus $(10^3 \text{ MW})/(1.02 \text{ MW}) = 981$ of them would be needed to equal the output of a 1 GW nuclear power plant.
ASSESS   Although the yield from wind energy is not as high as nuclear energy, wind has become one of the fastest-growing components of energy supply in the world.

69.   **INTERPRET**   In this problem we want to find the time it takes for the can to drain out all its water through its hole. An integral is needed since the water level is continuous, from 0 to height h.

DEVELOP   Let y be the height of the water above the bottom of the can, then $-dy/dt$ is the magnitude of the flow speed of the top surface of the water draining out (y decreases as a function of time). The continuity equation gives

$$-\frac{dy}{dt}A_0 = v_1 A_1 \quad \rightarrow \quad dt = -\left(\frac{A_0}{A_1}\right)v_1\,dy$$

where subscript 1 refers to the small hole in the bottom. For most of the time, $v_1 \approx \sqrt{2gy}$ (see Example 15.6) and we assume the top of the can is open).

EVALUATE   Carrying out the integration, we find the total time required to be

$$t = \int dt \approx -\int_h^0 \left(\frac{A_0}{A_1}\right)\frac{dy}{\sqrt{2gy}} = \frac{A_0}{A_1\sqrt{2g}}\,2\sqrt{y}\,\Big|_0^h = \frac{A_0}{A_1}\sqrt{\frac{2h}{g}}$$

ASSESS   This result is approximate since dy/dt cannot be neglected compared to v_1 when y is small. If we use Bernoulli's equation without this approximation, then

$$\frac{1}{2}\rho\left(\frac{dy}{dt}\right)^2 + \rho gy = \frac{1}{2}\rho v_1^2$$

since the pressure is atmospheric pressure at both the top of the can and the hole. Combining with the continuity equation gives

$$v_1 = \sqrt{2gy + (dy/dt)^2} = -\left(\frac{A_0}{A_1}\right)\frac{dy}{dt} \quad \rightarrow \quad \frac{dy}{dt} = -\sqrt{\frac{2gy}{(A_0/A_1)^2 - 1}}$$

Integration of this yields a more exact outflow time of $t = \sqrt{\frac{2h}{g}[(A_0/A_1)^2 - 1]}$.

71.   **INTERPRET**   Using the fact that the density and pressure in Earth's atmosphere are proportional: $\rho = p/h_0 g$, where h_0 is a constant, we want to find the atmospheric density as a function of height.

DEVELOP   The variation of pressure with height in the Earth's atmosphere follows from Equation 15.2 (with h replaced by $-h$, since height is positive upward whereas depth is positive downward). Thus, $dp = -\rho g\,dh$. If pressure and density are proportional (as given in Problem 70), then

$$dp = -\frac{p\,dh}{h_0}$$

Integrating the expression then yields pressure as a function of height h.

EVALUATE   **(a)** The above equation can be integrated from the surface values, $h = 0$ and p_0, to yield

$$\int_{p_0}^{p}\frac{dp}{p} = \ln\left(\frac{p}{p_0}\right) = -\int_0^h \frac{dh}{h_0} = -\frac{h}{h_0} \quad \rightarrow \quad p = p_0 e^{-h/h_0}$$

This is called the law of atmospheres; it applies exactly if the temperature is constant.

(b) The mass of atmosphere contained in a thin spherical shell of thickness dh, at height h, is

$$dm = \rho\,dV = \left(\rho_0 e^{-h/h_0}\right)4\pi(R_E + h)^2\,dh = 4\pi\rho_0(R_E + h)^2 e^{-h/h_0}\,dh$$

where R_E is the radius of the Earth and $R_E + h$ the radius of the shell. The mass of atmosphere below height h_1 is

$$M(h_1) = \int_0^{h_1} dm = 4\pi\rho_0 R_E^2 \int_0^{h_1}\left(1 + 2\frac{h}{R_E} + \frac{h^2}{R_E^2}\right)e^{-h/h_0}\,dh$$

The integrals can be evaluated easily enough with the use of the table of integrals in Appendix B. However, if $h_1/R_E \ll 1$, only the first term is important. (Even if h_1 is large, the exponential term is negligibly small for $h \ll h_0$ and none of the terms contribute significantly for large h.) To a good approximation, therefore

$$M(h_1) \approx 4\pi\rho_0 R_E^2 \int_0^{h_1} e^{-h/h_0}\,dh = 4\pi\rho_0 R_E^2 h_0 (1 - e^{-h_1/h_0})$$

The total mass of the atmosphere is approximately $M(\infty) = 4\pi\rho_0 R_E^2 h_0$, so the height bounding half the total mass is given by the equation $\frac{M}{2} = M(1 - e^{-h_1/h_0})$ or

$$h_1 = h_0 \ln 2 = (8.2 \text{ km})(0.693) = 5.7 \text{ km}$$

ASSESS   Since $\rho = p/h_0 g$, in terms of density, $\rho = \rho_0 e^{-h/h_0}$, where $\rho_0 = p_0/h_0 g$. Since $e^{-h_0 \ln 2/h_0} = \frac{1}{2}$, both p and ρ fall to half their surface values at a height $h_0 \ln 2 = (8.2 \text{ km})(0.693) = 5.7$ km, the same as the mass in part (**b**).

73.   **INTERPRET**   We are given the mass and radius of a sphere, and told how the density varies. Our goal is to find a parameter in the density equation. We will do this by integrating the density equation symbolically, and then setting the result equal to the known mass and solving for the unknown parameter.
DEVELOP   The equation for the density of the sphere is $\rho(r) = \rho_0 e^{r/R}$. The radius is $R = 0.10$ m and the mass is $M = 15$ kg. We will multiply $\rho(r)$ by dV to obtain the differential mass dm, then integrate this from 0 to R and set the result equal to M to find ρ_0.
Since the density depends only on r, we will do the integration in the form of spherical shells of thickness dr: $dV = 4\pi r^2 dr$.
EVALUATE

$$dm = \rho(r)dV = (\rho_0 e^{r/R})(4\pi r^2 dr) \rightarrow m = 4\pi\rho_0 \int_0^R r^2 e^{r/R} dr$$

$$\rightarrow m = 4\pi\rho_0 [re^{r/R}(2R^2 - 2rR + r^2)|_0^R] = 4\pi\rho_0 [(e-2)R^3]$$

$$\rightarrow \rho_0 = \frac{m}{4\pi R^3(e-2)} = 1660 \text{ kg/m}^3$$

ASSESS   This is very similar to a technique called "normalization," commonly used in statistical mechanics and quantum theory.

75.   **INTERPRET**   In Problem 44, we determined the force on a dam due to the water behind the dam. In this problem, we find the torque around the bottom edge of the same dam. We will use $F = pA$ and $\tau = yF$, as shown in the figure.
DEVELOP   The pressure varies with depth, according to $p(y) = \rho g(H - y)$. We will find the force $dF = pdA$, and thus the torque $d\tau = ydF$, from each horizontal strip across the dam. Integrating $d\tau$ gives us the total torque. The dam has width $w = 1500$ m and the water is $H = 95$ m deep. The density of water is $\rho = 1000$ kg/m³.
EVALUATE   The pressure $d\tau = ydF = y(p(y))dA = y(\rho g(H - y))(wdy)$. Integrate this from $y = 0$ to $y = H$.

$$\tau = \rho g w \int_0^H y(H - y)dy = \rho g w \left(\frac{H}{2}y^2 - \frac{y^3}{3}\right)\Big|_0^H = \rho g w \left(\frac{H^3}{2} - \frac{H^3}{3}\right) = \frac{1}{6}\rho g w H^3$$

$$\rightarrow \tau = 2.1 \times 10^{12} \text{ Nm}$$

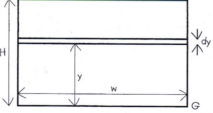

ASSESS   The units in our final equation are $\tau = \frac{kg}{m^3}\frac{m}{s^2}m\,m^3 = $ kg m/s² m = Nm, which works out correctly.

77.   **INTERPRET**   We find the expected density of a mix of immiscible liquids, and compare it with a measured density to see if the mix is what it should be. The density of a mix of liquids should be the total mass divided by the total volume.
DEVELOP   The "official" dressing is 1 part vinegar to 3 parts oil, measured by volume. So the dressing should have volume $4V$ and mass $m_{vinegar} + m_{oil} = \rho_{vinegar}V + 3\rho_{oil}V$. Calculate the density of this mix, and compare it with the measured density $\rho' = 0.97$ g/cm³. If the density of the sample is higher than it should be, then it has probably been diluted with water. The density of oil is $\rho_{oil} = 0.92$ g/cm³, and the density of vinegar is $\rho_{vinegar} = 1.0$ g/cm³.
EVALUATE

$$\rho = \frac{\rho_{vinegar}V + 3\rho_{oil}V}{4V} = \frac{\rho_{vinegar} + 3\rho_{oil}}{4} = 0.94 \text{ g/cm}^3$$

ASSESS   The dressing has been altered.

79.   **INTERPRET**   The question here is really "Is the water pressure sufficient to get the water to the top floor?" We'll look at it the other way: if there was a pipe full of water from the top to the bottom of the building, what would be the pressure at the bottom? If the pressure at the bottom of this hypothetical pipe is less than the measured pressure at the water heater, then there is enough pressure at the water heater to get hot water to the top floor.

DEVELOP   We will use $p = \rho h$, where $\rho = 62.2$ lb/ft^3 and $h = 33$ ft. (We don't use $p = \rho gh$, since we are given *weight* density rather than *mass* density.) The measured pressure is $p_m = 18$ lb/in$^2 \times \frac{144 \text{ in}^2}{\text{ft}^2} = 2600$ lb/ft^2, and the height is $h = 33$ ft.

EVALUATE $p = \rho h = 2050$ lb/ft^2. This pressure is less than the pressure at the heater, so there is sufficient water pressure to reach the top of the building.

ASSESS   A quicker way to solve this is to remember that atmospheric pressure (about 15 psi) can support a column of water about 32 feet high. This pressure (18 psi) is higher, so it can push water even higher.

TEMPERATURE AND HEAT

EXERCISES

Section 16.1 Heat, Temperature, and Thermodynamic Equilibrium

15. **INTERPRET**   This problem is about converting temperature from the Fahrenheit scale to the Celsius scale.
    DEVELOP   The two temperature scales are related by Equation 16.2:

$$T_F = \frac{9}{5}T_C + 32$$

EVALUATE   Solving the above equation for the Celsius temperature, we obtain

$$T_C = \frac{5}{9}(T_F - 32) = \frac{5}{9}(68 - 32) = 20°C$$

ASSESS   This is a useful result to remember, since 20°C, or 68 °F is a typical room temperature.

17. **INTERPRET**   Given both Fahrenheit and Celsius scales, we want to know when T_F and T_C are numerically equal.
    DEVELOP   The two temperature scales are related by Equation 16.2:

$$T_F = \frac{9}{5}T_C + 32$$

The condition that the readings are the same numerically is $T_F = (\frac{9}{5})T_C + 32 = T_C$.
EVALUATE   The above equation can be solved to give

$$T_C = -\frac{5}{4}(32) = -40 = T_F$$

ASSESS   This is the only temperature in which both scales yield the same reading: $-40°F = -40°C$.

19. **INTERPRET**   This problem is about converting temperature from the Celsius scale to the Fahrenheit scale.
    DEVELOP   The two temperature scales are related by Equation 16.2: $T_F = \frac{9}{5}T_C + 32$.
    EVALUATE   Solving the above equation for the Fahrenheit temperature, we obtain

$$T_F = \frac{9}{5}(39.1) + 32 = 102.4°F$$

ASSESS   The temperature is way above the normal body temperature of 98.6°F (or 37°C). Call the doctor immediately!

Section 16.2 Heat Capacity and Specific Heat

21. **INTERPRET**   We find the energy necessary to change the temperature of an object by a given amount. This involves the heat capacity of the object and the temperature.
    DEVELOP   We use the equation $Q = mc\Delta T$. The mass of the aluminum block is $m = 2.0$ kg, the specific heat is $c = 900$ J/kg·K, and the temperature change is $\Delta T = 18$ C°.
    EVALUATE $Q = mc\Delta T = 32,000$ J.
    ASSESS   The same value would be the heat released by the aluminum if it cooled 18 C°.

23. **INTERPRET**   The problem is about the average power output of the human body, given the information about the energy acquired from diet.
    DEVELOP   The energy gained from the diet is

$$\Delta E = (2 \times 10^6 \text{ cal})(4.184 \text{ J/cal}) = 8.37 \times 10^6 \text{ J}$$

EVALUATE   The average power output is

$$\overline{P} = \frac{\Delta E}{\Delta t} = \frac{8.37 \times 10^6 \text{ J}}{86,400 \text{ s}} = 96.9 \text{ W}$$

ASSESS   The average power output by the human body at rest is about 80 W, the same as a bright light bulb.

25.   **INTERPRET**   Given the energy it takes to warm the wrench, we want to find its heat capacity, as well as the specific heat of the metal.

    DEVELOP   The heat capacity of an object is given by $C = \frac{\Delta Q}{\Delta T}$, where ΔQ is the amount of heat transfer that results in a temperature change ΔT. As shown in Equation 16.3, the specific heat of a material is $c = C/m$.

    EVALUATE   (a) The average heat capacity is

$$C = \frac{\Delta Q}{\Delta T} = \frac{2.52 \text{ kJ}}{15 \text{ K}} = 168 \text{ J/K}$$

(b) The average specific heat of the material is

$$c = \frac{C}{m} = \frac{168 \text{ J/K}}{0.350 \text{ kg}} = 480 \text{ J/kg} \cdot \text{K}$$

ASSESS   The wrench is probably made of iron which has a specific of 448 J/kg · K.

Section 16.3 Heat Transfer

27.   **INTERPRET**   This problem is about converting heat loss expressed in Btu/h to SI units.

    DEVELOP   One Btu (British thermal unit) is equal to 1054 J, which is the amount of heat that's needed to raise the temperature of 1 lb of water from 63°F to 64°F.

    EVALUATE   The conversion to SI units is

$$1 \frac{\text{Btu}}{\text{h}} = (1 \text{ Btu/h})(1055 \text{ J/Btu})(1 \text{ h/3600 s}) = 0.293 \text{ W}$$

ASSESS   Our result shows that 1 W is about 3.4 Btu/h. The power output of air conditioners is commonly given in terms of Btu/h.

29.   **INTERPRET**   This problem is about the rate of heat conduction through the stove top.

    DEVELOP   We assume a steady flow of heat through the $A = 90 \text{ cm} \times 40 \text{ cm} = 0.36 \text{ m}^2$ area face, with no flow through the edges. The rate of heat flow is given by Equation 16.5:

$$H = -kA\frac{\Delta T}{\Delta x}$$

EVALUATE   From Table 16.2, we find the thermal conductivity of steel to be $k = 46$ W/m · K. Thus, the rate of heat conduction is

$$H = -kA\frac{\Delta T}{\Delta x} = -(46 \text{ W/m} \cdot \text{K})(0.36 \text{ m}^2)\frac{(295°\text{C} - 310°\text{C})}{0.0045 \text{ m}} = 55.2 \text{ kW}$$

ASSESS   The heat flow is positive, for x going from the inside of the stove to the outside, because the temperature gradient, $\Delta T/\Delta x$, is negative.

31.   **INTERPRET**   This problem is about the rate of heat conduction through the concrete slab.

    DEVELOP   We assume a steady flow of heat through the $A = 8.0 \text{ m} \times 12 \text{ m} = 96 \text{ m}^2$ area face, with no flow through the edges. The rate of heat flow is given by Equation 16.5:

$$H = -kA\frac{\Delta T}{\Delta x}$$

EVALUATE   From Table 16.2, we find the thermal conductivity of concrete to be $k = 1$ W/m · K. Thus, the rate of heat conduction is

$$H_{\text{floor}} = -kA\frac{\Delta T}{\Delta x} = -(1 \text{ W/m} \cdot \text{K})(96 \text{ m}^2)\frac{(293 \text{ K} - 283 \text{ K})}{0.23 \text{ m}} = -4.17 \text{ kW}$$

ASSESS   The energy loss through the floor by conduction is substantial. That's why carpeting can prevent heat loss and keeps the house warm during winter seasons.

33.   **INTERPRET**   This problem is about the $\mathcal{R}$-factors for 1-inch thicknesses of various materials.

DEVELOP   The $\mathcal{R}$-factor of a material is given by Equation 16.8:

$$\mathcal{R} = RA = \frac{\Delta x}{k}$$

where R is the thermal resistance and k is the thermal conductivity of a material having a thickness Δx.

EVALUATE   Using Table 16.2, with $k_{air} = 0.18$ Btu $\cdot$ in./h $\cdot$ ft^2 $\cdot$ F° for air, we have

$$\mathcal{R}_{air} = \frac{1 \text{ in.}}{0.18 \text{ Btu} \cdot \text{in./h} \cdot \text{ft}^2 \cdot \text{F}°} = 5.56 \text{ ft}^2 \cdot \text{F}° \cdot \text{h/Btu}$$

Similarly, with $k_{concrete} = 7$, $k_{fiberglass} = 0.29$, $k_{glass} = 5-6$, $k_{styrofoam} = 0.20$, and $k_{pine} = 0.78$ (all in units of Btu $\cdot$ in./h $\cdot$ ft^2 $\cdot$ F°), the $\mathcal{R}$-factors are

$$\mathcal{R}_{concrete} = \frac{1 \text{ in.}}{7 \text{ Btu} \cdot \text{in./h} \cdot \text{ft}^2 \cdot \text{F}°} = 0.143 \text{ ft}^2 \cdot \text{F}° \cdot \text{h/Btu}$$

$$\mathcal{R}_{fiberglass} = \frac{1 \text{ in.}}{0.29 \text{ Btu} \cdot \text{in./h} \cdot \text{ft}^2 \cdot \text{F}°} = 3.45 \text{ ft}^2 \cdot \text{F}° \cdot \text{h/Btu}$$

$$\mathcal{R}_{glass} = \frac{1 \text{ in.}}{(5-6) \text{ Btu} \cdot \text{in./h} \cdot \text{ft}^2 \cdot \text{F}°} = (0.167 - 0.2) \text{ ft}^2 \cdot \text{F}° \cdot \text{h/Btu}$$

$$\mathcal{R}_{styrofoam} = \frac{1 \text{ in.}}{0.20 \text{ Btu} \cdot \text{in./h} \cdot \text{ft}^2 \cdot \text{F}°} = 5 \text{ ft}^2 \cdot \text{F}° \cdot \text{h/Btu}$$

$$\mathcal{R}_{pine} = \frac{1 \text{ in.}}{0.78 \text{ Btu} \cdot \text{in./h} \cdot \text{ft}^2 \cdot \text{F}°} = 1.28 \text{ ft}^2 \cdot \text{F}° \cdot \text{h/Btu}$$

ASSESS   The $\mathcal{R}$-factor of a material is inversely proportional to the thermal conductivity. Good thermal insulators such as Styrofoam or wood have large $\mathcal{R}$-factors.

Section 16.4 Thermal-Energy Balance

35.   **INTERPRET**   This is an energy-balance problem. We are given the energy loss per time per degree of temperature difference, and the temperature difference. We wish to find the rate of energy loss, which is the power required to maintain the temperature.

DEVELOP   We multiply the energy loss rate per degree by the temperature difference in degrees.

EVALUATE $(14 \text{ W/°C}) \times (160°\text{C}) = 2.2$ kW.

ASSESS   2 kW is a reasonable power requirement for an oven.

37.   **INTERPRET**   This problem involves radiative heat loss, and the Stefan-Boltzmann law.

DEVELOP   We use $P = e\sigma A T^4$. The power is $P = 100$ W, the temperature is $T = 3000$ K, and $\sigma = 5.67 \times 10^{-8}$ W/m^2 $\cdot$ K^4. We wish to find the area A. We will assume that the emissivity is $e \approx 1$.

EVALUATE

$$P = e\sigma A T^4 \rightarrow A = \frac{P}{e\sigma T^4} = 2.2 \times 10^{-5} \text{ m}^2$$

ASSESS   This is about 20 square millimeters, which seems reasonable for the total area of a light bulb filament.

PROBLEMS

39.   **INTERPRET**   This problem is about finding the pressure at different temperatures, given its value at a reference temperature.

DEVELOP   The thermometric equation for an ideal constant-volume gas thermometer is

$$\frac{p}{T} = \frac{p_{ref}}{T_{ref}} \rightarrow p = \left(\frac{T}{T_{ref}}\right) p_{ref}$$

If we use the given values at the normal melting point of ice, then the pressure-temperature relation is given by

$$p = \left(\frac{T}{T_{\text{ref}}}\right) p_{\text{ref}} = \frac{T}{273.15\ \text{K}}(101\ \text{kPa})$$

EVALUATE   (a) When the temperature is the normal boiling point of water $T = 100°C = 373.15\ \text{K}$, the pressure is

$$p = \frac{373.15\ \text{K}}{273.15\ \text{K}}(101\ \text{kPa}) = 138\ \text{kPa}$$

(b) If the temperature is the normal boiling point of oxygen (90.2 K), then

$$p = \frac{90.2\ \text{K}}{273.15\ \text{K}}(101\ \text{kPa}) = 33.4\ \text{kPa}$$

(c) If the temperature is the normal boiling point of mercury (630 K), then

$$p = \frac{630\ \text{K}}{273.15\ \text{K}}(101\ \text{kPa}) = 233\ \text{kPa}$$

ASSESS   The linear relationship between p and T is depicted in Figure 16.3.

41.   **INTERPRET**   In this problem we are asked to calculate the boiling point of SO_2, given the height difference between the liquid levels in a constant-volume gas thermometer.

DEVELOP   The thermometric equation for an ideal constant-volume gas thermometer is

$$\frac{p}{T} = \frac{p_{\text{ref}}}{T_{\text{ref}}} \quad \rightarrow \quad p = \left(\frac{T}{T_{\text{ref}}}\right) p_{\text{ref}}$$

where T is measured in the Kelvin scale. Since the pressure in the constant-volume gas thermometer shown is proportional to h, the temperature of the boiling point of SO_2 is

$$T = T_3 \frac{p}{p_3} = T_3 \frac{h}{h_3}$$

EVALUATE   From the equation above, we find the boiling point of SO_2 to be

$$T = (273.16\ \text{K})\frac{57.8\ \text{mm}}{60.0\ \text{mm}} = 263\ \text{K} = -10.0°C$$

ASSESS   For a constant-volume gas thermometer, p/T is constant. Since pressure can be measured in mm of mercury ($p = \rho g h$) it is also true that h/T is constant.

43.   **INTERPRET**   This problem is about the amount of energy a body loses after running a marathon.

DEVELOP   The energy expended in running a marathon for a person with the given mass is

$$\Delta Q = (125\ \text{kcal/mi})(26.2\ \text{mi}) = 3.28 \times 10^3\ \text{kcal}$$

Knowing the amount of energy per gram of fats allows us to answer the question.

EVALUATE   Since typical fats contain about 9 kcal per gram, ΔQ is equivalent to the energy content of

$$\frac{3.28 \times 10^3\ \text{kcal}}{9\ \text{kcal/g}} = 364\ \text{g}$$

or about 13 oz of fat.

ASSESS   Running a marathon is a good way to burn the fat stored in the body.

45.   **INTERPRET**   We are interested in the energy needed to raise the temperature of a system. The problem involves specific heat.

DEVELOP   The energy ΔQ required to increase the temperature by ΔT is given by Equation 16.3: $\Delta Q = mc\Delta T$, where c is the specific heat and m is the mass of the material. The specific heats of some common materials can be found in Table 16.1.

EVALUATE   (a) When just the pan is heated, with $c_{\text{Cu}} = 386\ \text{J/kg} \cdot \text{K}$, the energy required is

$$\Delta Q = m_{\text{Cu}} c_{\text{Cu}}\ \Delta T = (0.8\ \text{kg})(386\ \text{J/kg} \cdot \text{K})(90\ \text{K} - 15\ \text{K}) = 23.2\ \text{kJ} = 5.54\ \text{kcal}$$

(b) If the pan contains water and both are heated between the same temperatures, we then have

$$\Delta Q = (m_{\text{Cu}} c_{\text{Cu}} + m_{\text{H}_2\text{O}} c_{\text{H}_2\text{O}})\ \Delta T = 23.2\ \text{kJ} + (1\ \text{kg})(4184\ \text{J/kg} \cdot \text{K})(75\ \text{K}) = 337\ \text{kJ} = 80.5\ \text{kcal}$$

(c) With $m_{Hg} = 4$ kg of mercury replacing the water,

$$\Delta Q = (m_{Cu}c_{Cu} + m_{Hg}c_{Hg})\,\Delta T = 23.2 \text{ kJ} + (4 \text{ kg})(140 \text{ J/kg} \cdot \text{K})(75 \text{ K}) = 65.2 \text{ kJ} = 15.6 \text{ kcal}$$

ASSESS   The energy required is proportional to the specific heat c. In this problem,

$$c_{Hg}\,(140 \text{ J/kg} \cdot \text{K}) < c_{Cu}\,(386 \text{ J/kg} \cdot \text{K}) < c_{H_2O}\,(4184 \text{ J/kg} \cdot \text{K})$$

47.   **INTERPRET**   We are asked about the time needed to raise the water temperature, given the power output of the microwave. The problem involves the specific heat of water.

DEVELOP   The energy ΔQ required to increase the temperature by ΔT is given by Equation 16.3: $\Delta Q = mc\Delta T$, where c is the specific heat and m is the mass of the material. The specific heats of some common materials can be found in Table 16.1.

EVALUATE   **(a)** With $c_{H_2O} = 4184$ J/kg $\cdot$ K and $\Delta T = 90$ K the energy required is

$$\Delta Q = m_{H_2O}c_{H_2O}\,\Delta T = \rho_{H_2O}Vc_{H_2O}\,\Delta T = (1000 \text{ kg/m}^3)(250 \times 10^{-6} \text{ m}^3)(4184 \text{ J/kg} \cdot \text{K})(90 \text{ K})$$
$$= 94{,}140 \text{ J}$$

Thus, the time required is $\Delta t = \frac{\Delta Q}{P} = \frac{94140 \text{ J}}{625 \text{ W}} = 151 \text{ s} = 2.51$ min.

ASSESS   A microwave oven is a fast and efficient means of heating up water.

49.   **INTERPRET**   We are given two ways to heat up water: by microwave oven with a paper cup, or by a pan on a stovetop. We'd like to know when it becomes quicker to use the stovetop and the rate of temperature change when this happens.

DEVELOP   The temperature rise per second is equal to the heat supplied per second (i.e., the power supplied if there are no losses) divided by the total heat capacity of the water and its container:

$$\frac{\Delta T}{\Delta t} = \frac{\Delta Q / C_{tot}}{\Delta t} = \frac{\bar{P}}{C_{tot}}$$

where $\bar{P} = \Delta Q / \Delta t$ is the average power supplied and $C_{tot} = C_{H_2O} + C_{cnt}$ is the total heat capacity, provided the water and container both have the same instantaneous temperature. This assumes that heat is supplied sufficiently slowly that the water and container share it and stay in instantaneous thermal equilibrium.

For the paper cup used in the microwave oven, $C_{cnt} \approx 0$, whereas for the pan used on the stove burner, $C_{cnt} = 1.4$ kJ/K. The rates of temperature rise are

$$\left(\frac{\Delta T}{\Delta t}\right)_{mw} = \frac{\bar{P}_{mw}}{C_{tot, mw}} = \frac{\bar{P}_{mw}}{C_{H_2O}}, \qquad \left(\frac{\Delta T}{\Delta t}\right)_{stove} = \frac{\bar{P}_{stove}}{C_{tot, stove}} = \frac{\bar{P}_{stove}}{C_{H_2O} + C_{cnt}}$$

EVALUATE   **(a)** The condition that $(\Delta T / \Delta t)_{stove} > (\Delta T / \Delta t)_{mw}$ implies

$$\frac{\bar{P}_{stove}}{C_{H_2O} + C_{cnt}} > \frac{\bar{P}_{mw}}{C_{H_2O}} \quad \rightarrow \quad \frac{\bar{P}_{stove}}{\bar{P}_{mw}} > \frac{C_{H_2O} + C_{cnt}}{C_{H_2O}} = 1 + \frac{C_{cnt}}{C_{H_2O}}$$

or

$$C_{H_2O} > \frac{\bar{P}_{mw}}{\bar{P}_{stove} - \bar{P}_{mw}}C_{cnt} = \left(\frac{625 \text{ W}}{1000 \text{ W} - 625 \text{ W}}\right)(1.4 \text{ kJ/K}) = 2.33 \text{ kJ/K}$$

Since $C_{H_2O} = m_{H_2O}c_{H_2O}$, where $c_{H_2O} = 4184$ J/kg $\cdot$ K, the minimum mass of water is $m_{H_2O} = 0.558$ kg.

(b) The rate of temperature rise when $m_{H_2O} = 0.558$ kg is

$$\frac{\Delta T}{\Delta t} = \left(\frac{\Delta T}{\Delta t}\right)_{mw} = \left(\frac{\Delta T}{\Delta t}\right)_{stove} = \frac{\bar{P}_{stove}}{C_{H_2O} + C_{cnt}} = \frac{1 \text{ kW}}{2.33 \text{ kJ/K} + 1.4 \text{ kJ/K}} = 0.268 \text{ J/s}$$

ASSESS   When the heat capacity of water, $C_{H_2O} = m_{H_2O}c_{H_2O}$, is small, the microwave is faster, whereas when C_{H_2O} is large, the stove burner is faster.

51. **INTERPRET**   Given the power output of the stove burner and the amount of time it takes to heat up the water, we want to know how much water is in the kettle. This problem involves specific heat.

DEVELOP   The energy supplied by the stove burner heats the kettle and water in it from 20°C to 100°C, with $\Delta T = 80$ K. If we neglect any losses of heat and the heat capacity of the burner, this energy is just the burner's power output times the time:

$$\Delta Q = \bar{P}\Delta t = (m_{H_2O}c_{H_2O} + m_K c_K)\Delta T$$

This equation can be used to solve for m_{H_2O}.

EVALUATE   Since all of these quantities are given except for the mass of the water, we can solve for m_{H_2O}:

$$m_{H_2O} = \frac{1}{c_{H_2O}}\left(\frac{\bar{P}\Delta t}{\Delta T} - m_K c_K\right) = \frac{1}{4184 \text{ J/kg} \cdot \text{K}}\left(\frac{(2 \text{ kW})(5.4 \times 60 \text{ s})}{80 \text{ K}} - (1.2 \text{ kg})(447 \text{ J/kg} \cdot \text{K})\right)$$
$$= 1.81 \text{ kg}$$

ASSESS   We find m_{H_2O} to be proportional to Δt. This makes sense because the more water in the kettle, the more time it takes to heat up the water.

53. **INTERPRET**   The object of interest is the car. The problem deals with transformation of energy from the kinetic energy of the car to the thermal energy of the brake disks.

DEVELOP   By energy conservation, the loss of kinetic energy of the car is equal to the thermal energy gained by the four brakes:

$$\Delta Q = \Delta K \quad \rightarrow \quad \frac{1}{2}M_{car}v^2 = 4m_{brake}c\,\Delta T$$

EVALUATE   From the equation above, with $v = 40$ km/h $= 11.1$ m/s, the change of temperature is

$$\Delta T = \frac{M_{car}v^2/2}{4m_{brake}c} = \frac{(1500 \text{ kg})(11.1 \text{ m/s})^2/2}{4(5 \text{ kg})(502 \text{ J/kg} \cdot \text{K})} = 9.22 \text{ K}$$

ASSESS   This is a big increase in temperature. The brakes can get very hot depending on how fast the car was moving initially.

55. **INTERPRET**   Our system consists of two materials, water and copper, which are initially at different temperatures. They are brought together and reach a thermal equilibrium. We want to find the mass of the copper.

DEVELOP   Let us assume that all the heat lost by the copper is gained by the water, with no heat transfer to the container or its surroundings. Then $-\Delta Q_{Cu} = \Delta Q_{H_2O}$ (as in Example 16.2), or

$$-m_{Cu}c_{Cu}(T - T_{Cu}) = m_{H_2O}c_{H_2O}(T - T_{H_2O})$$

The specific heats of copper and water can be found in Table 16.1.

EVALUATE   Expressing all the temperatures in the Kelvin scale and solving for m_{Cu}, one finds

$$m_{Cu} = \frac{m_{H_2O}c_{H_2O}(T - T_{H_2O})}{c_{Cu}(T_{Cu} - T)} = \frac{(1 \text{ kg})(4184 \text{ J/kg} \cdot \text{K})(298 \text{ K} - 293 \text{ K})}{(386 \text{ J/kg} \cdot \text{K})(573 \text{ K} - 298 \text{ K})} = 0.197 \text{ kg}$$

ASSESS   Since the water has much greater mass and higher specific heat, its temperature change is less compared to copper.

57. **INTERPRET**   This problem is about conductive heat flow. We want o find the heat flow rate along an iron rod.

DEVELOP   We assume a uniform variation of temperature along the length of the rod and no heat-flow through its sides. The heat flow rate is given by Equation 16.5:

$$H = -kA\frac{\Delta T}{\Delta x}$$

EVALUATE   Substituting the numerical values, we get

$$H = -kA\frac{\Delta T}{\Delta x} = -(80.4 \text{ W/m} \cdot \text{K})\pi(0.015 \text{ m})^2\frac{100 \text{ K}}{0.4 \text{ m}} = -14.2 \text{ W}$$

Here, the minus sign signifies a heat-flow from right to left in Fig. 16.16.

ASSESS   The flow rate H increases with the temperature gradient, $\Delta T/\Delta x$. The minus sign means that heat flows from the hotter end of the rod to the cooler end of the rod.

59.   **INTERPRET**   This problem is about heat loss through the insulation around the house and the amount of oil that must be burned to compensate for the loss.

DEVELOP   The difference in heat loss between $\mathcal{R}$-factors of 19 and 2.1, for a window area of $A = 40$ ft² and the given temperature difference, is

$$\Delta H = H_2 - H_1 = -A\,\Delta T\left(\frac{1}{\mathcal{R}_2} - \frac{1}{\mathcal{R}_1}\right)$$

A positive ΔH represents a greater heat-flow from inside the house to outside, or a loss of energy. Knowing ΔH allows us to determine the quantity of oil needed.

EVALUATE   The rate difference is

$$\Delta H = H_2 - H_1 = -A\Delta T\left(\frac{1}{\mathcal{R}_2} - \frac{1}{\mathcal{R}_1}\right)$$

$$= -(40 \text{ ft}^2)(15°\text{F} - 68°\text{F})\left(\frac{1}{2.1 \text{ ft}^2 \cdot \text{F}° \cdot \text{h/Btu}} - \frac{1}{19 \text{ ft}^2 \cdot \text{F}° \cdot \text{h/Btu}}\right)$$

$$= 898 \text{ Btu/h}$$

Over a winter month ($\Delta t = 30$ days $= 720$ h), the oil that would have to be consumed to compensate for this loss is

$$V = (898 \text{ Btu/h})(720 \text{ h})(1 \text{ gal}/10^5 \text{ Btu}) = 6.47 \text{ gal}$$

ASSESS   At \$2.20/gal, your monthly bill for the oil would be \$14.23. But this is just for this one window. If you take into account the entire house (roof, wall,…), the cost could get quite high.

61.   **INTERPRET**   This problem deals with heat transfer. Heat is lost through the insulation around the house, but also produced by the human body. We want to find the temperature inside the house at the steady state.

DEVELOP   The average thermal resistance of the house is given by $1/R = 370$ W/C°. Note that the thermal resistance R is defined as the reciprocal of the rate of heat-flow per degree of temperature difference (see Equation 16.7):

$$\frac{1}{R} = -\frac{H}{\Delta T}$$

In thermal energy balance (steady state), the power released by the people (owner plus guests) is equal to the rate of heat loss from the house (otherwise the house would heat up or cool down).

EVALUATE   At the steady state, we have

$$\bar{P} = |H| = \frac{|\Delta T|}{R} \quad \rightarrow \quad (41)(100 \text{ W}) = (370 \text{ W/C}°)\,|\Delta T|$$

or $|\Delta T| = 11.1$ C°. Since the outside temperature is $T_{\text{out}} = 12°\text{C}$, the temperature inside the house is $T_{\text{in}} = 12°\text{C} + |\Delta T| = 23.1°\text{C}$.

ASSESS   Increasing the number of guests at the party will increase the temperature in the house. In fact, each additional guest would raise the temperature inside the house by 0.27 °C. If nobody is in the house, then $\Delta T = 0$, and $T_{\text{in}} = T_{\text{out}}$.

63.   **INTERPRET**   This problem is about the radiation from a hot metal strip.

DEVELOP   The strip is in energy balance between the input power and the net power radiated (the only transfer mechanism available). Thus, according to Equation 16.9,

$$P_{\text{in}} = P_{\text{rad}} = e\sigma A\left(T_1^4 - T_2^4\right)$$

This equation allows us to determine the temperature of the strip, T_1.

EVALUATE   The total surface area of the strip is $A = 2\,[(0.5\ \text{cm}) \times (5\ \text{cm}) + (0.01\ \text{cm}) \times (5.5\ \text{cm})] = 5.11 \times 10^{-4}\ \text{m}^2$ (the edges contributing only 2%). Making the numerical substitutions, we find:

$$T_1 = \left(\frac{P_{in}}{e\sigma A} + T_2^4\right)^{1/4} = \left(\frac{50\ \text{W}}{(1)(5.67 \times 10^{-8}\ \text{W/m}^2 \cdot \text{K}^4)(5.11 \times 10^{-4}\ \text{m}^2)} + (300\ \text{K})^4\right)^{1/4}$$

$$= 1.15 \times 10^3\ \text{K}$$

(**b**) In this case, all the input power is transferred by conduction through the box,

$$P_{in} = H = -\frac{T_2 - T_1}{R}$$

Thus, the temperature of the strip is

$$T_1 = T_2 + RP_{in} = 300\ \text{K} + (8.0\ \text{K/W})(50\ \text{W}) = 700\ \text{K}$$

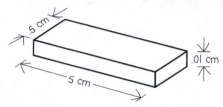

ASSESS   We get a higher temperature of the strip when heat transfer is caused by radiation than by conduction. Radiation dominates at high temperature because of its T^4 dependence.

65.   **INTERPRET**   Our system consists of two materials, water and an iron horseshoe, which are initially at different temperatures. They are brought together and reach a thermal equilibrium. We want to find the equilibrium temperature.

     DEVELOP   Let us assume that all the heat lost by the horseshoe is gained by the water, with no heat transfer to the container or its surroundings. Then $-\Delta Q_{Fe} = \Delta Q_{H_2O}$ (as in Example 16.2), or

$$-m_{Fe}c_{Fe}(T - T_{Fe}) = m_{H_2O}c_{H_2O}(T - T_{H_2O})$$

The specific heats of copper and water can be found in Table 16.1.

     EVALUATE   Solving for T, one finds

$$T = \frac{m_{Fe}c_{Fe}T_{Fe} + m_{H_2O}c_{H_2O}T_{H_2O}}{m_{Fe}c_{Fe} + m_{H_2O}c_{H_2O}}$$

$$= \frac{(1.1\ \text{kg})(0.107\ \text{kcal/kg} \cdot {}^\circ\text{C})(550{}^\circ\text{C}) + (15\ \text{kg})(1\ \text{kcal/kg} \cdot {}^\circ\text{C})(20{}^\circ\text{C})}{(1.1\ \text{kg})(0.107\ \text{kcal/kg} \cdot {}^\circ\text{C}) + (15\ \text{kg})(1\ \text{kcal/kg} \cdot {}^\circ\text{C})} = 24.1{}^\circ\text{C}$$

     ASSESS   The change of water temperature is $\Delta T_{H_2O} = T - T_{H_2O} = 24.1{}^\circ\text{C} - 20{}^\circ\text{C} = 4.1{}^\circ\text{C}$, while the change of temperature of the iron horseshoe is $|\Delta T_{Fe}| = 525.9{}^\circ\text{C}$. Since the water has much greater mass and higher specific heat, its temperature change is less compared to the horseshoe.

67.   **INTERPRET**   This problem is about the radiation from a burning log. We want to know its temperature.

     DEVELOP   If we neglect the radiation absorbed by the log from its environment (the fireplace brick, for example, does radiate heat to the room, but is probably at a temperature far less than red hot), then the net power radiated by the log is just that given by the Stefan-Boltzmann law (Equation 16.9):

$$P = e\sigma A T^4$$

Knowing the surface area of the log allows us to determine T.

     EVALUATE   The surface area of the log is

$$A = \pi dL + 2\pi R^2 = \pi d(L + R) = \pi(0.15\ \text{m})(0.65\ \text{m} + 0.075\ \text{m}) = 0.342\ \text{m}^2$$

Solving for T, we find

$$T = \left(\frac{P}{e\sigma A}\right)^{1/4} = \left(\frac{34 \times 10^3\ \text{W}}{(5.67 \times 10^{-8}\ \text{W/m}^2 \cdot \text{K}^4)(0.342\ \text{m}^2)}\right)^{1/4} = 1.15 \times 10^3\ \text{K}$$

ASSESS   When a burning log is glowing red hot, its temperature is above $1000°C$. If the temperature continues to rise, its color will then turn orange, and then yellow.

69.   **INTERPRET**   This problem is about the heat loss through various structural parts of the house via conduction.
    DEVELOP   We follow the approach outlined in Example 16.4. The heat flow rate is related to the $\mathcal{R}$-factor as

$$H = -kA\frac{\Delta T}{\Delta x} = -A\frac{\Delta T}{\Delta x/k} = -A\frac{\Delta T}{\mathcal{R}}$$

The window area here is $A_{window} = 10(2.5\text{ ft} \times 5.0\text{ ft}) = 125\text{ ft}^2$, and the wall area is 125 ft^2 less than in Example 16.4, or $A_{walls} = 1506\text{ ft}^2 - 125\text{ ft}^2 = 1381\text{ ft}^2$. Thus, the heat losses through these structural parts are:

$$|H|_{walls} = \left(\frac{1}{12.37}\frac{\text{Btu}}{\text{h}\cdot\text{ft}^2\cdot°\text{F}}\right)(1381\text{ ft}^2)(50°\text{F}) = 5583\text{ Btu/h}$$

$$|H|_{roof} = \left(\frac{1}{31.37}\frac{\text{Btu}}{\text{h}\cdot\text{ft}^2\cdot°\text{F}}\right)(1164\text{ ft}^2)(50°\text{F}) = 1855\text{ Btu/h}$$

$$|H|_{windows} = \left(\frac{1}{0.90}\frac{\text{Btu}}{\text{h}\cdot\text{ft}^2\cdot°\text{F}}\right)(125\text{ ft}^3)(50°\text{F}) = 6944\text{ Btu/h}$$

If we include the gain from the south windows, $4(12.5\text{ ft}^2)(30\text{ Btu/h}\cdot\text{ft}^2) = 1500\text{ Btu/h}$, the net rate of loss of energy from the entire house is

$$H_{total} = (5583 + 1855 + 6944 - 1500)\text{ Btu/h} = 12.88 \times 10^3\text{ Btu/h}$$

EVALUATE   (a) The monthly fuel bill is

$$(12.88 \times 10^3\text{ Btu/h})(24 \times 30\text{ h/mo})(1\text{ gal/}10^5\text{ Btu})(\$2.20/\text{gal}) = \$204/\text{mo}$$

(b) The solar gain from the south windows is worth

$$(1500\text{ Btu/h})(24 \times 30\text{ h/mo})(1\text{ gal/}10^5\text{ Btu})(\$2.20/\text{gal}) = \$23.76/\text{mo}$$

ASSESS   This is an expensive fuel bill. You probably would want to improve the insulation.

71.   **INTERPRET**   This problem is about radiation received by Pluto from the Sun. Treating Pluto as a blackbody, we want to find its average surface temperature.
    DEVELOP   Sunlight intensity varies with the inverse square of the distance from the Sun. Since at Earth's orbit the intensity is $I_{Earth} = 1.37\text{ kW/m}^2$, the intensity at Pluto's orbit is

$$I_{Pluto} = I_{Earth}\left(\frac{R_{SE}}{R_{PS}}\right)^2 = (1.37\text{ kW/m}^2)\left(\frac{1.50 \times 10^{11}\text{ m}}{5.9 \times 10^{12}\text{ m}}\right)^2 = 0.886\text{ W/m}^2$$

Let's assume that the effective solar intensity is only 70% of this: $I'_{Pluto} = 0.70 I_{Pluto} = 0.620\text{ W/m}^2$. The area of absorbing surface for Pluto is $A = \pi R_{Pluto}^2$, so the radiated power received by Pluto is $P = I'_{Pluto}A = I'_{Pluto}(\pi R_{Pluto}^2)$. The temperature of Pluto can be obtained using the Stefan-Boltzmann law.
    EVALUATE   The Stefan-Boltzmann law gives

$$P = e\sigma A_s T_{Pluto}^4 \rightarrow \sigma(4\pi R_{Pluto}^2)T_{Pluto}^4 = I'_{Pluto}(\pi R_{Pluto}^2)$$

where $A_s = 4\pi R_{Pluto}^2$ is the surface area of Pluto. Thus, the temperature of Pluto is

$$T_{Pluto} = \left(\frac{I'_{Pluto}}{4\sigma}\right)^{1/4} = \left(\frac{0.620\text{ W/m}^2}{4(5.67 \times 10^{-8}\text{ W/m}^2\cdot\text{K}^4)}\right)^{1/4} = 40.7\text{ K}$$

ASSESS   Our result is in good agreement with the mean temperature of 44 K for Pluto. The general expression for this estimate of a planet's equilibrium temperature is $T = (I/4\sigma)^{1/4}$, where $I = (r_E/r)^2 I_E(1-a)$, I_E is the solar constant and a the planet's albedo (percent sunlight reflected).

73.   **INTERPRET**   We find the change in Earth's average temperature if there is a 10% reduction in solar intensity. This is an energy balance problem.
    DEVELOP   We will use ratios. The normal solar intensity is S, the emissivity is e, and the current average temperature is $T = 287$ K. The power coming in from the sun is $P_{in} = AS = \pi R^2 S$, and the power lost to radiation is $P_{out} = e\sigma AT^4$. At equilibrium, these will balance.

EVALUATE $P_{in} = P_{out} \rightarrow \pi R_v^2 S = e\sigma (4\pi R_v^2)T^4 \rightarrow T = \sqrt[4]{\frac{S}{4e\sigma}}$, normally. The new temperature after the impact is

$T' = \sqrt[4]{\frac{S'}{4e\sigma}} = \sqrt[4]{\frac{0.90S}{4e\sigma}}$. The ratio of the two is $\frac{T'}{T} = \sqrt[4]{\frac{S'}{S}} = \sqrt[4]{0.90} \rightarrow T' = T\sqrt[4]{0.90} = 0.974T = 279.5$ K.

ASSESS   This change in temperature of nearly 8°C is a significant effect.

75. **INTERPRET**   How long will it take to heat a given amount of water with a fixed power supply? This is a heat capacity problem.

DEVELOP   The power available is $P = 1.42 \times 10^9$ J/s. The mass of water is $m = 5.4 \times 10^6$ kg, and the specific heat of water is $c = 4184$ J/kg·K. We need to change the water temperature $\Delta T = 340$ K. We find the energy this requires, and divide by the power to find the time.

EVALUATE $Q = mc\Delta T \rightarrow t = \frac{mc\Delta T}{P} = 5400$ s $= 1.5$ h

ASSESS   We have assumed that the pressure of the cooling system is maintained in such a way that the water remains liquid despite this high temperature. In a later chapter we will consider the heat required for *phase transitions* such as from liquid to vapor.

77. **INTERPRET**   This is a heat capacity problem, but with a heat capacity that changes with temperature. We solve for heat Q by integrating over T.

DEVELOP   The mass is given as $m = 40$ g, and $dQ = mcdT$. We substitute $c = 31\left(\frac{T}{343\,K}\right)^3$ J/(g K), and integrate from $T_1 = 10$ K to $T_2 = 25$ K.

EVALUATE

$$Q = \int_{T_1}^{T_2} m\left(\frac{31T}{(343\,K)}\right)^3 \frac{J}{g\,K}\,dT = m\left(\frac{31}{(343\,K)}\right)^3 \frac{J}{g\,K}\int_{T_1}^{T_2}T^3 dT = m\left(\frac{31}{(343\,K)}\right)^3 \frac{J}{g\,K}\left[\frac{1}{4}T^4\right]_{T_1}^{T_2}$$

$$\rightarrow Q = \frac{m}{4}\left(\frac{31}{(343\,K)}\right)^3 \frac{J}{g\,K}\left[T_2^4 - T_1^4\right] = 281\text{ J}$$

ASSESS   At more normal temperatures, the specific heat of copper is $c = 386$ J/kgK, so the heat required to change the temperature of 40 grams of copper by 15°C would be $Q = 232$ J. Our answer is probably in the right ballpark, then.

79. **INTERPRET**   We derive Newton's law of cooling, and apply it to the time it takes a house to freeze.

DEVELOP   We start with $dQ = mcdT$ and $-\frac{dQ}{dt} = -kA\frac{\Delta T}{\Delta x} = -\frac{\Delta T}{R}$. The heat capacity of the house is given as $mc = 6.5 \times 10^6$ J/K, and $R = 6.67$ mK/W. Use $\Delta T = (T_0 - T)$, where $T_0 = -15°C = 258$ K is the outside temperature and T is the temperature of the house. Solve for the time at which $T = 0°C = 273$ K.

EVALUATE

$$mcdT = \frac{T_0 - T}{R}dt \rightarrow mcR\frac{dT}{T_0 - T} = dt$$

$$mcR\int_{T_i}^{T_f}\frac{dT}{T_0 - T} = \int_0^t dt = t$$

$$t = mcR\ln\left(\frac{T_0 - T_i}{T_0 - T_f}\right) = 36.7 \times 10^3 \text{ s} = 10 \text{ h}$$

ASSESS   This answer seems reasonable for the situation described.

EXERCISES

Section 17.1 Gases

19. **INTERPRET** We are dealing with an ideal gas. We are given the pressure, temperature, and volume, and want to find the number of gas molecules.

    DEVELOP We shall use the ideal-gas law, $pV = NkT$, given in Equation 17.1, to find the number of molecules.

    EVALUATE From Equation 17.1, we have

    $$N = \frac{PV}{kT} = \frac{(180 \text{ kPa})(8.5 \times 10^{-3} \text{ m}^3)}{(1.38 \times 10^{-23} \text{ J/K})(350 \text{ K})} = 3.17 \times 10^{23}$$

    ASSESS One mole has $N_A = 6.02 \times 10^{23}$ molecules. Thus, we have about 0.53 mole of molecules in the system.

21. **INTERPRET** We are dealing with an ideal gas. We want to know how volume changes with temperature.

    DEVELOP To compare different states of an ideal gas, it is often convenient to express Equation 17.1 as a ratio:

    $$\frac{P_1 V_1}{P_2 V_2} = \frac{N_1 T_1}{N_2 T_2}$$

    In this problem, the pressure is constant, and if no gas escapes or enters, then N is also constant. Therefore, the above equation becomes

    $$\frac{V_1}{V_2} = \frac{T_1}{T_2}$$

    where T is in the Kelvin scale.

    EVALUATE (a) If $T_2 = 100°C = 373 \text{ K}$ and $T_1 = 200°C = 473 \text{ K}$ then

    $$V_1 = \left(\frac{473 \text{ K}}{373 \text{ K}} \right) V_2 = 1.27 \, V_2$$

    (b) If $T_1 = 2T_2$, then $V_1 = 2V_2$.

    ASSESS The fact that $V \sim T$ for a given mass of ideal gas at constant pressure is known as the law of Charles and Gay-Lussac.

23. **INTERPRET** We treat air molecules as ideal gas. Given the pressure, temperature, and volume, we want to find the number of air molecules.

    DEVELOP We shall use the ideal-gas law, $pV = NkT$, given in Equation 17.1, to find the number of molecules.

    EVALUATE The number of air molecules is

    $$N = \frac{PV}{kT} = \frac{(10^{-10} \text{ Pa})(10^{-3} \text{ m}^3)}{(1.38 \times 10^{-23} \text{ J/K})(273 \text{ K})} = 2.65 \times 10^7$$

    ASSESS One mole has $N_A = 6.02 \times 10^{23}$ molecules. Thus, we have about 4.4×10^{-17} mole of molecules in the system.

25. **INTERPRET** In this problem we want to compare the speeds of two different molecules that are at different temperatures.

DEVELOP The thermal speed of a molecule is given by Equation 17.4:

$$v_{th} = \sqrt{\frac{3kT}{m}}$$

where T is the temperature and m is the mass of the gas molecule.

EVALUATE Comparing the thermal speeds for H_2 ($m_{H_2} = 2u$) and SO_2 ($m_{SO_2} = 64u$) at the given temperatures, we find

$$\frac{v_{th}(H_2)}{v_{th}(SO_2)} = \sqrt{\frac{T_{H_2}}{T_{SO_2}} \frac{m_{SO_2}}{m_{H_2}}} = \sqrt{\frac{75\ K}{350\ K} \frac{64u}{2u}} = 2.62$$

So, the hydrogen is faster.

ASSESS The thermal speed of a gas molecule is proportional to $\sqrt{T}$ and inversely proportional to $\sqrt{m}$.

Section 17.2 Phase Changes

27. **INTERPRET** This problem is about melting, and it involves heat of fusion. We want to identify the substance in question.

DEVELOP Using Equation 17.5, we find the heat of transformation to be

$$L = \frac{Q}{m} = \frac{200\ J}{0.008\ kg} = 25\ kJ/kg$$

Knowing this property allows us to identify what the substance is.

EVALUATE The closest figure in Table 17.1 is the heat of fusion for lead ($L_f = 25$ kJ/kg). So we identify the substance as lead.

ASSESS Heat of fusion, L_f, is one of the chemical properties of a material. So knowing L_f allows us to identify what the material is.

29. **INTERPRET** This problem is about phase change of CO_2. The object of interest is the heat that must be extracted for solidification to take place.

DEVELOP The heat of transformation in this process can be calculated by using Equation 17.5, $Q = mL_s$.

EVALUATE Equation 17.5 gives $Q = mL_s = (0.25\ kg)(573\ kJ/kg) = 143$ kJ.

ASSESS This is the heat that must be extracted to turn CO_2 gas into frozen CO_2, or dry ice. To revert back to the gaseous state, the same amount of heat must be absorbed.

31. **INTERPRET** We treat air as an ideal gas. We are given the pressure and temperature, and we want to find the density of gas.

DEVELOP We shall use the ideal-gas law, $pV = nRT$, given in Equation 17.2, to find the density of air molecules.

EVALUATE The molar density implied by the ideal gas law (which is a good approximation for air under the conditions stated in the problem) is

$$\frac{n}{V} = \frac{p}{RT} = \frac{300\ kPa}{(8.314\ J/K \cdot mol)(273\ K + 34\ K)} = 118\ mol/m^3$$

ASSESS The result implies that one mole of air molecules under this condition occupies about 8.47 L of volume. This can be compared with the 22.4 L at the standard temperature (273 K) and pressure (101 kPa).

Section 17.3 Thermal Expansion

33. **INTERPRET** We identify this as a thermal expansion problem. We know the initial volume, the material, and the change in temperature: we need to find the new volume.

DEVELOP The change in volume is given by $\beta = \frac{\Delta V}{V \Delta T}$. The initial volume is $v = 1.00$ L. The material is ethyl alcohol, which has a volume coefficient of expansion $\beta = 75 \times 10^{-5}$ K^{-1}, and the change in temperature is $\Delta T = -18$ K.

EVALUATE $\Delta V = \beta V \Delta T = -0.0135\ L.$ The new volume of liquid is $V' = (1.000\ L) + \Delta V = 0.9865\ L.$

ASSESS   This is enough to observe in your own refrigerator: Seal a volume of liquid at room temperature, put it in the refrigerator, and observe the deformation of the container when it cools.

35. **INTERPRET**   This problem deals with thermal expansion of a steel washer. The quantity of interest is the diameter of the washer, so the relevant quantity is the coefficient of linear expansion, α.

    DEVELOP   The coefficient of linear expansion is defined as (see Equation 17.7):

    $$\alpha = \frac{\Delta L/L}{\Delta T}$$

    For steel, its value is (see Table 17.2) $\alpha = 12 \times 10^{-6}\ \mathrm{K^{-1}}.$ This is the equation we shall use to solve for $\Delta T.$

    EVALUATE   From the equation above, we get

    $$\Delta T = \frac{\Delta L}{\alpha L} = \frac{(9.55\ \mathrm{mm} - 9.52\ \mathrm{mm})}{(12 \times 10^{-6}\ \mathrm{K^{-1}})(9.52\ \mathrm{mm})} = 263\ \mathrm{K}$$

    Since the initial temperature is $0°C,$ we must heat up the washer to $263°C.$

    ASSESS   Since α is very small, a large increase in temperature only results in a small increase of the washer's diameter.

PROBLEMS

37. **INTERPRET**   The system of interest is the solar corona which we treat as ideal gas. The quantity of interest is the number density of air molecules.

    DEVELOP   The number density implied by the ideal gas law (Equation 17.1) is

    $$pV = NkT \quad \rightarrow \quad \frac{N}{V} = \frac{p}{kT}$$

    EVALUATE   Applying the above equation to the solar corona, we obtain

    $$\left(\frac{N}{V}\right)_{corona} = \frac{p}{kT} = \frac{3 \times 10^{-2}\ \mathrm{Pa}}{(1.38 \times 10^{-23}\ \mathrm{J/K})(2 \times 10^{6}\ \mathrm{K})} = 1.09 \times 10^{15}\ \mathrm{m^{-3}}$$

    On the other hand, for air at STP, the particle density is

    $$\left(\frac{N}{V}\right)_{STP} = \frac{p}{kT} = \frac{1.013 \times 10^{5}\ \mathrm{Pa}}{(1.38 \times 10^{-23}\ \mathrm{J/K})(273\ \mathrm{K})} = 2.69 \times 10^{25}\ \mathrm{m^{-3}}$$

    which is about 2.5×10^{10} times greater than $(N/V)_{corona}.$

    ASSESS   As expected, the air density is a function of altitude in both the solar and terrestrial atmospheres.

39. **INTERPRET**   The object of interest is the cylinder compressed with air. We are given the pressure, temperature, and volume, and want to find the number of air molecules.

    DEVELOP   We shall treat the air as an ideal gas (although somewhat risky at 180 atm) and use the ideal-gas law, $pV = nRT,$ given in Equation 17.2, to find the number of molecules. The volume of the cylinder is

    $$V = \pi(d/2)^2 h = \pi(0.10\ \mathrm{m})^2(1.0\ \mathrm{m}) = 0.01\ \pi\ \mathrm{m^3}$$

    EVALUATE   (a) Applying the ideal-gas law gives

    $$n = \frac{pV}{RT} = \frac{(180\ \mathrm{atm})(1.013 \times 10^{5}\ \mathrm{Pa/atm})(0.01\pi\ \mathrm{m^3})}{(8.314\ \mathrm{J/K \cdot mol})(293\ \mathrm{K})} = 235\ \mathrm{mol}$$

    where we have used $T = 20°C = 293\ \mathrm{K}$ as the room temperature.

    (b) If the pressure is $p' = 1\ \mathrm{atm},$ then the volume would be

    $$V' = \frac{nRT}{p'} = \left(\frac{p}{p'}\right)V = \left(\frac{180\ \mathrm{atm}}{1\ \mathrm{atm}}\right)(0.01\pi\ \mathrm{m^3}) = 5.65\ \mathrm{m^3}$$

    ASSESS   When temperature is held constant, $pV = $ constant for an ideal gas. Therefore, decreasing the pressure increases the volume.

41. **INTERPRET**   The object of interest is the flask filled with air, which we treat as an ideal gas. We explore the effect of changing temperature and pressure.

DEVELOP   When the flask is immersed in boiling water, its volume is kept fixed. Therefore, the ideal-gas law, $pV = nRT$, gives

$$\frac{p_2}{p_1} = \frac{T_2}{T_1}$$

The initial conditions of the gas are $p_1 = 1$ atm, $V_1 = 3$ L, and $T_1 = 293$ K. Using the ideal-gas law, we find the number of molecules to be

$$n_1 = \frac{p_1 V_1}{RT_1} = \frac{(1\ \text{atm})(1.013 \times 10^5\ \text{Pa/atm})(3 \times 10^{-3}\ \text{m}^3)}{(8.314\ \text{J/K} \cdot \text{mol})(293\ \text{K})} = 0.125\ \text{mol}$$

When the flask is opened at $T_2 = 373$ K, the pressure decreases to $p_2 = 1$ atm, so the quantity of gas remaining is

$$n_2 = \frac{p_1 V_1}{RT_2} = \left(\frac{T_1}{T_2}\right) n_1$$

EVALUATE   **(a)** From the equation above, we find the maximum pressure reached in the flask to be

$$p_2 = \left(\frac{T_2}{T_1}\right) p_1 = \left(\frac{373\ \text{K}}{293\ \text{K}}\right)(1\ \text{atm}) = 1.27\ \text{atm}$$

(b) After opening the flask the quantity of gas left is

$$n_2 = \left(\frac{T_1}{T_2}\right) n_1 = \left(\frac{293\ \text{K}}{373\ \text{K}}\right)(0.125\ \text{mol}) = 0.0980\ \text{mol}$$

Therefore, the amount that escaped was $\Delta n = n_1 - n_2 = 0.0268$ mol.

(c) The pressure of the remaining gas is

$$p_3 = \frac{n_2 RT_1}{V_1} = \left(\frac{n_2}{n_1}\right) p_1 = \left(\frac{0.098\ \text{mol}}{0.125\ \text{mol}}\right)(1\ \text{atm}) = 0.786\ \text{atm}$$

ASSESS   Pressure is proportional to the number of molecules in the volume. After some gas molecules have escaped, the pressure goes down.

43. **INTERPRET**   This problem is about melting, and it involves heat of fusion. We want to know how much ice would melt by exploding a 1-megaton nuclear bomb.

DEVELOP   A 1-megaton nuclear device releases about $\Delta Q = 4.16 \times 10^{15}$ J of energy. The heat of fusion for water is $L_f = 334$ kJ/kg.

EVALUATE   Using Equation 17.5, we find that this amount of energy is capable of melting

$$m = \frac{Q}{L_f} = \frac{4.16 \times 10^{15}\ \text{J}}{334\ \text{kJ/kg}} = 1.25 \times 10^{10}\ \text{kg}$$

of ice at the normal melting point of 0°C.

ASSESS   The amount of ice melted by this explosion is huge! The impact could be disastrous.

45. **INTERPRET**   This problem is about melting, so heat of fusion is involved. We want to know how much energy is needed to melt a given quantity of ice.

DEVELOP   The total mass of ice is

$$m = \rho V = \rho A h = (917\ \text{kg/m}^3)(8.2 \times 10^{10}\ \text{m}^2)(1.3\ \text{m}) = 9.78 \times 10^{13}\ \text{kg}$$

The amount of heat required for melting can be calculated by using Equation 17.5, $Q = L_f m$, where $L_f = 334$ kJ/kg for ice.

EVALUATE   Substituting the values, the heat required to melt at 0°C is

$$Q = mL_f = (9.78 \times 10^{13}\ \text{kg})(334\ \text{kJ/kg}) = 3.26 \times 10^{19}\ \text{J}$$

(b) The average power absorbed is

$$\bar{P} = \frac{Q}{t} = \frac{(3.26 \times 10^{19}\ \text{J})}{(21\ \text{d})(86400\ \text{s/d})} = 1.80 \times 10^{13}\ \text{W}$$

ASSESS   The average intensity is

$$\bar{I} = \frac{\bar{P}}{A} = \frac{1.80 \times 10^{13}\ \text{W}}{8.2 \times 10^{10}\ \text{m}^2} = 220\ \text{W/m}^2$$

On the other hand, the average intensity from direct sunlight is about $\bar{I}_s = 1000\ \text{W/m}^2$. So the energy source for the melting process is most likely to have come from sunlight. (Note that not all the energy from the sunlight is absorbed by the ice.)

47. **INTERPRET**   This problem is about melting, so heat of fusion is involved. We want to know how long it takes to melt a given quantity of ice.

DEVELOP   The absorbed power which contributes to the melting process is

$$\bar{P}_{\text{absorbed}} = 0.75\bar{I}_s A$$

where $\bar{I}_s = 200\ \text{W/m}^2$ and A is the area of the lake. On the other hand, the heat needed for melting can be calculated by using Equation 17.5, $Q = L_f m$, where $L_f = 334\ \text{kJ/kg}$ for ice. The mass of the ice is $m = \rho V = \rho(Ah)$, where ρ is the density of ice and h is the thickness.

EVALUATE   The amount of time required to melt the ice is

$$t = \frac{Q}{\bar{P}_{\text{absorbed}}} = \frac{L_f m}{0.75\bar{I}_s A} = \frac{L_f \rho A h}{0.75\bar{I}_s A} = \frac{L_f \rho h}{0.75\bar{I}_s} = \frac{(335\ \text{kJ/kg})(917\ \text{kg/m}^3)(0.5\ \text{m})}{(0.75)(200\ \text{W/m}^2)} = 1.02 \times 10^6\ \text{s} = 11.8\ \text{d}$$

ASSESS   Our result of 11.8 days for the ice in the lake to melt completely sounds reasonable.

49. **INTERPRET**   This problem involves raising the temperature of ice to the melting point and then changing the phase, so both specific heat and heat of fusion are involved.

DEVELOP   The energy needed to raise the temperature is given by Equation 16.3, $\Delta Q = mc\Delta T$. Equation 17.5, $Q = Lm$, gives the energy for phase change.

EVALUATE   Adding up the two energies, we obtain

$$Q = mc_{\text{ice}}\Delta T + mL_f = (10\ \text{kg})[(2.05\ \text{kJ/kg} \cdot \text{K})(10\ \text{K}) + 334\ \text{kJ/kg}] = 0.205\ \text{MJ} + 3.34\ \text{MJ} = 3.545\ \text{MJ}$$

ASSESS   About 94% of the energy is actually used to melt the ice, only 6% for raising the temperature.

51. **INTERPRET**   The problem deals with both raising the temperature of water and vaporizing it. So specific heat and heat of vaporization are involved.

DEVELOP   The energy needed to raise the temperature of water is given by Equation 16.3, $\Delta Q = mc\Delta T$. Equation 17.5, $Q = Lm$, gives the energy for phase change.

EVALUATE   Adding up both contributions, the energy required to heat the water to 100°C and boil it away completely is $Q = m(c\Delta T + L_v)$ which is supplied at a given rate $P = Q/t$. Thus, the amount of time required is

$$t = \frac{Q}{P} = \frac{m(c\Delta T + L_v)}{P} = \frac{\rho V(c\Delta T + L_v)}{P}$$

$$= \frac{(10^3\ \text{kg/m}^3)(420\ \text{m}^3)[(4.184\ \text{kJ/kg} \cdot \text{K})(100°\text{C} - 20°\text{C}) + 2257\ \text{kJ/kg}]}{200\ \text{MW}} = 5.44 \times 10^3\ \text{s}$$

$$= 90.7\ \text{min} = 1.51\ \text{h}$$

ASSESS   Failsafe cooling systems are crucial for preventing nuclear meltdown.

53. **INTERPRET**   Our system consists of both ice and warm water. We want to know how much ice is needed to bring the water to 0°C.

DEVELOP   Assume that the only heat transfer is between the punch and the ice. To cool the water to 0°C, the amount of heat that must be extracted is (using Equation 16.3)

$$\Delta Q = mc\Delta T = (16\ \text{kg})(4.184\ \text{kJ/kg} \cdot \text{K})(25°\text{C} - 0°\text{C}) = 1.67\ \text{MJ}$$

This amount of heat is used to melt ice at 0°C.

EVALUATE   To bring the water temperature down to 0°C, the minimum mass of ice needed is

$$m = \frac{\Delta Q}{L_f} = \frac{1.67 \text{ MJ}}{334 \text{ kJ/kg}} = 5.01 \text{ kg}$$

ASSESS   Actually the punch would be diluted with 5.01 kg of melt-water. So to reduce the dilution, sufficient ice at a temperature below 0°C is needed.

55.   **INTERPRET**   This problem involves both a temperature rise and a phase change. The objects of interest are water and aluminum.

DEVELOP   The energy needed to raise the temperature of the aluminum block is given by Equation 16.3, $\Delta Q = mc\Delta T$. On the other hand, Equation 17.5, $Q = Lm$, gives the heat released by the steam before condensing to water.

EVALUATE   The heat required to raise the temperature of the block is

$$\Delta Q_{Al} = m_{Al} c_{Al} \Delta T_{Al} = (40 \text{ kg})(0.90 \text{ kJ/kg} \cdot \text{K})(100°C - 50°C) = 1800 \text{ kJ}$$

The heat released by a quantity of condensing steam at 100°C is

$$Q_{H_2O} = m_{H_2O} L_v = m_{H_2O} (2257 \text{ kJ/kg})$$

If just this heat is transferred to the block, as suggested, then $Q_{H_2O} = \Delta Q_{Al}$, and the mass of the steam is

$$m_{H_2O} = \frac{Q_{H_2O}}{L_v} = \frac{\Delta Q_{Al}}{L_v} = \frac{1800 \text{ kJ}}{2257 \text{ kJ/kg}} = 0.798 \text{ kg}$$

ASSESS   The fact that $L_v = 2257$ kJ/kg means that for each 1 kg of steam undergoing condensation, 2257 kJ of heat is released. This is more than enough to raise the temperature of the aluminum block in the problem. So our result of 0.798 kg is reasonable.

57.   **INTERPRET**   This problem involves both a temperature rise and a phase change. The object of interest is the water in the microwave oven.

DEVELOP   The energy needed to raise the temperature of the water is given by Equation 16.3, $\Delta Q = mc\Delta T$. Equation 17.5, $Q = Lm$, gives the heat needed to vaporize water.

In 20 minutes, $Q = (0.5 \text{ kW})(20 \times 60 \text{ s}) = 600$ kJ of heat energy is transferred to the water (if we ignore energy absorbed by a container or lost to the surroundings). The energy consumed in raising the water's temperature to the normal boiling point is

$$Q_1 = mc \, \Delta T = (0.3 \text{ kg})(4.184 \text{ kJ/kg} \cdot \text{K})(100°C - 20°C) = 100 \text{ kJ}$$

Therefore $Q' = Q - Q_1 = 500$ kJ is left to vaporize some of the water.

EVALUATE   Using Equation 17.5, the amount of water vaporized is

$$m' = \frac{Q'}{L_v} = \frac{500 \text{ kJ}}{2257 \text{ kJ/kg}} = 221 \text{ g}$$

Therefore, only $\Delta m = m - m' = 300$ g $- 221$ g $= 78.7$ g of boiling water (or less than 3 oz) is all that remains.

ASSESS   The excess heat from the microwave oven vaporizes the water. This is precisely what causes your food to dry out when you heat it in the microwave oven for too long.

59.   **INTERPRET**   This problem involves heat transfer between ice and water.

DEVELOP   Assume that all the heat lost by the water is gained by the ice. The temperature of the water drops and that of the ice rises. If either reaches 0°C, a change of phase occurs, freezing or melting, depending on which reaches 0°C first.

To cool to 0°C, the amount of heat the water would lose is

$$\Delta Q_{water} = m_{water} c_{water} \, \Delta T_{water} = (1 \text{ kg})(4.184 \text{ kJ/kg} \cdot \text{K})(5 \text{ K}) = 20.9 \text{ kJ}$$

On the other hand, to warm to 0°C, the ice would gain

$$\Delta Q_{ice} = m_{ice} c_{ice} \Delta T_{ice} = (1 \text{ kg})(2.05 \text{ kJ/kg} \cdot \text{K})(40 \text{ K}) = 82.0 \text{ kJ}$$

Evidently, the water reaches 0°C first.

EVALUATE The amount of heat transfer during the change of phase is $\Delta Q_{ice} - \Delta Q_{water} = 82.0 \text{ kJ} - 20.9 \text{ kJ} = 61.1 \text{ kJ}$. Therefore, the amount of water that freezes is

$$\Delta m = \frac{61.1 \text{ kJ}}{334 \text{ kJ/kg}} = 0.183 \text{ kg}$$

The final mixture is at 0°C and contains 1.183 kg of ice and $1 \text{ kg} - 0.183 \text{ kg} = 0.818 \text{ kg}$ of water.

ASSESS The equilibrium temperature in this case is 0°C where water and ice coexist.

61. **INTERPRET** This problem is about thermal expansion. Since it involves volume, the relevant quantity is the coefficient of volume expansion, β, whose value can be used to identify the substance in question.

DEVELOP As in Example 17.5, we assume that the thermal expansion of the cylinder is negligible compared to that of the liquid. Then the entire change in volume is due to the liquid. The amount of volume change can be calculated from Equation 17.6, $\beta = (\Delta V/V)/\Delta T$.

EVALUATE Substituting the values given in the problem statement, we have

$$\beta = \frac{\Delta V/V}{\Delta T} = \frac{(-75 \text{ mL})/(2000 \text{ mL})}{-50 \text{ K}} = 75 \times 10^{-5} \text{ K}^{-1}$$

From Table 17.2, we see that this matches the coefficient for ethyl alcohol.

ASSESS The thermal expansion coefficient for ethyl alcohol is much greater than the material the cylinder is made of. So our assumption is justified.

63. **INTERPRET** This problem is about thermal expansion. Since it involves volume, the relevant quantity is the coefficient of volume expansion, β.

DEVELOP From Equation 17.6, $\beta = (\Delta V/V)/\Delta T$, the volume change due to an increase in temperature is

$$\Delta V = \frac{\beta \Delta T}{V}$$

where β is the coefficient of volume expansion, and V is the volume at $T_0 = 10°C$. Therefore, to avoid any spill, we want the volume of the gasoline to be

$$V_{total} = V + \Delta V = V(1 + \beta \, \Delta T) = 60 \text{ L}$$

when the temperature is $T_1 = 25°C$.

EVALUATE The change of temperature is $\Delta T = 25°C - 10°C = 15 \text{ K}$, and $\beta = 95 \times 10^{-5} \text{ K}^{-1}$. Therefore, the volume of the gasoline at $T_0 = 10°C$ is

$$V = \frac{V_{total}}{1 + \beta \, \Delta T} = \frac{60 \text{ L}}{1 + (95 \times 10^{-5} \text{ K}^{-1})(15 \text{ K})} = 59.2 \text{ L}$$

ASSESS Gasoline expands when being heated. So, you want to leave some room in the container to account for the volume change.

65. **INTERPRET** This problem involves both heat conduction and phase change. The object of interest is the block of ice.

DEVELOP Using Equation 16.7, we find that a temperature difference of $\Delta T = 0°C - 20°C = -20 \text{ K}$ causes a steady heat flow of

$$H = -\frac{\Delta T}{R} = \frac{20 \text{ K}}{0.12 \text{ K/W}} = 167 \text{ W}$$

into the refrigerator. On the other hand, the amount of heat the block of ice can absorb while melting is

$$Q = mL_f = 15 \text{ kg} (334 \text{ kJ/kg}) = 5.01 \text{ MJ}$$

The amount of time it takes for the ice to melt is then $t = Q/H$ since H is the rate of energy loss.

EVALUATE At the above rate, the melting process would take

$$t = \frac{Q}{H} = \frac{5.01 \text{ MJ}}{167 \text{ W}} = 3.01 \times 10^4 \text{ s} = 8.35 \text{ h}$$

ASSESS The block of ice lasts for quite a long time! Note that increasing the thermal resistance R will decrease H, and hence increase t.

67.   **INTERPRET**   This problem is about proving that the coefficient of volume expansion of an ideal gas at constant pressure is the inverse temperature in the Kelvin scale.

DEVELOP   As mentioned in the text following Equation 17.6, β is defined in general as

$$\beta = \frac{dV/V}{dT} = \frac{1}{V}\frac{dV}{dT}$$

The ideal-gas law is given by Equation 17.1. The two equations can be combined to give the proof.

EVALUATE   For an ideal gas at constant pressure, $V = NkT/p$. This gives $dV/dT = Nk/p$. Substituting the equation into the above expression for β gives

$$\beta = \frac{1}{V}\frac{dV}{dT} = \frac{1}{V}\frac{Nk}{p} = \frac{Nk}{NkT} = \frac{1}{T}$$

ASSESS   The unit of β is K^{-1}, which is the same as the inverse temperature in the Kelvin scale.

69.   **INTERPRET**   The problem is about the volume of water at a given temperature, given its coefficient of volume expansion as a function of temperature.

DEVELOP   In Problem 68, water's coefficient of volume expansion in the temperature range from 0°C to about 20°C is given approximately by $\beta = a + bT + cT^2$, where T is in Celsius and $a = -6.43 \times 10^{-5}°C^{-1}$, $b = 1.70 \times 10^{-5}°C^{-2}$, and $c = -2.02 \times 10^{-7}°C^{-3}$. Thus, the volume as a function of temperature is

$$\frac{dV}{V} = \beta dT = (a + bT + cT^2)dT$$

Integrating the expression and imposing the boundary condition allows us to find the volume of water at 12°C.

EVALUATE   Integrating the expression above gives

$$\int_{V_1}^{V_2}\frac{dV}{V} = \ln\left(\frac{V_2}{V_1}\right) = \int_{T_1}^{T_2}\beta dT = \int_{T_1}^{T_2}(a + bT + cT^2)dT = a(T_2 - T_1) + \frac{1}{2}b(T_2^2 - T_1^2) + \frac{1}{3}c(T_2^3 - T_1^3)$$

For $T_1 = 0°C$, $T_2 = 12°C$, and coefficients a, b and c given above, the right hand side is 3.36×10^{-4}. With $V_1 = 1.00000$ L, exponentiation gives

$$V_2 = V_1 e^{3.36 \times 10^{-4}} = 1.00034 \text{ L}$$

ASSESS   The fraction of volume change is $(V_2 - V_1)/V_1 = 0.00034$.

71.   **INTERPRET**   This problem deals with thermal expansion of a brass pendulum. The quantity of interest is its length, so the relevant quantity is the coefficient of linear expansion, α.

DEVELOP N swings of a pendulum clock produce a time reading of $t = N\tau$ where $\tau = 2\pi\sqrt{L/g}$ is the period. If the clock is accurate at $T_1 = 20°C$, at some other temperature T_2, the length of the pendulum becomes $L_2 = L_1(1 + \alpha \Delta T)$. Therefore, the ratio of the periods is

$$\frac{\tau_2}{\tau_1} = \sqrt{\frac{L_2}{L_1}} = \sqrt{1 + \alpha \Delta T}$$

The error in the clock is the difference in its time-readings for N swings at T_2, relative to T_1:

$$\Delta t = t_2 - t_1 = N\tau_2 - N\tau_1 = N\tau_1\left(\frac{\tau_2}{\tau_1} - 1\right)$$

This is the equation we shall solve to find the time it takes for the clock to err by 1 minute.

EVALUATE   Substituting the ratio of the periods into the expression for Δt gives

$$\Delta t = t_1\left(\sqrt{1 + \alpha \Delta T} - 1\right) = t_1\left(1 + \frac{1}{2}\alpha \Delta T + \cdots - 1\right) \approx \frac{1}{2}t_1\alpha \Delta T$$

where we have used Taylor-series expansion (see Appendix A) since $\alpha\Delta T \ll 1$. The time for an error of 1 min to accumulate is therefore

$$t_1 = \frac{\Delta t}{\frac{1}{2}\alpha\,\Delta T} = \frac{(-1\ \text{min})}{\frac{1}{2}(19\times10^{-6}\ \text{K}^{-1})(18°C - 20°C)} = 5.26\times10^4\ \text{min} = 36.5\ \text{d}$$

where we used α for brass from Table 17.2. Since $L_2 < L_1, \tau_2 < \tau_1$ so $\Delta t = t_2 - t_1 < 0$, and the clock at 18°C is fast.

ASSESS   The period of a pendulum, $\tau = 2\pi\sqrt{L/g}$, increases with its length L. Due to thermal expansion, the pendulum length at 20°C is greater than that at 18°C, and consequently, its period is longer.

73.   **INTERPRET**   This problem involves pressure, temperature, and volume. We will assume that the ideal gas law applies, and find the volume of one mole under the given conditions.

DEVELOP   The ideal gas law is $pV = nRT$, where $R = 8.314$ J/K $\cdot$ mol. The pressure is given as 9.1×10^6 Pa, $P = 90$ ATM $=$ and the temperature is $T = 730$ K. We want to find the volume of one mole of gas, and see if this volume is less than 1 L.

EVALUATE

$$pV = nRT \rightarrow V = \frac{nRT}{p} = 6.7\times10^{-4}\ \text{m}^3 = 0.67\ L$$

ASSESS   This design will work, as one liter contains more than a mole at this pressure and temperature

75.   **INTERPRET**   This is a heat problem. Our goal is to find the time required to vaporize a given amount of ice with a constant-power microwave oven.

DEVELOP   We will assume that all of the microwave oven's power goes into the ice, and use the equations for heat of fusion, then heat for the temperature change, and finally heat of vaporization: $Q = L_f m + mc\Delta T + L_v m$, where $L_f = 334$ kJ/kg, $c = 4184$ J/kg $\cdot$ K, $L_v = 2257$ kJ/kg, $\Delta T = 100$ K, and $m = 0.500$ kg. The microwave provides $P = 500$ J/s, so we will divide the heat by the power to find the time. We will compare this time with the claimed value of 10 minutes.

EVALUATE

$$Q = m(L_f + c\Delta T + L_v) \rightarrow t = \frac{Q}{p} = \frac{m}{p}(L_f + c\Delta T + L_v)$$
$$\rightarrow t = 3000\ \text{s} = 50\ \text{min}$$

ASSESS   If this microwave vaporizes 0.500 kg of ice in 10 minutes, then it is putting out about 5 times as much power as it should. Either there is something wrong with the oven, or the claim is not correct.

18

HEAT, WORK, AND THE FIRST LAW OF THERMODYNAMICS

EXERCISES

Section 18.1 The First Law of Thermodynamics

15. **INTERPRET**   We identify the system as the water in the insulated container. The problem is about work done to raise the temperature of a system. The first law of thermodynamics is involved.

    DEVELOP   Since the container is perfectly insulated thermally, no heat enters or leaves the water in it. Thus, $Q = 0$ and the first law of thermodynamics in Equation 18.1 gives $\Delta U = Q - W = -W$. The change in the internal energy of the water is determined from its temperature rise, $\Delta U = mc\,\Delta T$ (see comments in Section 16.1 on internal energy).

    EVALUATE   The work done on the water is

    $$W = -\Delta U = -mc\,\Delta T = -(1 \text{ kg})(4.184 \text{ kJ/kg} \cdot \text{K})(7 \text{ K}) = -29.3 \text{ kJ}$$

    ASSESS   The negative sign signifies that work was done on the water.

17. **INTERPRET**   We identify the system as the gas that undergoes expansion. The problem is about the change of internal energy of a system and involves the first law of thermodynamics.

    DEVELOP   The heat added to the gas is $Q = Pt = (40 \text{ W})(25 \text{ s}) = 1000 \text{ J}$. In addition, the amount of work it does on its surrounding is $W = 750 \text{ J}$. The change in internal energy can be found by using the first law of thermodynamics given in Equation 18.1.

    EVALUATE   Using Equation 18.1 we find

    $$\Delta U = Q - W = 1000 \text{ J} - 750 \text{ J} = 250 \text{ J}$$

    ASSESS   Since $\Delta U > 0$, we conclude that the internal energy has increased.

19. **INTERPRET**   This problem is about heat and mechanical energy, which are related by the first law of thermodynamics. The system is the automobile engine.

    DEVELOP   Since we are dealing with rates, we make use of Equation 18.2:

    $$\frac{dU}{dt} = \frac{dQ}{dt} - \frac{dW}{dt}$$

    If we assume that the engine system operates in a cycle, then $dU/dt = 0$. The engine's mechanical power output can then be calculated once the heat output is known.

    EVALUATE   The above conditions yield $(dQ/dt)_{\text{out}} = 68 \text{ kW}$ and $(dW/dt) = 0.17(dQ/dt)_{\text{in}}$. Equation 18.2 then gives

    $$\frac{dW}{dt} = \frac{dQ}{dt} = \left(\frac{dQ}{dt}\right)_{\text{in}} - \left(\frac{dQ}{dt}\right)_{\text{out}} = \frac{1}{0.17}\frac{dW}{dt} - \left(\frac{dQ}{dt}\right)_{\text{out}}$$

    or

    $$\frac{dW}{dt} = \frac{(dQ/dt)_{\text{out}}}{(0.17)^{-1} - 1} = \frac{68 \text{ kW}}{(0.17)^{-1} - 1} = 13.9 \text{ kW}$$

    ASSESS   We find the mechanical power output dW/dt to be proportional to the heat output, $(dQ/dt)_{\text{out}}$. In addition, dW/dt also increases with the percentage of the total energy released in burning gasoline that ends up as mechanical work.

Section 18.2 Thermodynamic Processes

21.   **INTERPRET**   The expansion of the ideal gas involves two stages: an isochoric (constant-volume) process and an isobaric (constant-pressure) process. We are asked to find the total work done by the gas.

DEVELOP   For an isochoric process, $\Delta V = 0$ and $W = 0$. On the other hand, for an isobaric process, the work done is $W = p\Delta V$.

EVALUATE   Path AC is isochoric, so $W_{AC} = 0$. Similarly, path CB is isobaric, so the work done during this stage is

$$W_{CB} = p_2(V_2 - V_1) = 2p_1(2V_1 - V_1) = 2p_1V_1$$

Thus, the total work done is $W_{ACB} = W_{AC} + W_{CB} = 0 + 2p_1V_1 = 2p_1V_1$.

ASSESS   In the pV diagram, Fig. 18.19, the area under AC is zero, and that under CB, a rectangle, is $2p_1V_1$. The work done by the gas is the area under the pV curve.

23.   **INTERPRET**   The constant temperature of 300 K indicates that the process is isothermal.

DEVELOP   We assume the gas to be ideal and apply the ideal-gas law given in Equation 17.1: $pV = NkT$. For an isothermal process, $T = $ constant, we obtain $p_1V_1 = p_2V_2$. The total work done by the gas can be calculated using Equation 18.4:

$$W_{12} = nRT \ln\left(\frac{V_2}{V_1}\right)$$

EVALUATE   (a) For the isothermal expansion process, the volume increases by a factor of

$$\frac{V_2}{V_1} = \frac{p_1}{p_2} = \frac{100 \text{ kPa}}{75 \text{ kPa}} = \frac{4}{3}$$

(b) Using Equation 18.4, the work done by the gas is

$$W_{12} = nRT \ln\left(\frac{V_2}{V_1}\right) = (0.3 \text{ mol})(8.314 \text{ J/mol} \cdot \text{K})(300 \text{ K})\ln\left(\frac{4}{3}\right) = 215 \text{ J}$$

ASSESS   Since $V_2 > V_1$, we find the work to be positive, $W_{12} > 0$. This makes sense because the gas inside the balloon must do positive work to expand outward.

25.   **INTERPRET**   The thermodynamic process here is adiabatic, with no heat flowing between the system (the gas) and its environment.

DEVELOP   In an adiabatic process, $Q = 0$, and the first law of thermodynamics becomes $\Delta U = -W$. The temperature and volume are related by Equation 18.11b:

$$TV^{\gamma-1} = \text{constant}$$

EVALUATE   From the equation above, we have

$$T_1V_1^{\gamma-1} = T_2V_2^{\gamma-1} \quad \rightarrow \quad \frac{V_2}{V_1} = \left(\frac{T_1}{T_2}\right)^{1/(\gamma-1)}$$

Thus, for the temperature to double, the volume change is

$$\frac{V_2}{V_1} = \left(\frac{T_1}{T_2}\right)^{1/(\gamma-1)} = \left(\frac{1}{2}\right)^{1/(1.4-1)} = 0.177$$

ASSESS   We see that increasing the temperature along the adiabat is accompanied by a volume decrease. In addition, since $pV^{\gamma} = $ constant, the final pressure is also increased:

$$p_2 = p_1\left(\frac{V_1}{V_2}\right)^{\gamma} = p_1\left(\frac{1}{0.177}\right)^{1.4} = 11.3p_1$$

Section 18.3 Specific Heats of an Ideal Gas

27.   **INTERPRET**   The problem is about the specific heat of a mixture of gases. We want to know what fraction of the molecules is monatomic.

DEVELOP   The internal energy of a mixture of two ideal gases is

$$U = f_1 N \bar{E}_1 + f_2 N \bar{E}_2$$

where f_1 is the fraction of the total number of molecules, N, of type 1, and $\bar{E}_1$ is the average energy of a molecule of type 1, etc. Classically, $\bar{E} = g(\frac{1}{2}kT)$, where g is the number of degrees of freedom. The molar specific heat at constant volume is

$$C_V = \frac{1}{n}\frac{dU}{dT} = \frac{N_A}{N}\frac{d}{dT}\left(f_1 N g_1 \frac{1}{2}kT + f_2 N g_2 \frac{1}{2}kT \right) = \frac{1}{2}R(f_1 g_1 + f_2 g_2)$$

Suppose that the temperature range is such that $g_1 = 3$ for the monatomic gas, and $g_2 = 5$ for the diatomic gas, as discussed in Section 18.3. Then

$$C_V = \frac{1}{2}R(3f_1 + 5f_2) = R(2.5 - f_1)$$

where $f_2 = 1 - f_1$ since the sum of the fractions of the mixture is one. Now, C_V can also be specified by the ratio

$$\gamma = \frac{C_P}{C_V} = \frac{C_V + R}{C_V} = 1 + \frac{R}{C_V} \quad \rightarrow \quad C_V = \frac{R}{\gamma - 1}$$

Equating the two expressions allows us to solve for f_1.

EVALUATE   Solving, we find $2.5 - f_1 = \frac{1}{\gamma-1} = \frac{1}{0.52}$, or $f_1 = 57.7\%$.

ASSESS   From the equation above, we see that the specific-heat ratio can be written as

$$\gamma = 1 + \frac{1}{2.5 - f_1}$$

In the limit where all the gas molecules are monatomic, $f_1 = 1$, and $\gamma = 1.67$. On the other hand, if all the molecules are diatomic, then $f_1 = 0$ and the specific-heat ratio is $\gamma = 1.4$. The equation yields the expected results in both limits.

29.   **INTERPRET**   The thermodynamic process is adiabatic, and we want to know the temperature change when work is done on the gas.

DEVELOP   In an adiabatic process, $Q = 0$ so $\Delta U = -W$. From Equation 18.6, $\Delta U = nC_V \Delta T$, the change in temperature is

$$\Delta T = \frac{\Delta U}{nC_V} = \frac{-W}{nC_V}$$

Note that the change in internal energy is equal to the negative of the work done *by* the gas, or equivalently, the work done *on* the gas. If the work done per mole *on* the gas is $(-W)/n = 2.5$ kJ/mol, then $\Delta T = (2.5 \text{ kJ/mol})/C_V$.

EVALUATE   (a) For an ideal monatomic gas, $C_V = \frac{3}{2}R = \frac{3}{2}(8.314 \text{ J/mol} \cdot \text{K})$, so $\Delta T = 200$ K.
(b) For an ideal diatomic gas (with five degrees of freedom), $C_V = \frac{5}{2}R$ so $\Delta T = 120$ K.

ASSESS   Since the diatomic gas has a greater specific heat C_V, its temperature change is less than that of the monatomic gas.

PROBLEMS

31.   **INTERPRET**   The constant temperature of 440 K indicates that the process is isothermal.

DEVELOP   We apply the ideal-gas law given in Equation 17.1: $pV = NkT$. For an isothermal process, $T = $ constant, we obtain $p_1 V_1 = p_2 V_2$. Since $\Delta U = 0$ for an isothermal process, the heat absorbed is equal to the total work done by the gas (Equation 18.4):

$$Q = W = nRT \ln\left(\frac{V_2}{V_1} \right)$$

EVALUATE   (a) Using the equation above, the heat absorbed is $Q = W = 3.3$ kJ.

(b) Equation 18.4 gives

$$n = \frac{W}{RT \ln(V_2/V_1)} = \frac{3.3 \text{ kJ}}{(8.314 \text{ J/mol} \cdot \text{K})(440 \text{ K}) \ln 10} = 0.392 \text{ mol}$$

ASSESS   The heat absorbed by the gas is equal to the work done by the gas on its surrounding as it expands, and there is no change in temperature.

33.   **INTERPRET**   We take the air inside the spherical bubble to behave like an ideal gas at constant temperature. So the process is isothermal.

DEVELOP   We apply the ideal-gas law given in Equation 17.1: $pV = NkT$. For an isothermal process, $T = $ constant, we obtain $p_1 V_1 = p_2 V_2$. Since the volume of a spherical bubble of diameter d is $V = 4\pi(d/2)^3/3 = \pi d^3/6$, the relation between the diameter and the pressure can be written as

$$p_1\left(\frac{\pi d_1^3}{6}\right) = p_2\left(\frac{\pi d_2^3}{6}\right) \rightarrow \frac{d_2}{d_1} = \left(\frac{p_1}{p_2}\right)^{1/3}$$

EVALUATE   (a) Using the equation above, we find the diameter at the maximum pressure to be

$$d_2 = \left(\frac{p_1}{p_2}\right)^{1/3} d_1 = \left(\frac{(80 + 760) \text{ mm of Hg}}{(125 + 760) \text{ mm of Hg}}\right)^{1/3} (1.52 \text{ mm}) = 1.49 \text{ mm}$$

(b) The work done *on* the air is the negative of Equation 18.4, or

$$W_{\text{on air}} = -nRT \ln\left(\frac{V_2}{V_1}\right) = -p_1 V_1 \ln\left(\frac{p_1}{p_2}\right) = p_1 V_1 \ln\left(\frac{p_2}{p_1}\right)$$

$$= (840 \text{ mm of Hg})\left(\frac{101.3 \text{ kPa}}{760 \text{ mm of Hg}}\right)\frac{\pi}{6}(1.52 \text{ mm})^3 \ln\left(\frac{885 \text{ mm of Hg}}{885 \text{ mm of Hg}}\right)$$

$$= 10.7 \ \mu\text{J}$$

ASSESS   Positive work is done by the blood in compressing the air bubble.

35.   **INTERPRET**   The thermodynamic process here is adiabatic, with no heat flowing between the system (the gas) and its environment.

DEVELOP   In an adiabatic process, $Q = 0$, and the first law of thermodynamics becomes $\Delta U = -W$. The pressure and volume are related by Equation 18.11a: $PV^\gamma = $ constant. This implies

$$p_1 V_1^\gamma = p_2 V_2^\gamma \rightarrow \frac{p_2}{p_1} = \left(\frac{V_1}{V_2}\right)^\gamma$$

EVALUATE   Taking the natural logarithm on both sides of the above to solve for γ, we obtain

$$\ln\left(\frac{p_2}{p_1}\right) = \gamma \ln\left(\frac{V_1}{V_2}\right) \rightarrow \gamma = \frac{\ln(p_2/p_1)}{\ln(V_1/V_2)} = \frac{\ln 2.55}{(\ln 2)} = 1.35$$

ASSESS   The value of γ indicates that gas consists of polyatomic molecules.

37.   **INTERPRET**   The problem involves a cyclic process. The three processes that make up the cycle are: isothermal (AB), isochoric (BC), and isobaric (CA).

DEVELOP   Along the isotherm AB where $T = $ constant, we have $p_A V_A = p_B V_B$. For an isothermal process, the work done by the gas is (Equation 18.4):

$$W = Q = nRT \ln\left(\frac{V_2}{V_1}\right)$$

EVALUATE   (a) If AB is an isotherm, with $V_A = 5$ L and $V_B = 1$ L, then the ideal gas law gives

$$P_B = \left(\frac{V_A}{V_B}\right)P_A = \left(\frac{5}{1}\right)(60 \text{ kPa}) = 300 \text{ kPa}$$

(b) The work done *by* the gas in the isothermal process *AB* is

$$W_{AB} = nRT_A \ln\left(\frac{V_B}{V_A}\right) = p_A V_A \ln\left(\frac{V_B}{V_A}\right) = (300 \text{ J}) \ln\left(\frac{1}{5}\right) = -483 \text{ J}$$

The process *BC* is isochoric and $W_{BC} = 0$. Similarly, the process *CA* is isobaric so

$$W_{CA} = p_A(V_A - V_C) = (60 \text{ kPa})(5 \text{ L} - 1 \text{ L}) = 240 \text{ J}$$

The total work done by the gas is

$$W_{ABCA} = W_{AB} + W_{BC} + W_{CA} = -483 \text{ J} + 0 + 240 \text{ J} = -243 \text{ J}$$

The work done *on* the gas is the negative of this, that is, $W_{\text{on gas}} = 243 \text{ J}$.

ASSESS   Since the process is cyclic, the system returns to its original state, there's no net change in internal energy, and $\Delta U = 0$. This implies that $Q = W_{ABCA} = -243 \text{ J}$. That is, 243 J of heat must come *out* of the system.

39.   **INTERPRET**   We identify the thermodynamic process here as adiabatic compression.

DEVELOP   In an adiabatic process, $Q = 0$, and the first law of thermodynamics becomes $\Delta U = -W$. The temperature and volume are related by Equation 18.11b:

$$TV^{\gamma-1} = \text{constant}$$

From the equation above, we obtain

$$T_1 V_1^{\gamma-1} = T_2 V_2^{\gamma-1} \quad \rightarrow \quad T_2 = T_1\left(\frac{V_1}{V_2}\right)^{\gamma-1}.$$

where V_1/V_2 is the compression ratio (for *T* and *V* at maximum compression).

EVALUATE   Substituting the values given, we have

$$T_2 = T_1\left(\frac{V_1}{V_2}\right)^{\gamma-1} = (303 \text{ K})(8.5)^{0.4} = 713 \text{ K} = 440°\text{C}$$

(*Note: T* appearing in the gas laws is the absolute temperature.)

ASSESS   The higher the compression ratio V_1/V_2, the greater the temperature at the maximum compression, and hence a higher thermal efficiency.

41.   **INTERPRET**   We identify the thermodynamic process here as adiabatic compression.

DEVELOP   In an adiabatic process, $Q = 0$, and the first law of thermodynamics becomes $\Delta U = -W$. The temperature and volume are related by Equation 18.11b:

$$TV^{\gamma-1} = \text{constant}$$

From the equation above, we obtain

$$T_1 V_1^{\gamma-1} = T_2 V_2^{\gamma-1} \quad \rightarrow \quad T_2 = T_1\left(\frac{V_1}{V_2}\right)^{\gamma-1}$$

where V_1/V_2 is the compression ratio (for *T* and *V* at maximum compression). In addition, since $pV^\gamma = \text{constant}$, the final pressure is $p_2 = p_1(V_1/V_2)^\gamma$.

EVALUATE   **(a)** Substituting the values given in the problem statement, we find the air temperature at the maximum compression to be

$$T_2 = T_1\left(\frac{V_1}{V_2}\right)^{\gamma-1} = (320 \text{ K})(10.2)^{0.4} = 810 \text{ K}$$

(b) The corresponding pressure is

$$p_2 = p_1(V_1/V_2)^\gamma = (101.3 \text{ kPa})(10.2)^{1.4} = 2.62 \text{ MPa} = 25.8 \text{ atm}$$

ASSESS   The higher the compression ratio V_1/V_2, the greater the temperature and pressure at the maximum compression, and hence, a higher thermal (fuel) efficiency.

43.   **INTERPRET**   This problem is to explore how different ways of adding heat (isothermal, isochoric, or isobaric) affects the final temperature of the system.

DEVELOP   In an isothermal process, the temperature T is kept constant. With $\Delta U = 0$, the first law of thermodynamics gives $W = Q$. In an isochoric process, $\Delta V = 0$, and $W = 0$. First law of thermodynamics gives $Q = \Delta U = nC_V\Delta T$. Finally, in an isobaric process, $\Delta p = 0$ and

$$Q = nC_p\Delta T = n(C_V + R)\Delta T$$

These are the equations we shall use to solve for ΔT and W in each case.

EVALUATE   **(a)** When heat is added isothermally, T is constant, so the final temperature is $T_2 = 300$ K. Since $\Delta U = 0$, $W = Q = 1.5$ kJ.

(b) From the above, we see that in an isochoric process, $W = 0$ and

$$\Delta T = \frac{Q}{nC_V} = \frac{1.5 \text{ kJ}}{(2 \text{ mol})(\frac{5}{2}R)} = 36.1 \text{ K}$$

Therefore, $T_2 = 300$ K $+ \Delta T = 336$ K.

(c) In an isobaric process,

$$\Delta T = \frac{Q}{nC_p} = \frac{Q}{n(C_V + R)} = \frac{1.5 \text{ kJ}}{(2 \text{ mol})(\frac{5}{2}R + R)} = 25.8 \text{ K}$$

and $T_2 = 326$ K. The work done is

$$W = p\Delta V = nR\Delta T = \frac{R}{C_p}Q = \frac{R}{(7R/2)}Q = 429 \text{ J}$$

ASSESS   Comparing all three cases, we find

$$\Delta T : \text{ isothermal} < \text{isobaric} < \text{isochoric}$$
$$W: \text{ isochoric } < \text{isobaric} < \text{isothermal}$$

The results agree with that illustrated in Table 18.1.

45.   **INTERPRET**   The problem involves a cyclic process, and we identify three separate stages of the cycle: adiabatic, isochoric, and isothermal.

DEVELOP   In an adiabatic process (AB), $Q = 0$, and the first law of thermodynamics becomes $\Delta U = -W$. The pressure and volume are related by Equation 18.11a: $pV^\gamma = $ constant. This implies

$$p_A V_A^\gamma = p_B V_B^\gamma \;\rightarrow\; p_B = \left(\frac{V_A}{V_B}\right)^\gamma p_A$$

Point C lies on an isotherm with A, so the ideal-gas law (Equation 17.1) yields

$$p_C = \frac{p_A V_A}{V_C}$$

EVALUATE   **(a)** From the equation above, the pressure at point B is

$$p_B = \left(\frac{V_A}{V_B}\right)^\gamma p_A = (250 \text{ kPa})\left(\frac{1}{3}\right)^{1.67} = 39.9 \text{ kPa}$$

(b) The pressure at point C is $p_C = p_A \frac{V_A}{V_C} = (250 \text{ kPa})(\frac{1}{3}) = 83.3$ kPa.

(c) The net work done *by* the gas is $W_{ABCA} = W_{AB} + W_{BC} + W_{CA}$. W_{AB} is for an adiabatic process (Equation 18.12) and equals

$$W_{AB} = \frac{p_A V_A - p_B V_B}{\gamma - 1} = \frac{(250 \text{ kPa})(1 \text{ m}^3) - (39.9 \text{ kPa})(3 \text{ m}^3)}{0.67} = 194 \text{ kJ}$$

W_{BC} is for an isochoric process and equals zero. Finally, W_{CA} is for an isothermal process (Equation 18.4) and equals

$$W_{CA} = nRT_A \ln\left(\frac{V_A}{V_C}\right) = (250 \text{ kJ})\ln\left(\frac{1}{3}\right) = -275 \text{ kJ}$$

Thus,

$$W_{ABCA} = W_{AB} + W_{BC} + W_{CA} = 194 \text{ kJ} + 0 + (-275 \text{ kJ}) = -80.2 \text{ kJ}$$

The work done *on* the gas is the negative of this.

ASSESS Since the process is cyclic, the system returns to its original state, there's no net change in internal energy, and $\Delta U = 0$. This implies that $Q = W_{ABCA} = -80.2$ kJ. That is, 80.2 kJ of heat must come *out* of the system.

47. **INTERPRET** We find the work done in a given heat cycle. In each part of the cycle, the work done is the area under the p-V curve.

DEVELOP The gas is is taken through four parts of a cycle.
(a) It is heated at constant volume until the pressure is doubled.
(b) It is compressed adiabatically until its volume is $\frac{1}{4}$ the initial value.
(c) It is cooled at constant volume to a temperature of 300 K.
(d) It is expanded isothermally until it returns to the original state.

Parts **(a)** and **(c)** do no work, as the volume is constant.

The work in part **(b)** is the area under an adiabatic curve, $W_b = \frac{p_1 V_1 - p_2 V_2}{\gamma - 1}$. In order to use this, we will need the final pressure, which we can obtain by using $pV^\gamma = \text{constant}$.

The work in part **(d)** is the area under an isothermal curve, $W_d = nRT \ln(\frac{V_2}{V_1})$. We will need to know the number of moles n, which we can obtain from $pV = nRT$.

The total work is the sum $W = W_b + W_d$.

We are told that the gas is an ideal gas with $\gamma = 1.4$, it has a volume $V_1 = 4.0$ L $= 4 \times 10^{-3}$ m^3 at $T = 300$ K, and $p_1 = 100$ kPa.

EVALUATE The pressure during part **(b)** changes from $2p_1$ to $(2p_1)V_1^\gamma = p_2 V_2^\gamma \rightarrow p_2 = 2p_1(\frac{V_1}{V_2})^\gamma = 2p_1(4)^\gamma$. The work for part **(b)** is then $W_b = \frac{2p_1 V_1 - 2p_1 4^\gamma (\frac{1}{4} V_1)}{\gamma - 1} = \frac{2p_1 V_1}{\gamma - 1}(1 - 4^{\gamma - 1})$.

The number of moles of gas is $n = \frac{pV}{RT}$, and the work for part **(d)** is $W_d = nRT \ln(\frac{V_f}{V_i}) = (\frac{p_1 V_1}{RT_1}) RT_1 \ln(\frac{V_1}{\frac{1}{4}V_1}) = p_1 V_1 \ln(4)$.

The total work is then $W_b + W_d = \frac{2p_1 V_1}{\gamma - 1}(1 - 4^{\gamma - 1})W_d + p_1 V_1 \ln(4) = p_1 V_1 \left[\frac{2(1 - 4^{\gamma - 1})}{\gamma - 1} + \ln 4\right] = -928$ J.

ASSESS The work done is negative. If we draw a p-V diagram of this motion, we can see that the area under the curve is negative, since the gas goes around the cycle "counterclockwise."

49. **INTERPRET** The problem involves a cyclic process, and we identify three separate stages of the cycle: adiabatic, isochoric, and isothermal.

DEVELOP In an adiabatic process $(AB), Q = 0$, and the first law of thermodynamics becomes $\Delta U = -W$. The pressure and volume are related by Equation 18.11a: $pV^\gamma = \text{constant}$, and the work done by the gas is

$$W_{AB} = \frac{p_A V_A - p_B V_B}{\gamma - 1}$$

Since $\Delta V_{BC} = 0$ for the isochoric process, $W_{BC} = 0$. Similarly, for an isothermal process, the work done is (Equation 18.4)

$$W_{CA} = nRT_A \ln\left(\frac{V_A}{V_C}\right)$$

The minimum volume attained is V_B, which is the volume at the end of the adiabatic compression. Note that $V_B = V_C$ since $\Delta V_{BC} = 0$.

EVALUATE The solutions below are presented in the reverse order.

(c) The same individual processes applied in this problem are in the same order as in Example 18.4, so the pV diagram looks just like Figure 18.14, except that $V_A = 25$ L and $P_B = 3P_A = 3(50 \text{ kPa})$.

(b) The minimum volume attained can be found from the adiabatic law (Equation 18.11a):

$$V_C = V_B = V_A \left(\frac{p_A}{p_B}\right)^{1/\gamma} = (25 \text{ L})\left(\frac{1}{3}\right)^{1/1.67} = 12.9 \text{ L}$$

(a) The work done *on* the gas is the negative of the work done *by* the gas:

$$W_{\text{on gas}} = -W_{ABCA} = -(W_{AB} + W_{BC} + W_{CA}) = -\frac{p_A V_A - p_B V_B}{\gamma - 1} - 0 - p_A V_A \ln\left(\frac{V_A}{V_C}\right)$$

$$= -\frac{(50 \text{ kPa})(25 \text{ L}) - (150 \text{ kPa})(12.9 \text{ L})}{1.67 - 1} - (50 \text{ kPa})(25 \text{ L})\ln\left(\frac{25 \text{ L}}{12.9 \text{ L}}\right)$$

$$= 211 \text{ J}$$

since AB is adiabatic, BC is isochoric, and CA is isothermal.

ASSESS   Since the process is cyclic, the system returns to its original state, there's no net change in internal energy, and $\Delta U = 0$. This implies that $Q = W_{ABCA} = -211$ J. That is, 211 J of heat must come *out* of the system.

51.   **INTERPRET**   The volume of the flask does not change, so we identify the thermodynamic process as isochoric.
DEVELOP   In an isochoric process, $\Delta V = 0$, and $W = 0$. The first law of thermodynamics (Equation 18.1) gives $Q = \Delta U = nC_V \Delta T$.
EVALUATE   Substituting the values given, we find the heat that must be added to raise the temperature to be

$$Q = nC_V \Delta T = \left(\frac{p_0 V_0}{RT_0}\right)(2.5R)\Delta T = \frac{(100 \text{ kPa})(5 \text{ L})}{R(273 \text{ K})}(2.5R)(20 \text{ K}) = 91.6 \text{ J}$$

where we have used the ideal gas law to find n.
ASSESS   Adding the heat raises the air temperature inside the flask. Since the volume does not change, by ideal-gas law, the pressure must also go up. The final pressure can be calculated as

$$p = \frac{nRT}{V_0} = \frac{nRT_0}{V_0}\frac{T}{T_0} = p_0 \frac{T}{T_0} = (100 \text{ kPa})\frac{293 \text{ K}}{273 \text{ K}} = 107 \text{ kPa}$$

53.   **INTERPRET**   The thermodynamic process here involves two stages: isothermal compression followed by an adiabatic compression.
DEVELOP   During the first stage, since the gas is compressed isothermally, $\Delta T = 0$, there is no change in the temperature of the system, but the volume is reduce from V_0 to $V_1 = V_0/3$. During the next stage of adiabatic compression, $Q = 0$, and the first law of thermodynamics becomes $\Delta U = -W$. The temperature and volume are related by Equation 18.11b:

$$TV^{\gamma-1} = \text{constant}$$

which gives

$$T_1 V_1^{\gamma-1} = T_2 V_2^{\gamma-1} \quad \rightarrow \quad T_2 = T_1 \left(\frac{V_1}{V_2}\right)^{\gamma-1}$$

The final temperature is T_2.
EVALUATE   Substituting the values given, we find T_2 to be

$$T_2 = T_1\left(\frac{V_1}{V_2}\right)^{\gamma-1} = (273 \text{ K})\left(\frac{V_0/3}{V_0/5}\right)^{(7/5)-1} = (273 \text{ K})\left(\frac{5}{3}\right)^{0.4} = 335 \text{ K}$$

ASSESS   Since $TV^{\gamma-1} = $ constant for an adiabatic process, the compression results in an increase of the final temperature.

55.   **INTERPRET**   The problem involves a cyclic process, and we identify three separate stages of the cycle: isochoric (AB), isobaric (BC), and isothermal (CA).
DEVELOP   The work done *by* the gas in each segment of the cycle is summarized in Table 18.1. For the isochoric process (path AB), $\Delta V = 0$, and $W = 0$. For the isobaric process (path BC), the work done is

$$W_{BC} = p_B(V_C - V_B)$$

Finally, for the isothermal process (CA), the work done is (Equation 18.4)

$$W_{CA} = nRT_A \ln\left(\frac{V_A}{V_C}\right)$$

EVALUATE   (a) Using the equations above, we obtain

$$W_{AB} = 0 \qquad \text{(isochoric)}$$

$$W_{BC} = p_B(V_C - V_B) = (250 \text{ kPa})(1 \text{ L} - 5 \text{ L}) = -1000 \text{ J} \qquad \text{(isobaric)}$$

$$W_{CA} = nRT_A \ln\left(\frac{V_A}{V_C}\right) = p_A V_A \ln\left(\frac{V_A}{V_C}\right) = (50 \text{ kPa})(5 \text{ L})\ln(5) = 402 \text{ J} \qquad \text{(isothermal)}$$

Adding up all the contributions, the net work done *by* the gas is

$$W_{ABCA} = W_{AB} + W_{BC} + W_{CA} = 0 - 1000 \text{ J} + 402 \text{ J} = -598 \text{ J}$$

The net work done *on* the gas is simply the negative of this, $W_{\text{on gas}} = -W_{ABCA} = 598$ J.

(b) With *V* held constant, the heat transferred is

$$Q_{AB} = nC_V \Delta T = \frac{nR(T_B - T_A)}{\gamma - 1} = \frac{p_B V_B - p_A V_A}{\gamma - 1} = \frac{(250 \text{ kPa} - 50 \text{ kPa})(5 \text{ L})}{1.4 - 1} = 2.50 \text{ kJ}$$

Since $Q_{AB} > 0$, heat is transferred into the gas.

ASSESS   At constant volume, the gas must be heated in order to raise its pressure. In the above, we have used $C_V = R/(\gamma - 1)$. The equation can be derived as follows:

$$\gamma = \frac{C_P}{C_V} = \frac{C_V + R}{C_V} = 1 + \frac{R}{C_V} \quad \rightarrow \quad C_V = \frac{R}{\gamma - 1}$$

57. **INTERPRET**   We identify the thermodynamic process as adiabatic expansion of a gas mixture.

DEVELOP   In an adiabatic process (AB), $Q = 0$, and the pressure and volume are related by Equation 18.11a: $pV^\gamma = $ constant. This implies

$$p_1 V_1^\gamma = p_2 V_2^\gamma \quad \rightarrow \quad \frac{p_2}{p_1} = \left(\frac{V_1}{V_2}\right)^\gamma$$

Taking the natural logarithm on both sides of the above to solve for γ, we obtain

$$\ln\left(\frac{p_2}{p_1}\right) = \gamma \ln\left(\frac{V_1}{V_2}\right) \quad \rightarrow \quad \gamma = \frac{\ln(p_2/p_1)}{\ln(V_1/V_2)} = \frac{\ln(1/3)}{\ln(1/2)} = \frac{-\ln(3)}{-\ln(2)} = 1.58$$

To find the fraction of the molecules that are argon, we use the result from Exercise 27:

$$f_1 = 2.5 - \frac{1}{\gamma - 1}$$

EVALUATE   Substituting $\gamma = 1.58$ into the equation above gives

$$f_{Ar} = 2.5 - \frac{1}{\gamma - 1} = 2.5 - \frac{1}{1.58 - 1} = 0.790 = 79.0\%$$

ASSESS   In the limit where all the gas molecules are monatomic, $f_1 = 1$, and $\gamma_{\text{monatomic}} = 1.67$. On the other hand, if all the molecules are diatomic, then $f_1 = 0$ and the specific-heat ratio is $\gamma_{\text{diatomic}} = 1.4$. Our ratio of $\gamma = 1.58$ is closer to 1.67. This implies that the gas mixture is predominantly monatomic.

59. **INTERPRET**   This problem is about melting, and it involves heat of fusion. The source of energy is the mechanical energy of the rock.

DEVELOP   The mechanical energy of the rock (originally gravitational potential energy) melted the ice (changed its internal energy) and no heat energy was transferred ($Q = 0$).

EVALUATE   From the first law of thermodynamics, we have

$$-W = \Delta U \quad \rightarrow \quad m_{\text{rock}} gh = m_{\text{ice}} L_f$$

Thus, we find the height from which the rock is dropped to be

$$h = \frac{m_{ice}L_f}{m_{rock}g} = \frac{(0.0063 \text{ kg})(334 \text{ J/g})}{(8.5 \text{ kg})(9.8 \text{ m/s}^2)} = 25.3 \text{ m}$$

ASSESS In this problem, the rock did positive work on the ice-water system. Therefore, the work done *by* the system is negative, $W < 0$.

61. **INTERPRET** We identify the thermodynamic process as isobaric, with pressure being kept at 1.0 atm.

DEVELOP Since the process is isobaric ($p = 1$ atm), the work done is $W = p\Delta V$. The initial volume of 1 g of water at 50°C is approximately $V_1 = 1$ mL, while the final volume of 1 g of steam at 200°C (from the ideal gas law) is

$$V_2 = \frac{nRT_2}{p} = \frac{(\frac{1}{18} \text{ mol})(8.314 \text{ J/mol} \cdot \text{K})(473 \text{ K})}{1.013 \times 10^5 \text{ Pa}} = 2.16 \text{ L}$$

On the other hand, the heat added (at constant pressure) is

$$Q = mc_{water}(100°C - 50°C) + mL_v + nC_p(200°C - 100°C)$$

which includes raising the temperature of the water, changing its state to steam, and raising the temperature of the steam.

EVALUATE (a) Substituting the values given, we find the work done by the system to be

$$W = p\Delta V = p(V_2 - V_1) = (1 \text{ atm})(2.16 - 0.001)\text{L}(101.3 \text{ J/L} \cdot \text{atm}) = 218 \text{ J}$$

(b) Using $C_p = C_V + R = 5.3R$, and values from Tables 16.1 and 17.1, we find

$$Q = (1 \text{ g})(4.184 \text{ J/g} \cdot \text{K})(50 \text{ K}) + 2257 \text{ J} + \left(\frac{1}{18}\text{mol}\right)(5.3)(8.314 \text{ J/mol} \cdot \text{K})(100 \text{ K})$$

$$= 0.209 \text{ kJ} + 2.257 \text{ kJ} + 0.245 \text{ kJ} = 2.711 \text{ kJ}$$

ASSESS Most of the heat added goes to changing the phase of the system from water to steam. By first law of thermodynamics, the internal energy of the system has increased by $\Delta U = Q - W$.

63. **INTERPRET** We identify our system as the horizontal piston-cylinder system containing an ideal gas. We want to demonstrate that the piston undergoes simple harmonic motion when slightly displaced.

DEVELOP Since the piston-cylinder system is horizontal, we do not need to consider the force of gravity on the piston. At equilibrium, the pressure forces from inside and outside the piston are equal, so the gas pressure at the equilibrium position of the piston is p_0. We also assume that the gas temperature at equilibrium is T_0, so $p_0V_0 = nRT_0$, where $V_0 = Ax_0$ is the volume at equilibrium. When the piston is displaced from its equilibrium position by an amount Δx (positive to the right), the horizontal force on it is $F = pA - p_0A = (p - p_0)A$, and Newton's second law gives an acceleration of

$$\frac{d^2(\Delta x)}{dt^2} = \frac{(p - p_0)A}{M}$$

For isothermal expansions and compressions of the gas,

$$pV = p_0V_0 \quad \rightarrow \quad pA(x_0 + \Delta x) = p_0Ax_0$$

For small displacements $\Delta x \ll x_0$, the pressure is

$$p = p_0\frac{x_0}{x_0 + \Delta x} = p_0\frac{1}{1 + (\Delta x/x_0)} \approx p_0[1 - (\Delta x/x_0)]$$

(see the binomial approximation in Appendix A). Substituting the expression for p into Newton's second law equation allows us to show that the piston executes simple harmonic motion.

EVALUATE Combining the two equations yields

$$\frac{d^2(\Delta x)}{dt^2} = \frac{(p - p_0)A}{M} = -\frac{p_0A}{M}\frac{\Delta x}{x_0} = -\omega^2\Delta x$$

where $\omega^2 = p_0 A/Mx_0$. This is the equation for simple harmonic motion of the piston, about its equilibrium position x_0, with angular frequency $\omega = \sqrt{p_0 A/Mx_0}$. Since $p_0 V_0 = p_0 Ax_0 = nRT_0$, we may eliminate x_0 to obtain

$$\omega = \frac{p_0 A}{\sqrt{MnRT_0}}$$

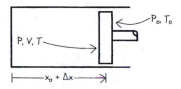

ASSESS   In order for the gas temperature to remain constant, as assumed above, heat must flow into and out of the gas. This requires time, so the motion of the piston must be very slow. If the motion is rapid (or if the cylinder is thermally insulated), there is no time for heat transfer in the gas and the expansions and compressions are adiabatic. In this case,

$$pV^\gamma = p_0 V_0^\gamma \quad \rightarrow \quad pA^\gamma(x_0 + \Delta x)^\gamma = p_0 A^\gamma x_0^\gamma$$

or $p = p_0(1 + \Delta x/x_0)^{-\gamma}$. For small displacements $\Delta x \ll x_0$, we have $p \approx p_0(1 - \gamma\Delta x/x_0)$ and

$$\frac{d^2(\Delta x)}{dt^2} = \frac{(p - p_0)A}{M} = -\frac{\gamma p_0 A}{M}\frac{\Delta x}{x_0} = -\omega^2 \Delta x$$

where $\omega^2 = \gamma p_0 A/Mx_0$. This represents simple harmonic motion with

$$\omega = \sqrt{\gamma p_0 A/Mx_0} = p_0 A\sqrt{\gamma/MnRT_0}$$

65.   **INTERPRET**   We identify the thermodynamic process as adiabatic with $Q = 0$.

DEVELOP   In an adiabatic process, the temperature and pressure are related by Equation 18.11a: $pV^\gamma = $ constant. The equation can be rewritten as

$$pV^\gamma = p^{1-\gamma}(pV)^\gamma = p^{1-\gamma}(nRT)^\gamma$$

which implies that $p^{1-\gamma}T^\gamma = $ constant. The work done *by* the air is

$$W = \frac{p_i V_i - p_f V_f}{\gamma - 1}$$

EVALUATE   (a) Using the equation above, the final temperature is

$$T_f = T_i\left(\frac{p_f}{p_i}\right)^{(\gamma-1)/\gamma} = (262\text{ K})\left(\frac{86.5\text{ kPa}}{62.0\text{ kPa}}\right)^{(1.4-1)/\gamma} = 288\text{ K} = 15.2°\text{C}$$

(b) The work done on one cubic meter of air (the negative of Equation 18.12) is

$$W_{\text{on gas}} = -\frac{p_i V_i - p_f V_f}{\gamma - 1} = \frac{(p_f V_f - p_i V_i)}{\gamma - 1} = \frac{p_i V_i}{\gamma - 1}\left[\left(\frac{p_f}{p_i}\right)^{(\gamma-1)/\gamma} - 1\right]$$

$$= \frac{(62.0\text{ kPa})(1\text{ m}^3)}{1.4 - 1}\left[\left(\frac{86.5\text{ kPa}}{62.0\text{ kPa}}\right)^{(1.4-1)/1.4} - 1\right] = 15.5\text{ kJ}$$

where we used Equation 18.11a to eliminate V_f.

ASSESS   The final volume is

$$V_f = V_i\left(\frac{p_i}{p_f}\right)^{1/\gamma} = (1\text{ m}^3)\left(\frac{62.0\text{ kPa}}{86.5\text{ kPa}}\right)^{1/1.4} = 0.788\text{ m}^3$$

which is less than the initial volume V_i. So, positive work is done on the air to compress it.

67.   **INTERPRET**   The problem involves a cyclic process, and we identify four separate stages of the cycle: adiabatic compression, isobaric expansion, adiabatic expansion, and isochoric cooling.

DEVELOP   From Table 18.1, the work done and heat absorbed during each of the four processes comprising the diesel cycle is:

$$1 \rightarrow 2 \text{ (adiabatic)} \quad W_{12} = (p_1 V_1 - p_2 V_2)/(\gamma - 1), \ Q_{12} = 0$$

$$2 \rightarrow 3 \text{ (isobaric)} \quad W_{23} = p_2 (V_3 - V_2), \ Q_{23} = nC_P (T_3 - T_2) \equiv Q_h$$

$$3 \rightarrow 4 \text{ (adiabatic)} \quad W_{34} = (p_3 V_3 - p_4 V_4)/(\gamma - 1), \ Q_{34} = 0$$

$$4 \rightarrow 1 \text{ (isovolumic)} \quad W_{41} = 0, \ Q_{41} = nC_V (T_1 - T_4) \equiv -Q_c$$

For the whole cycle, the work done is

$$W = \frac{p_1 V_1 - p_2 V_2 + p_3 V_3 - p_4 V_4}{(\gamma - 1)} + p_3 V_3 - p_2 V_2 = \frac{\gamma(p_3 V_3 - p_2 V_2) - p_4 V_4 + p_1 V_1}{(\gamma - 1)}$$

while the heat added can be written as

$$Q_h = nC_P (T_3 - T_2) = \frac{\gamma(P_3 V_3 - P_2 V_2)}{\gamma - 1}$$

where we used the ideal gas law, $nT = PV/R$, and

$$\frac{C_P}{R} = \frac{C_P}{C_P - C_V} = \frac{\gamma}{\gamma - 1}$$

Therefore, the work done by the system can be rewritten as

$$W = Q_h - \frac{(P_4 V_4 - P_1 V_1)}{(\gamma - 1)} = Q_h - \frac{(P_4 V_4 - P_1 V_1)Q_h}{\gamma(P_3 V_3 - P_2 V_2)}$$

The efficiency is

$$e = \frac{W}{Q_h} = 1 - \frac{(P_4 V_4 - P_1 V_1)}{\gamma(P_3 V_3 - P_2 V_2)}$$

EVALUATE   The adiabatic law can now be used to express every product in terms of $P_2 V_2$ and the compression and cutoff ratios: $P_2 = P_3$, $V_1 = V_4$, $V_1/V_2 = r$, $V_3/V_2 = r_c$, $P_1 V_1^\gamma = P_2 V_2^\gamma$, and $P_3 V_3^\gamma = P_4 V_4^\gamma$, so

$$p_1 V_1 = p_2 V_2 r^{1-\gamma}$$

$$p_3 V_3 = p_2 V_2 r_c$$

$$p_4 V_4 = p_2 V_2 r^{1-\gamma} r_c^\gamma$$

Using these expressions, we find the efficiency to be

$$e = 1 - \frac{r^{1-\gamma}\left(r_c^\gamma - 1\right)}{\gamma(r_c - 1)}$$

ASSESS   The efficiency of the Diesel engine is always less than one. Diesel engines are more efficient than gasoline engines.

69.   **INTERPRET**   We derive the equivalent of $W = nRT \ln(\frac{V_2}{V_1})$ for a *non*-ideal gas that includes Van der Waals forces. We are given the equation relating pressure and volume for a Van der Waals gas.

DEVELOP   The equation relating pressure and volume for a Van der Waals gas is given as $[p + a(\frac{n}{V})^2](V - nb) = nRT$. We solve this for pressure, and integrate to find the work.

EVALUATE

$$\left[p + a\left(\frac{n}{V}\right)^2\right](V - nb) = nRT \rightarrow p(V - nb) + a\left(\frac{n}{V}\right)^2 (V - nb) = nRT$$

$$\rightarrow p = \frac{1}{V - nb}\left[nRT - a\left(\frac{n}{V}\right)^2 (V - nb)\right] = \frac{nRT}{V - nb} - a\left(\frac{n}{V}\right)^2$$

Integrate this to find work:

$$W = \int_{V_1}^{V_2} p\,dV = \int_{V_1}^{V_2}\left[\frac{nRT}{V-nb} - a\left(\frac{n}{V}\right)^2\right]dV = \left[nRT\ln(V-nb) + \frac{an^2}{V}\right]_{V_1}^{V_2}$$
$$\rightarrow W = nRT\ln\left(\frac{V_2-nb}{V_1-nb}\right) + an^2\left(\frac{1}{V_2} - \frac{1}{V_1}\right)$$

ASSESS This simplifies to $W = nRT\ln(\frac{V_2}{V_1})$ if a and b are zero.

71. **INTERPRET** We calculate the rate $\frac{dT}{dy}$ at which air cools as it rises, approximating the process as adiabatic.

DEVELOP The hydrostatic equation is $\frac{dp}{dh} = \rho g = -\frac{dp}{dy}$, and for adiabatic processes $nC_v dT = -p\,dV$. We differentiate the ideal gas law $pV = nRT$, with the previous substitution, to find the relationship between dT and dp. Finally we use the hydrostatic equation to relate dp to dy and thus find $\frac{dT}{dy}$. The molecular weight of air is given as $M = 0.029$ kg/mol, and $\gamma = 1.4$.

EVALUATE

$$pV = nRT \rightarrow nRdT = pdV + Vdp = -nC_v dT + Vdp = nRdT$$
$$\rightarrow Vdp = n(R + C_v)dT = nC_p dT$$

Volume is $V = \frac{m}{\rho} = \frac{nM}{\rho}$, so $\frac{M}{\rho}dp = C_p dT$. From $\rho g = -\frac{dp}{dy}$, we have $dp = -\rho g dy$, so $\frac{M}{\rho}(-\rho g dy) = C_p dT \rightarrow \frac{dT}{dy} = -\frac{Mg}{C_p}$.

$$C_p = \frac{\gamma R}{\gamma - 1}, \text{ so } \frac{dT}{dy} = -\frac{Mg}{\gamma R}(\gamma - 1) = -0.0098 \text{ K/m}$$

ASSESS This value, –9.8 K/km, is the approximate change in temperature as you gain altitude.

73. **INTERPRET** We find the pressure and volume of air within a diving bell, given that the temperature varies in such a way that the pressure and volume vary according to a given equation.

DEVELOP The equation relating pressure and temperature is given as $p = p_0\sqrt{\frac{V_0}{V}}$, where $V_0 = 17$ m³ and $p_0 = 1$ atm. The value of $p_{max} = 1.5$ atm. We can plug these values into the equation, and see what is the resulting value of V. If it's more than $V_{min} = 8.66$ m³, the design is ok.

EVALUATE Solve the given equation for V: $p = p_0\sqrt{\frac{V_0}{V}} \rightarrow V = \frac{p_0^2 V_0}{p^2} = 7.56$ m³.

ASSESS The design needs work. Note also that since the pressure appears only in a ratio, we do not need to convert to SI units.

75. **INTERPRET** We find the heat required to heat a volume of gas at constant volume.

DEVELOP The molar specific heat of air at this range of temperatures is $C_v = 2.5R$. The temperature change is from 0°C to 20°C, and the initial volume at 0°C is 5.0 L. The heat required will be $Q = nC_v\Delta T$, so if we find the number of moles n using $pV = nRT$, we can find Q. We will assume that the initial pressure is $p_i = 101.3$ kPa.

EVALUATE

$$pV = nRT \rightarrow n = \frac{pV}{RT}. \quad Q = nC_v\Delta T = \frac{p_i V_i}{RT_i}(2.5R)\Delta T = 93 \text{ J}$$

ASSESS Note that we are using the ideal-gas law, so it is important to use absolute temperature and volume in m^3.

19

THE SECOND LAW OF THERMODYNAMICS

EXERCISES

Sections 19.2 and 19.3 The Second Law of Thermodynamics and Its Applications

15. **INTERPRET** This problem is about the thermal efficiency of a heat engine.

    DEVELOP If it were a reversible engine, its efficiency would be that of a Carnot engine, given by Equation 19.3:

    $$e_{\text{Carnot}} = 1 - \frac{T_c}{T_h}$$

    EVALUATE Substituting the values given in the problem, we obtain

    $$e_{\text{Carnot}} = 1 - \frac{T_c}{T_h} = 1 - \frac{2.7 \text{ K}}{5600 \text{ K}} = 1 - 4.82 \times 10^{-4} \approx 99.95\%$$

    ASSESS The engine efficiency is almost 100%. This is too good to be true.

17. **INTERPRET** This problem is about a Carnot engine that operates via the Carnot cycle.

    DEVELOP The efficiency of an engine, by definition is,

    $$e = \frac{W}{Q_h}$$

    where W and Q_h are the work done and heat absorbed per cycle.

    EVALUATE (a) From the equation above, the efficiency of the engine is

    $$e = \frac{W}{Q_h} = \frac{350 \text{ J}}{900 \text{ J}} = 38.9\%$$

    (b) W and Q_h are related to the heat rejected per cycle by the first law of thermodynamics (since ΔU per cycle is zero). The relation is

    $$Q_c = Q_h - W = 900 \text{ J} - 350 \text{ J} = 550 \text{ J}$$

    (c) For a Carnot engine operating between two temperatures, $T_h/T_c = Q_h/Q_c$, so

    $$T_h = T_c \left(\frac{Q_h}{Q_c} \right) = (283 \text{ K}) \left(\frac{900 \text{ J}}{550 \text{ J}} \right) = 463 \text{ K} = 190°\text{C}$$

    ASSESS The maximum temperature T_h is greater than T_c, as our calculation confirms. Note that Carnot's theorem applies to the ratio of absolute temperatures.

19. **INTERPRET** This problem is about the work done by a refrigerator to freeze water. Heat of transformation is involved in the phase change.

    DEVELOP The amount of heat that must be extracted in order to freeze the water is

    $$Q_c = mL_f = (0.67 \text{ kg})(334 \text{ kJ/kg}) = 224 \text{ kJ}$$

    The work consumed by the refrigerator while extracting this heat is given by Equation 19.4, $W = Q_c/\text{COP}$.

    EVALUATE Substituting the values given, we obtain

    $$W = \frac{Q_c}{\text{COP}} = \frac{224 \text{ kJ}}{4.2} = 53.3 \text{ kJ}$$

ASSESS   A COP of 4.2 means that each unit of work can transfer 4.2 units of heat from inside the refrigerator. A smaller COP would mean that more work is required to freeze the water.

Section 19.4 Entropy and Energy Quality

21.   **INTERPRET**   This problem asks for the entropy increase after heating up water.

DEVELOP   For a substance with constant specific heat (in this case at constant pressure), $dQ = mcdT$, and the change in entropy is

$$\Delta S = \int_1^2 \frac{dQ}{T} = mc \int_{T_1}^{T_2} \frac{dT}{T} = mc \ln\left(\frac{T_2}{T_1}\right)$$

EVALUATE   Substituting the values given, we have

$$\Delta S = mc \ln\left(\frac{T_2}{T_1}\right) = (0.25 \text{ kg})(4.184 \text{ kJ/kg} \cdot \text{K}) \ln\left(\frac{368 \text{ K}}{283 \text{ K}}\right) = 275 \text{ J/K}$$

ASSESS   The final entropy of the system has increased.

23.   **INTERPRET**   This problem is about the entropy increase associated with melting a block of lead.

DEVELOP   For a change of phase at constant temperature,

$$\Delta S = \frac{\Delta Q}{T} = \frac{mL_f}{T}$$

EVALUATE   From the above equation, we find the mass of lead to be

$$m = \frac{T \Delta S}{L_f} = \frac{(600 \text{ K})(900 \text{ J/K})}{24.7 \text{ kJ/kg}} = 21.9 \text{ kg}$$

ASSESS   As expected, the mass of the block is proportional to the change in entropy.

PROBLEMS

25.   **INTERPRET**   This problem is about a Carnot engine, its work, efficiency and power output.

DEVELOP   For a cyclic operation, the change in internal energy is zero, $\Delta U = 0$. From the first law of thermodynamics, we have $W = Q_h - Q_c$. Once the work W is known, its efficiency can be obtained as $e = W/Q_h$. For a Carnot engine, $Q_c/Q_h = T_c/T_h$.

EVALUATE   (a) The work done by the engine during each cycle is

$$W = Q_h - Q_c = 890 \text{ J} - 470 \text{ J} = 420 \text{ J}$$

(**b**) The efficiency of the engine is

$$e = \frac{W}{Q_h} = \frac{420 \text{ J}}{890 \text{ J}} = 47.2\%$$

(**c**) For a Carnot engine, $T_c = \left(\frac{Q_c}{Q_h}\right) T_h = \left(\frac{470 \text{ J}}{890 \text{ J}}\right)(550 \text{ K}) = 290 \text{ K}$. Note that $T_c < T_h$, as expected.

(**d**) The mechanical power output of the engine is

$$P_{out} = \frac{W}{t} = \frac{(420 \text{ J/cycle})}{22 \text{ cycles/s}} = 9.24 \text{ kW}$$

ASSESS   The efficiency of the engine can also be calculated using Equation 19.3:

$$e = 1 - \frac{T_c}{T_h} = 1 - \frac{290 \text{ K}}{550 \text{ K}} = 47.2\%$$

27.   **INTERPRET**   This problem is about a power plant, its rate of energy extraction, efficiency, and the highest temperature it attains.

DEVELOP   From Equation 16.3, $\Delta Q = mc\Delta T$, we see that in order to raise the temperature of the cooling water by 8.5 K, heat must be exhausted to it at a rate of

$$\frac{dQ_c}{dt} = c\left(\frac{dm}{dt}\right)\Delta T = (4.184 \text{ kJ/kg} \cdot \text{K})(2.8 \times 10^4 \text{ kg/s})(8.5 \text{ K}) = 996 \text{ MW}$$

We take this to be all the heat rejected by the power plant. Since the work output, dW/dt, is also given, the heat input to the plant (extracted from its fuel) is

$$\frac{dQ_h}{dt} = \frac{dQ_c}{dt} + \frac{dW}{dt}$$

In terms of the rates, the efficiency of the plant is $e = (dW/dt)/(dQ_h/dt)$. If we consider the plant to operate like a Carnot engine, then its highest temperature can be calculated using $Q_c/Q_h = T_c/T_h$.

EVALUATE   (a) Substituting the values given, we obtain

$$\frac{dQ_h}{dt} = \frac{dQ_c}{dt} + \frac{dW}{dt} = 996 \text{ MW} + 750 \text{ MW} = 1.75 \text{ GW}$$

(b) The plant's efficiency (from the definition of efficiency in terms of rates) is

$$e = \frac{dW/dt}{dQ_h/dt} = \frac{750 \text{ MW}}{1.75 \text{ GW}} = 43.0\%$$

(c) With the assumption that the plant operates like an ideal Carnot engine, then

$$\frac{T_h}{T_c} = \frac{Q_h}{Q_c} = \frac{dQ_h/dt}{dQ_c/dt} = \frac{1.75 \text{ GW}}{996 \text{ MW}} = 1.75$$

(Note that the energy rate per cycle and the energy rate per second are proportional.) If $T_c = 15°C = 288$ K, then

$$T_h = 1.75 T_c = 1.75(288 \text{ K}) = 505 \text{ K} = 232°C$$

ASSESS   The actual highest temperature would be somewhat greater than this, because the actual efficiency is always less than the Carnot efficiency.

29.   **INTERPRET**   This problem is about a power plant, and we are interested in its rate of using cooling water.
DEVELOP   For a cyclic operation, the change in internal energy is zero, $\Delta U = 0$. From the first law of thermodynamics, we have $W = Q_h - Q_c$. Therefore, the total rate at which heat is exhausted by all power plants is

$$\frac{dQ_c}{dt} = \frac{d}{dt}(Q_h - W) = \frac{dW}{dt}\left(\frac{1}{e} - 1\right) = (2 \times 10^{11} \text{ W})\left(\frac{1}{33\%} - 1\right) = 4 \times 10^{11} \text{ W}$$

The mass rate of flow at which water could absorb this amount of energy, with only a 5°C temperature rise, is given by:

$$\frac{dQ_c}{dt} = \frac{dm}{dt} c_{\text{water}} \Delta T$$

The equation can be solved to give the mass rate.
EVALUATE   Solving the equation above, we obtain

$$\frac{dm}{dt} = \frac{4 \times 10^{11} \text{ W}}{(4184 \text{ J/kg} \cdot \text{K})(5°C)} = 1.91 \times 10^7 \text{ kg/s}$$

or about 1 Mississippi (a self-explanatory unit of river flow).
ASSESS   In order to absorb the power output of 2×10^{11} W with only 5°C increase of temperature, we expect the mass flow rate to be large.

31.   **INTERPRET**   This problem is about a freezer. The highest and the lowest temperatures are $T_h = 32°C = 305$ K and $T_c = 0°C = 273$ K, respectively.
DEVELOP   The coefficient of performance (COP) of a reversible freezer is given by Equation 19.4:

$$\text{COP} = \frac{Q_c}{W} = \frac{Q_c}{Q_h - Q_c} = \frac{T_c}{T_h - T_c}$$

Once the COP is known, we can then solve for Q_c, and the amount of water the freezer can freeze in one hour is $m = Q_c/L_f$.
EVALUATE   (a) The COP of the freezer is $\text{COP} = \frac{T_c}{T_h - T_c} = \frac{273 \text{ K}}{305 \text{ K} - 273 \text{ K}} = 8.53$.
(b) The heat extracted in one hour is

$$Q_c = \text{COP } W = 8.53 \times (12 \text{ kWh}) = 369 \text{ MJ}$$

The amount of water it could freeze is $m = \frac{Q_c}{L_f} = \frac{369 \text{ MJ}}{334 \text{ kJ/kg}} = 1.10 \times 10^3$ kg.

ASSESS Typical freezers have a COP lower than 8.53. Thus, more electrical energy is needed to freeze the same amount of water.

33. **INTERPRET** This problem is about a Carnot engine, and we want to know its efficiency and the highest temperature it attains.

DEVELOP From Equation 16.3, $\Delta Q = mc\Delta T$, we see that in order to raise the temperature of the cooling water by ΔT, heat must be exhausted to the engine at a rate of

$$\frac{dQ_c}{dt} = c\left(\frac{dm}{dt}\right)\Delta T$$

For a cyclic operation, the change in internal energy is zero, $\Delta U = 0$. From the first law of thermodynamics, we have $W = Q_h - Q_c$, or $Q_h = Q_c + W$. Therefore, the rate of heat input to the engine is

$$\frac{dQ_h}{dt} = \frac{dQ_c}{dt} + \frac{dW}{dt}$$

The efficiency of the engine is $e = \frac{dW/dt}{dQ_h/dt}$.

EVALUATE (a) The rate of heat exhausted by the engine is

$$\frac{dQ_c}{dt} = c\left(\frac{dm}{dt}\right)\Delta T = (4.184 \text{ kJ/kg} \cdot \text{K})(3.2 \text{ kg/s})(28°C - 23°C) = 66.9 \text{ kW}$$

Since $dW/dt = 150$ kW, the rate of heat input is

$$\frac{dQ_h}{dt} = \frac{dQ_c}{dt} + \frac{dW}{dt} = 66.9 \text{ kW} + 150 \text{ kW} = 216.9 \text{ kW}$$

Thus, the efficiency of the engine is

$$e = \frac{dW/dt}{dQ_h/dt} = \frac{150 \text{ kW}}{216.9 \text{ kW}} = 69.1\%$$

(b) For a Carnot engine with $T_c = \frac{1}{2}(23°C + 28°C) = 298.5$ K, its highest temperature is

$$T_h = \frac{T_c}{1-e} = \frac{298.5 \text{ K}}{1-0.691} = 967 \text{ K} = 694°C$$

ASSESS The efficiency of an engine is proportional to $T_h - T_c$, the temperature difference between the highest and the lowest temperatures. Since an efficiency of 69.1% is rather high, we expect the temperature difference to be large. In this case, it is $T_h - T_c = 967$ K $- 298.5$ K $= 668.5$ K.

35. **INTERPRET** This problem is about a refrigerator. We want to find its efficiency, given the heat extracted and the work output.

DEVELOP The heat extracted from the water is

$$Q_c = mc\Delta T = \rho Vc\Delta T = (10^3 \text{ kg/m}^3)(4 \times 10^{-3} \text{ m}^3)(4.184 \text{ kJ/kg} \cdot \text{K})(9°C - 1°C) = 134 \text{ kJ}$$

On the other hand, the work input is $W = Pt = (130 \text{ W})(4 \times 60 \text{ s}) = 31.2$ kJ. The coefficient of performance (COP) of the refrigerator can be calculated using Equation 19.4:

$$\text{COP} = \frac{Q_c}{W}$$

EVALUATE (a) Substituting the values given, the actual COP is

$$\text{COP} = \frac{Q_c}{W} = \frac{134 \text{ kJ}}{31.2 \text{ kJ}} = 4.29$$

(b) For a reversible refrigerator operating between $T_c = 274$ K (1°C) and $T_h = 298$ K (25°C), the COP is

$$\text{COP}_{rev} = \frac{T_c}{T_h - T_c} = \frac{274 \text{ K}}{298 \text{ K} - 274 \text{ K}} = 11.4$$

Thus, we see that the COP of the refrigerator is about 38% of the maximum possible COP_{rev}.

ASSESS The maximum possible COP is attained when the refrigerator cycle is reversible. The COP of a real refrigerator can never exceed COP_{rev}.

37.   **INTERPRET**   We analyze the COP required to save money with a heat pump, considering the cost of oil and of electricity. We will do this by calculating the cost of the heat gained by both the oil-burning heater and the electric heat pump.

DEVELOP   The Coefficient of Performance (COP) is the relationship between the heat sent to the cold reservoir and the work done. We set the heat Q_c to be the same for both heating mechanisms, and solve for COP. The cost of oil is $\$_{oil} = \frac{\$1.75}{30 \text{ kWh}} = \$0.0583 \text{ kWh}^{-1}$, and the cost of electricity is $\$_{electricity} = \$0.165 \text{ kWh}^{-1}$.

EVALUATE   The cost of electricity is more by a factor of $\frac{16.5}{5.83} = 2.83$. So in order to be cost-effective, the heat pump must have a COP of greater than 2.83, since $COP = \frac{Q_c}{W} \rightarrow Q_c = W \times COP$.

ASSESS   Most heat pumps have a COP much higher than this value, so it's probably a good idea to switch.

39.   **INTERPRET**   We are asked to find the COP, power usage, and operating cost compared to that of an oil-burning heater of a heat pump. We will assume that the heat pump is a Carnot heat pump.

DEVELOP   The maximum COP of a heat pump is $COP_{max} = \frac{T_c}{T_h - T_c} = \frac{Q_c}{W}$. We are given the temperatures involved as $T_c = 10°C = 283 \text{ K}$ and $T_h = 70°C = 343 \text{ K}$. The power consumption requested is proportional to W, and the rate at which heat is supplied is proportional to $Q_c = 20 \text{ kW}$. The cost of electricity is $\$_{electricity} = \$0.155 \text{ kWh}^{-1}$ and the cost of oil is $\$_{oil} = \frac{\$1.95 \text{ /gal}}{30 \text{ kWh/gal}} = \$0.065 \text{ kWh}^{-1}$.

EVALUATE

(a) $COP_{max} = \frac{T_c}{T_h - T_c} = 4.72$

(b) $COP = \frac{Q_c}{W} \rightarrow W = \frac{Q_c}{COP} \rightarrow \frac{W}{t} = \frac{Q_c}{t \times COP}$. The heat per time is 20 kW, so the power is $\frac{W}{t} = 4.24 \text{ kW}$.

(c) The cost of running the oil heater, per hour, is $\$_{oil} \times 20 \text{ kWh} = \$1.30$. The cost of running the heat pump to do the same job is $\$_{electricity} \times 4.24 \text{ kWh} = \$0.66$.

ASSESS   The cost of running the heat pump is just over half as much as running an oil-burning heater.

41.   **INTERPRET**   Our engine cycle consists of four paths, two of which are isochoric and two are isobaric.

DEVELOP   Label the states in Fig. 19.21 A, B, C, D going clockwise from the upper left corner. The work done and the heat absorbed during the isobaric segments AB and CD are

$$W_{AB} = p_A(V_B - V_A) = (6 \text{ atm})(6 \text{ L} - 2 \text{ L}) = 24 \text{ L} \cdot \text{atm}$$
$$W_{CD} = p_C(V_D - V_C) = (3 \text{ atm})(2 \text{ L} - 6 \text{ L}) = -12 \text{ L} \cdot \text{atm}$$

and

$$Q_{AB} = nC_P(T_B - T_A) = n\left(\frac{5}{2}R\right)\left(\frac{p_B V_B}{nR} - \frac{p_A V_A}{nR}\right) = \frac{5}{2}(36 - 12) \text{ L} \cdot \text{atm} = 60 \text{ L} \cdot \text{atm}$$

$$Q_{CD} = nC_P(T_D - T_C) = \frac{5}{2}(p_D V_D - p_C V_C) = \frac{5}{2}(6 - 18) \text{ L} \cdot \text{atm} = -30 \text{ L} \cdot \text{atm}$$

where we have assumed an ideal monatomic gas.

For the isochoric segments, we have

$$Q_{BC} = nC_V(T_C - T_B) = \frac{3}{2}(18 - 36) \text{ L} \cdot \text{atm} = -27 \text{ L} \cdot \text{atm}$$

$$Q_{DA} = nC_V(T_A - T_D) = \frac{3}{2}(12 - 6) \text{ L} \cdot \text{atm} = 9 \text{ L} \cdot \text{atm}$$

and $W_{BC} = W_{DA} = 0$. The net heat added for one cycle is

$$Q = Q_{AB} + Q_{BC} + Q_{CD} + Q_{DA} = (60 - 27 - 30 + 9) \text{ L} \cdot \text{atm} = (12 \text{ L} \cdot \text{atm})(101.3 \text{ J/L} \cdot \text{atm})$$
$$= 1.22 \text{ kJ}$$

and the net work done is $W = (24 + 0 - 12 + 0) \text{ L} \cdot \text{atm} = 12 \text{ L} \cdot \text{atm} = 1.22 \text{ kJ}$. Note that the first law of thermodynamics, applied to a cyclic process, requires that $W = Q$.

EVALUATE   (a) Since the heat absorbed is $Q_+ = (60 + 9) \text{ L} \cdot \text{atm} = 69 \text{ L} \cdot \text{atm}$, the efficiency is

$$e = \frac{W}{Q_+} = \frac{12}{69} = 17.4\%$$

(b) The maximum and minimum temperatures are $T_B = p_B V_B/nR$ and $T_D = p_D V_D/nR$, so the efficiency of a Carnot engine operating between these temperatures is

$$e_{\text{Carnot}} = 1 - \frac{T_D}{T_B} = 1 - \frac{p_D V_D}{p_B V_B} = 1 - \frac{6}{36} = 83.3\%$$

This is not a contradiction of Carnot's theorem, because the given engine does not operate between two heat reservoirs at fixed temperatures.

ASSESS   The efficiency of a real engine is always less or equal to that of a Carnot engine.

43.   **INTERPRET**   This problem is about the increase in entropy as the ice is melted and heated up.

DEVELOP   During the melting process at $T_1 = 0°C$, the change in entropy is

$$\Delta S_1 = \frac{\Delta Q}{T_1} = \frac{mL_f}{T_1}$$

During the warming to $T_2 = 15°C$, the change in entropy is (see Exercise 21) $\Delta S_2 = mc\ln(T_2/T_1)$.

EVALUATE   The total change in entropy is

$$\Delta S = \Delta S_1 + \Delta S_2 = m\left[\frac{L_f}{T_1} + c\ln\left(\frac{T_2}{T_1}\right)\right]$$

$$= (9.4 \times 10^4 \text{ kg})\left[\frac{334 \text{ kJ/kg}}{273 \text{ K}} + (4.184 \text{ kJ/kg} \cdot \text{K})\ln\left(\frac{288 \text{ K}}{273 \text{ K}}\right)\right] = 136 \text{ MJ/K}$$

ASSESS   As expected, the entropy change is positive in both melting and heating processes.

45.   **INTERPRET**   This problem is about entropy change with varying temperature while keeping pressure fixed.

DEVELOP   From the first law of thermodynamics $(dQ = dU + dW)$ and the properties of an ideal gas $(dU = nC_V dT$ and $PV = nRT)$, an infinitesimal entropy change is

$$dS = \frac{dQ}{T} = nC_V\frac{dT}{T} + \frac{P}{T}dV = nC_V\frac{dT}{T} + nR\frac{dV}{V}$$

When the pressure is constant, ideal-gas law gives $V_2/V_1 = T_2/T_1$.

EVALUATE   Substituting the second equation into the first one yields

$$\Delta S = nC_V\ln\left(\frac{T_2}{T_1}\right) + nR\ln\left(\frac{V_2}{V_1}\right) = n(C_V + R)\ln\left(\frac{T_2}{T_1}\right) = nC_p\ln\left(\frac{T_2}{T_1}\right)$$

ASSESS   The same expression can also be obtained by using $dQ = nC_p dT$ at constant pressure. Note that $\Delta S > 0$ if $T_2 > T_1$, as expected.

47.   **INTERPRET**   This problem is about the entropy change of the universe in an irreversible heat transfer.

DEVELOP   The house loses heat at $T_{\text{house}} = 20°C = 293$ K at a rate of $dQ/dt = 30$ kW, so

$$\left(\frac{dS}{dt}\right)_{\text{house}} = -\frac{1}{T_{\text{house}}}\frac{dQ}{dt} = -\frac{1}{293 \text{ K}}\frac{dQ}{dt}$$

The surroundings gain heat at $T_{\text{surroundings}} = -10°C = 263$ K at the same rate, so

$$\left(\frac{dS}{dt}\right)_{\text{surroundings}} = \frac{1}{T_{\text{surroundings}}}\frac{dQ}{dt} = \frac{1}{263 \text{ K}}\frac{dQ}{dt}$$

The sum of the two terms is the rate of the change of entropy of the universe.

EVALUATE   Adding both terms gives

$$\left(\frac{dS}{dt}\right)_{\text{universe}} = \left(\frac{dS}{dt}\right)_{\text{house}} + \left(\frac{dS}{dt}\right)_{\text{surroundings}} = \frac{dQ}{dt}\left(-\frac{1}{T_{\text{house}}} + \frac{1}{T_{\text{surroundings}}}\right)$$

$$= (30 \text{ kW})\left(-\frac{1}{293 \text{ K}} + \frac{1}{263 \text{ K}}\right) = 11.7 \text{ J/K} \cdot \text{s}$$

ASSESS   The entropy in the house decreases due to loss of heat, while the entropy in the outdoor surrounding goes up. The overall entropy of the universe increases, in accordance with the second law of thermodynamics.

49.   **INTERPRET**   This problem asks about the energy quality during a thermodynamic process during which entropy has increased.

DEVELOP   The change in entropy during the adiabatic free expansion process is

$$\Delta S = nR \ln\left(\frac{V_2}{V_1}\right)$$

Therefore, the energy made unavailable is $E_{\text{unavailable}} = T\Delta S$.

EVALUATE   Substituting the values given in the problem statement, the energy that becomes unavailable to do work, in the free expansion of an ideal gas (T remains constant), is

$$E_{\text{unavailable}} = T\Delta S = nRT \ln\left(\frac{V_2}{V_1}\right) = (8.7 \text{ mol})(8.314 \text{ J/mol} \cdot \text{K})(450 \text{ K})\ln(10) = 74.9 \text{ kJ}$$

ASSESS   This is the work that could have been recovered from a reversible isothermal expansion. However, due to the irreversible nature of the process, we give up the possibility of extracting this work.

51.   **INTERPRET**   This problem asks for the entropy change of the pan-water system, when thermal equilibrium has been reached.

DEVELOP   Assume all the heat lost by the pan is gained by the water. The equilibrium temperature is given by Equation 16.4, or

$$T_{\text{eq}} = \frac{(2.4 \text{ kg})(900 \text{ J/kg} \cdot \text{K})(428 \text{ K}) + (3.5 \text{ kg})(4184 \text{ J/kg} \cdot \text{K})(288 \text{ K})}{(2.4 \text{ kg})(900 \text{ J/kg} \cdot \text{K}) + (3.5 \text{ kg})(4184 \text{ J/kg} \cdot \text{K})} = 306 \text{ K}$$

Using the result of Exercise 21, the change in entropy for the pan is

$$\Delta S_{\text{pan}} = m_{\text{pan}} c_{\text{pan}} \ln\left(\frac{T_{\text{eq}}}{T_{\text{pan}}}\right)$$

Similarly, the change in entropy for the water is

$$\Delta S_{\text{water}} = m_{\text{water}} c_{\text{water}} \ln\left(\frac{T_{\text{eq}}}{T_{\text{water}}}\right)$$

The sum of these two terms is the change of entropy of the pan-water system.

EVALUATE   The entropy change of the pan and water together is

$$\Delta S = \Delta S_{\text{pan}} + \Delta S_{\text{water}} = m_{\text{pan}} c_{\text{pan}} \ln\left(\frac{T_{\text{eq}}}{T_{\text{pan}}}\right) + m_{\text{water}} c_{\text{water}} \ln\left(\frac{T_{\text{eq}}}{T_{\text{water}}}\right)$$

$$= (2.4 \text{ kg})(900 \text{ J/kg} \cdot \text{K}) \ln\left(\frac{306 \text{ K}}{428 \text{ K}}\right) + (3.5 \text{ kg})(4184 \text{ J/kg} \cdot \text{K})\ln\left(\frac{306 \text{ K}}{288 \text{ K}}\right) = 163 \text{ J/K}$$

ASSESS   The entropy change for the pan is negative, while that of the water is positive. The total entropy change is positive, in accordance with the second law of thermodynamics.

53.   **INTERPRET**   This problem is about the efficiency of an engine operating between two temperatures.

DEVELOP   We take the engine to be a reversible one, operating between the two given temperatures ($T_h = 420$ K and $T_c = $ normal ice point $= 273$ K). The efficiency can then be computed using Equation 19.3, $e_{\text{Carnot}} = 1 - T_c/T_h$.

EVALUATE   **(a)** Substituting the values given in the problem statement, we find the efficiency to be

$$e_{\text{Carnot}} = 1 - \frac{T_c}{T_h} = 1 - \frac{273 \text{ K}}{420 \text{ K}} = 35.0\%$$

(b) The total heat the block of ice can absorb, as it melts at 273 K, is

$$Q = mL_f = (10^3 \text{ kg})(334 \text{ kJ/kg}) = 334 \text{ MJ}$$

Then the melt-water temperature will rise and the engine's efficiency will drop. While running at the original efficiency, the engine exhausts heat at the rate

$$\frac{dQ_c}{dt} = \frac{dQ_h}{dt} - \frac{dW}{dt} = \left(\frac{1}{e_{\text{Carnot}}} - 1\right)\frac{dW}{dt} = \left(\frac{1}{0.35} - 1\right)(8.5 \text{ kW}) = 15.8 \text{ kW}$$

(combine the first law with the definition of efficiency). Thus, it can operate between the original temperatures for a time

$$t = \frac{Q}{dQ_c/dt} = \frac{334 \text{ MJ}}{15.8 \text{ kW}} = 2.12 \times 10^4 \text{ s} = 5.88 \text{ h}$$

ASSESS For real engines in which $e < e_{\text{Carnot}}$, heat is exhausted at a greater rate. This shortens the duration in which the engine can maintain its efficiency.

55. **INTERPRET** This problem is about the efficiency of the Otto cycle as a function of the compression ratio.
     DEVELOP Using the adiabatic and ideal gas laws, Equations 18.11a and b, and 17.2, the relationships between the temperatures, volumes, and pressures are given by (see Problem 54)

$$\frac{T_3}{T_4} = \left(\frac{V_4}{V_3}\right)^{\gamma-1} = r^{\gamma-1} = \left(\frac{V_1}{V_2}\right)^{\gamma-1} = \frac{T_2}{T_1}$$

$$\frac{p_3}{p_4} = \left(\frac{V_4}{V_3}\right)^{\gamma} = \left(\frac{V_1}{V_2}\right)^{\gamma} = \frac{p_2}{p_1}$$

and

$$\frac{T_3}{T_2} = \frac{p_3}{p_2} = 3 = \frac{p_4}{p_1} = \frac{T_4}{T_1}$$

where $r = V_1/V_2$ is the compression ratio. In terms of T_1, the temperatures are $T_2 = r^{\gamma-1}T_1$, $T_4 = 3T_1$, and $T_3 = r^{\gamma-1}T_4 = 3 \times r^{\gamma-1}T_1$.
Since no heat is transferred on the adiabatic segments, the heat input is

$$Q_h = Q_{23} = nC_V(T_3 - T_2) = nC_V r^{\gamma-1}(3-1)T_1 = 2 \times r^{\gamma-1}nC_V T_1$$

and the heat exhaust is

$$Q_c = -Q_{41} = -nC_V(T_1 - T_4) = 2nC_V T_1$$

EVALUATE (a) The efficiency of the Otto cycle is

$$e_{\text{Otto}} = 1 - \frac{Q_c}{Q_h} = 1 - r^{1-\gamma}$$

For a Carnot engine, the efficiency is

$$e_{\text{Carnot}} = 1 - \frac{T_c}{T_h} = 1 - \left(\frac{p_2}{p_3}\right)r^{1-\gamma} = 1 - \frac{1}{3}r^{1-\gamma}$$

(b) For small $r(r \to 1)$, $e_{\text{Otto}} \to 0$ and $e_{\text{Carnot}} \to 2/3$, while for large $r(r \to \infty)$, e_{Otto} and $e_{\text{Carnot}} \to 1$. A plot of the efficiencies as a function of r is shown on the right for $\gamma = 1.4$ and $1 \leq r \leq 15$. (The shape of the curves depends on the value of γ, of course.)

ASSESS Our result shows that $e_{\text{Otto}} < e_{\text{Carnot}}$, as expected. Note that the compression ratio r cannot be very large in practice, or pre-ignition results.

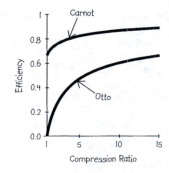

57. **INTERPRET**   Find the maximum efficiency of a power plant, given the temperature range of its cycle. We will calculate the Carnot efficiency, and compare this with the actual efficiency.

    DEVELOP   The high temperature $T_h = 950°F = 783$ K. The low temperature is $T_c = 90°F = 305$ K. The Carnot efficiency is given by $e = 1 - \frac{T_c}{T_h}$.

    EVALUATE   The maximum efficiency is $e = 1 - \frac{T_c}{T_h} = 61\%$.

    ASSESS   The actual efficiency of this plant is given as 25%, which is considerably lower due (at least in part) to having to evaporate moisture out of the wood-chip fuel.

59. **INTERPRET**   This problem deals with a Carnot engine whose heat reservoir temperature varies with time.

    DEVELOP   In time dt, the engine extracts heat $dQ_h = -mcdT_h$ from the block, and does work $dW = Pdt$. Since it is a Carnot engine,

    $$dW = e_{max}dQ_h = \left(\frac{T_h - T_c}{T_h}\right)(-mcdT_h) = Pdt$$

    The power is also assumed to be proportional to $T_h - T_c$, so the equation becomes

    $$-mc\frac{dT_h}{T_h} = \frac{P_0 dt}{T_{h0} - T_c}$$

    Integrating over the expression then yields T_h as a function of time. For (b), we note that the power output becomes zero when $T_h = T_c$.

    EVALUATE   (a) Integrating from $t = 0$ and T_{h0} to t and T_h, we obtain

    $$\int_{T_{h0}}^{T_h} \frac{dT_h'}{T_h'} = \ln\left(\frac{T_h}{T_{h0}}\right) = -\int_0^t \frac{P_0 dt}{mc(T_{h0} - T_c)} = \frac{-P_0 t}{mc(T_{h0} - T_c)}$$

    or $T_h(t) = T_{h0}e^{-P_0 t/mc(T_{h0}-T_c)}$.

    (b) The power output is zero for $T_h = T_c$. This occurs at time

    $$t_0 = (mc/P_0)(T_{h0} - T_c)\ln(T_{h0}/T_c)$$

    ASSESS   We find the instantaneous temperature of the hot block to decrease exponentially with time. At $t = 0$, $T_h(0) = T_{h0}$. However, for $P_0 t \gg mc(T_{h0} - T_c)$, T_h becomes very small. Note that the expression for P was originally assumed to be valid for $T_h \geq T_c$, or for times $t \leq t_0$. If we allow $T_h < T_c$, or $t > t_0$, then $dW = Pdt < 0$ becomes work input to an "engine" which acts like a refrigerator cooling the block.

61. **INTERPRET**   We find the change in entropy for a sample of copper at low temperatures, where the specific heat changes with temperature.

    DEVELOP   The specific heat of copper at low temperatures is given as $c = 31\left(\frac{T}{343\,K}\right)^3$ J·g·K. Also $dS = \frac{dQ}{T}$, and $dQ = mcdT$. We find the change in entropy by integrating from $T = 25$ K to $T = 10$ K. The mass of the copper is $m = 40$ g.

    EVALUATE

    $$dS = \frac{dQ}{T} \rightarrow \Delta S = \int_{T_1}^{T_2} \frac{mc}{T}dT = \int_{T_1}^{T_2} \frac{m\left(31(\frac{T}{343})^3\right)}{T}dT = \frac{31m}{343^3}\int_{T_1}^{T_2} T^2 dt$$

    $$\rightarrow \Delta S = \frac{31m}{343^3}\left[\frac{1}{3}\left(T_2^3 - T_1^3\right)\right] = -0.15 \text{ J/kg}$$

    ASSESS   This change is negative because the temperature goes down. Somewhere in this cooling process, the entropy of something else must have gone up by more than 0.15 J/kg.

63. **INTERPRET**   We find the entropy change of a mass of steam as it cools while being used to drive a heat engine. We also calculate the entropy change of the cold reservoir, and the total amount of work that the engine can do. We will use the definition of entropy and the amount of energy that becomes unavailable to do work.

DEVELOP   We calculate the change in entropy of the steam in two parts. We use $\Delta S_1 = -\frac{Q_{condensation}}{T_{steam}}$ for the change in entropy as the steam condenses, and then $\Delta S_2 = mc\ln\left(\frac{T_{ice}}{T_{steam}}\right)$ for the change in entropy of the water as it cools. To find the change in entropy of the reservoir, we use $\Delta S_r = \frac{E}{reservoir}$, where E is the heat that flows into the reservoir. During the process of cooling the steam, an amount of energy $E = T_{min}\Delta S$ becomes unavailable to do work, so ideally the rest of the energy becomes work: $W = Q_1 - E$. The mass of steam at $T_h = 373$ K is $m = 50$ kg. The temperature of the cold reservoir is $T_c = 273$ K. The latent heat of vaporization for the water is $L_v = 2256 \times 10^3$ J/kg, and the specific heat of water is $c = 4190$ J/kg·K.

EVALUATE

(a) The entropy change of the steam (and water) is $\Delta S_s = \Delta S_1 + \Delta S_2 = -\frac{mL_v}{373\,K} + mc\ln\left(\frac{273\,K}{373\,K}\right) = -368$ kJ/K.

(b) The entropy change of the reservoir is $\Delta S_r = \frac{E}{T_r} = \frac{-T_{min}\Delta S_s}{T_r} = -\Delta S_s = 368$ kJ/K.

(c) The work that the engine can do is $W = (mL_v + mc\Delta T) - E = 33$ MJ.

ASSESS   The amount of heat that becomes unavailable to do work is approximately 100 MJ, nearly three times the work done!

65. **INTERPRET**   We calculate how long a block of ice will last if we use it as a dump for the waste heat for an engine. We will use the Carnot efficiency and the work done by the engine to find the rate of heat wasted to the ice, and use the latent heat of fusion of the ice to find the rate at which the ice melts.

DEVELOP   The efficiency of the engine is given by $e = 1 - \frac{T_c}{T_h}$, so the work done by the engine is $W = Q_h e$ and we see that the waste heat is $Q_w = Q_h - W$. The latent heat of fusion of ice is $L_f = 334 \times 10^3$ J/kg. The power output of the engine is $P = 8.5$ kW. The temperatures involved are $T_h = 420$ K and $T_c = 273$ K, and the initial mass of ice is $m = 1000$ kg.

EVALUATE   We first calculate the efficiency: $e = 1 - \frac{273}{420} = 0.35$. The rate at which heat is used is then

$$P_h = \frac{Q_h}{t} = \frac{W}{et} = 24.3 \text{ kW},$$

and heat is wasted at a rate of

$$P_w = \frac{Q_w}{t} = \frac{Q_h}{t} - W = 15.8 \text{ kW}.$$

From $Q = mL_f$, we have

$$\frac{Q_w}{t} = \frac{m}{t}L_f \rightarrow t = \frac{mL_f}{P_w} = 21.2 \text{ s} = 5.88 \text{ hours}.$$

ASSESS   The ice does not last through the day.